THE

Eddie Bauer

GUIDE TO

FAMILY CAMPING

THE Eddie Bauer GUIDE TO FAMILY CAMPING

ARCHIE SATTERFIELD
and EDDIE BAUER

ILLUSTRATIONS BY TED RAND

ADDISON-WESLEY PUBLISHING COMPANY
READING, MASSACHUSETTS · MENLO PARK, CALIFORNIA
LONDON · AMSTERDAM · DON MILLS, ONTARIO · SYDNEY

Library of Congress Cataloging in Publication Data

Satterfield, Archie.
The Eddie Bauer guide to family camping.

Includes index.
1. Camping—Equipment and supplies.
2. Camping.
3. Family recreation.
I. Title.

GV191.76.S27 1982 688.7'654 82-13923
ISBN 0-201-07776-0
ISBN 0-201-07777-9 (pbk.)

ABCDEFGHIJ-DO-85432

Design by Val Paul Taylor

INTRODUCING THE *Eddie Bauer* OUTDOOR LIBRARY

Eddie Bauer has been serving the needs of outdoor enthusiasts for three generations. Since 1920, we have been dedicated to developing, testing, and manufacturing the finest in apparel and gear for outdoor adventures. Our aim has been to make outdoor adventures more enjoyable.

Now, sixty-two years after we began, we are answering the needs of another generation of outdoor activists. *Eddie Bauer* Outdoor Guides like this one on Family Camping will help newcomers get started. Others will contain up-to-date, tested information to make your outdoor excursions safe, warm, dry, and comfortable.

CONTENTS

ACKNOWLEDGMENTS

Books of this nature are always a team effort, and many individuals and companies have been very helpful in the research, reading drafts of the manuscript and bringing additional material to my attention.

Several members of the Eddie Bauer, Inc., staff have devoted many hours to this book, and it all began with a series of conversations with Jack Quinlan, assistant to the president. David V. Rudd, vice-president of marketing, has been the overall editor for the project, and much of the structure came from his suggestions. He in turn brought in other men and women from the company with their own areas of expertise. These include Jim Wheat, Ken Wherry, Bob Murphy, Marilyn Siehl, and Cort Green, recently retired from the staff.

Much of the material resulted from my own experiences camping with my family from the time the children were infants to the present, on trips all over the American West and in Alaska and the Yukon. Some of those trips were made through the cooperation of Kampgrounds of America (KOA) and Winnebago Industries, Inc.

For information on new developments in camping equipment, I have relied on the literature from Eddie Bauer, Inc., suppliers, and a selection of books, pamphlets, and product-information releases from them. Other material came from the following books:

Roughing It Easy (I and II) by Dian Thomas. New York: Warner Books, 1974, 1976.

Packrat Papers. Lynnwood, Washington. Signpost Publications, 1973, 1977.

Old Fashioned Dutch Oven Cookbook by Don Holm. Caldwell, Idaho: Caxton, 1969.

Medicine for Mountaineering by James A. Wilkerson, M.D. Seattle: The Mountaineers, 1967.

Map and Compass by Bjorn Kjellstrom. New York: Charles Scribner's Sons, 1976.

The Camping Trailer Handbook. Wichita, Kansas: Coleman Co., 1974.

One Pot Meals by Margaret Gin. San Francisco: 101 Productions, 1976.

The Picnic Gourmet by Joan Hemingway and Connie Maticich. New York: Random House, 1975, 1977.

Elegant Meals with Inexpensive Meats. San Francisco: Ortho Books, 1978.

Trailside Cookery by Russ Mohney. Harrisburg, Pennsylvania: Stackpole, 1976.

Campground and Trailer Park Guide. Chicago: Rand McNally and Co. Revised annually.

National Park Guide by Michael Frome. Chicago: Rand McNally and Co. Revised annually.

National Forest Guide by Len Hilts. Chicago: Rand McNally and Co., 1976.

And finally, four people were largely responsible for getting me involved in this projected series of the Eddie Bauer Outdoor Library: Richard V. Sawyer, who heard about it through the writers' grapevine and told me; Eddie Bauer, who was instrumental in my being selected as writer; Doe Coover, the Addison-Wesley senior editor, whose frequent trips to the Northwest have kept the book progressing professionally and whose format for the series has the simplicity of genius; and Dominick Abel, our literary agent, who fitted all the pieces together.

In spite of all this generous assistance from experts, some errors may find their way into the book like rain seeping through a poncho seam. If so, the fault is mine and not that of those who helped.

Archie Satterfield

FOREWORD

I am firmly convinced that one of the greatest benefits of wilderness camping as a form of recreation is that it promotes longevity. A dozen or so companions of my youth, both boys and girls, who joined me on wilderness treks have continued their keen interest in nature, have lived clean lives, and are now octogenarians like me and enjoy the same good health my wife, Christine, and I do.

During my fifty-five years of outdoor outfitting, I have always had a keen interest in those who traded with me, and I've been impressed with certain vital differences in their behavior.

Those who took advantage of every opportunity to get away from city life to enjoy outdoor treks—fishing, hunting, hiking, photography, or simply nature and clean outdoor living—have by and large enjoyed their retirement in good health.

In sharp contrast are those who told me, as they acquired costly outdoor equipment, that they were dedicated to first gaining financial security without taking time off for outdoor trips and waiting for their retirement years to enjoy outdoor recreation. Too many of these people soon died after retirement or suffered ill health, only to have their costly equipment brought back to me by their heirs for estate appraisal and usually disposal.

This is an example of what not to do. These people had denied themselves the enjoyment of the healthful outdoor life they had longed for, which would have prepared them for their retirement years.

I have been very fortunate in this respect. My business has also been my hobby, and when I opened my own shop it was like starting a lifelong vacation. I field-tested every piece of clothing and equipment that I sold so that people knew if it was in Eddie Bauer's store, he had personally put it to a rugged test. If I didn't trust equipment, it wasn't stocked. If I needed equipment that wasn't available elsewhere, I developed it myself.

This need for better equipment led to my invention of quilted goose-down clothing, followed by quilted goose-down sleeping bags. These inventions led to saving hundreds if not thousands of lives of armed forces personnel, including pilots and crewmen during World War II.

Throughout all the years I operated my own business, my employees and I were always out on camping trips, combining our recreation with field-testing equipment. If we said a good down coat would keep you warm in a storm on a glacier, we knew from personal experience that it would.

Family camping has been so rewarding to me and my many friends and customers that I'm confident this book will encourage other parents to take up family camping and introduce their children to this clean, healthful form of outdoor recreation.

Eddie Bauer

PART I
INTO THE GREAT OUTDOORS

INTRODUCTION

THE JOYS OF FAMILY CAMPING

If you have ever camped in the woods near a lake or stream, for the rest of your life you will remember the sound of the wind in the trees, the feel of the sun warm on your back, and the smell of coffee and bacon mingling with the woodsmoke. You will remember those evenings when the silence and stillness are punctuated only by the slap of a beaver's tail on the calm water and the fireflies trying to light up the night.

If you are lucky, you will wake before first light and lie in your cozy sleeping bag and watch the black sky slowly turn pale as the stars disappear, one by one, through false dawn into sunrise.

Although most campers speak of the adventures on camping trips, it is often the quiet little events you will remember for years to come. You may never discuss them because you consider these memories personal, but you will remember the smells of fresh leaves mingled with decaying wood, the patterns of sunlight filtering through the forest, the sounds of running water and the wind, birds calling and fish rising in the morning and evening.

The physical and mental benefits of camping are many, and few other forms of recreation offer more to strengthen the family bond. The distractions of young love, neighborhood bullies, stereophonic music, television, refrigerator raids, telephones ringing at dinnertime—all these recede once the family car is parked and you are setting up camp.

Camping offers peace and quiet far from the normal routine of our lives. It gives us an opportunity to get exercise in natural rather than urban, downtown-club settings. It puts us in a position of learning about nature firsthand rather than from books or television specials.

Children take to camping as naturally as they do to building sand castles on the beach. Household chores, a dreary form of undeserved punishment at home, are seen in a new light at the campground. Putting up tents and learning how to build fires and operate campstoves is great fun. Going to fetch a bucket of water isn't a chore as much as it is a small adventure.

It is a rare child who can't find something to do on a camping trip and to keep happily occupied almost every minute of the day. Infants will go anywhere their parents take them and will be happy if they're kept dry, warm, and well fed. Older children are so inventive and curious about new surroundings that they often resemble perpetual motion machines. A stick may be just a stick to an adult, but to a child it is a spear, a magic wand, a territorial boundary, a pencil for drawing pictures in the dirt, a cooking implement for hot dogs and marshmallows, a bow, or an arrow.

Perhaps more than any other family activity, camping adds pleasant experiences to a child's warehouse of memories, and happy adults are usually those who have pleasant memories from their childhood. These camping memories become an important part of an entire family's heritage, as much as the family scrapbooks children and grandchildren love to look through all their lives.

Just as world travelers bring back artwork, furnishings, and decorations from foreign countries, campers often make it a habit to bring home something to remind them of each trip. Some collect items for Christmas use, such as fir or pine cones to fashion into wreaths or to hang as ornaments. A handful of smooth river rocks that change colors when wet can be placed in a terrarium. Pieces of worm-carved wood and shells or sand dollars are prizes to be found on many beaches. Wild flowers can be pressed in a journal or diary with notes of where and when they were picked.

All these things, plus the simple pleasures of being outdoors in healthful surroundings, add to the family memories we cherish throughout our lives.

During recent years we have heard and read a great deal about the importance of taking up lifetime sports as opposed to the more strenuous team sports reserved for the young. Camping and its subsidiary activities are an ideal lifetime sport because they can be enjoyed throughout one's entire life, from infancy to old age. Camping can be as strenuous or as sedentary as you wish, from wilderness expeditions and mountain climbing to an overnight trip at a nearby campground.

This book, then, is a frank attempt to encourage more families to go camping. Although outdoor recreation has become one of the major leisure activities since the late 1940s, there still are millions of people who for various reasons have never spent a night in a tent. Except for strolls in city parks or along shores or nature trails in national parks, their experiences with nature have been limited to seeing it through their car windows.

Although being outdoors is a universal urge, many noncampers avoid it because they have had a bad experience on an earlier outing, or they are afraid they will have a bad experience. Another reason is the confusion over what kind of equipment is needed and the fear that acquiring this basic equipment requires an enormous investment.

Camping is a form of recreation that hardly knows social or racial barriers. Although you will frequently find the inevitable equipment snobs in some campgrounds, still it is a great social leveler, and you will find business tycoons and salaried wage earners shoulder to shoulder around the campfire.

Yet anyone can camp—young or experienced or neophyte. The out of doors isn't an alien environment to be feared. It is our natural habitat. And with today's equipment and camping conveniences, camping can be a pleasurable and beneficial experience for all. While it is true that campers run the risk of getting wet, chilled or otherwise uncomfortable, the proper clothing and other equipment will keep such problems to a minimum. Those who have gone on a few family camping trips learn to take such slight discomforts in stride and think of them as a reasonable trade-off for the pleasure of being outdoors and the closeness such trips almost invariably bring to families.

Because it is so popular, camping offers a wide variety of degrees of roughing it. You can go deep into the wilderness with everything you need in a backpack, or you can travel from campground to campground in a luxurious motorhome with all the amenities of home aboard. Between these extremes is a wide choice of how you enjoy the outdoors.

It is beyond doubt the most popular form of outdoor recreation in North America, and many people who enjoy outdoor trips don't really think of themselves as campers. They may be dedicated bird watchers or hunters or fishermen, and when they say they're going out on a trip, they seldom mention that camping is part of the program.

So this book has been written with the novice in mind, those who enjoy being outdoors but cannot find a book basic enough to provide answers to their questions. Since camping has become so enormously popular in recent years, the beginner has been ignored in favor of the more experienced people who know the basics. It is hoped this book will fill that gap.

CHAPTER 1

CAMPING IN AMERICAN HISTORY

The history of North America is rife with great adventures we would now call camping trips, wilderness treks or expeditions: Daniel Boone's steady westward push through Cumberland Gap to his final stop in Missouri; the great Canadian voyageurs who gradually pushed modern civilization across Canada to the Pacific and north to the islands of the Arctic Ocean; the farmers who floated their crops down the Mississippi to New Orleans and walked part of the way home on the Natchez Trace; the meat hunters who went out for days at a time to gather food for their families; the trappers and traders along the Ohio, Mississippi, and Missouri rivers.

The most dramatic examples of camping involved the great expeditions mounted during the nation's infancy when America was attempting to stretch its boundaries from the Atlantic to the Pacific. The best known of them all is the Lewis and Clark Expedition of 1804-6. No other undertaking was better planned or better carried out; they were gone two years, traveling in new and treacherous country with all the dangers of weather, terrain, disease, and unknown Indian tribes using equipment that today would make a dedicated camper shudder to contemplate. Yet when they returned to St. Louis from the mouth of the Columbia River, they had lost only one man, and he died of a ruptured appendix which would have killed him had he been in a Boston hospital at the time; the first successful surgery for appendicitis was still several years away.

The basic camping equipment

changed little from that period until the middle of this century. Firearms improved, boats and canoes had motors available, trains and automobiles enabled people to get into the wilderness quicker, but camping equipment was accepted as heavy and bulky.

But Americans have always been a restless people, and the European peasant who had his definite place in society back home came to America and found that he could let his imagination and ingenuity thrive. This freedom, the constant boom atmosphere in America, plus the most beautiful geography in the world kept Americans constantly interested in camping and made them seek ways to make it simpler and easier to enjoy. The outdoor life has always been one of the bonuses of living in America.

In many respects, camping in those early years was simpler because the population was so thin that camps could be set up almost anywhere. Those going some distance off the roads and railroads could use their team of horses and farm wagon or strings of pack animals to carry their equipment.

Some of the most famous campers during the late nineteenth century were the wealthy and members of European royalty who spent fortunes coming to America to visit the great wilderness west of the Mississippi. They went on safaris in Africa and expeditions in America, carrying every possible convenience with them, including gourmet cooks, servants, musical instruments, desks, bedroom sets, and vast collections of firearms.

For the rest, camping gradually became a form of recreation with short trips for hunting and fishing. Families would load up some kitchen equipment, a few blankets, some canvas, and put it all in the family wagon and go down to the nearby creek for a day or two of fishing and visiting with other friends on the same outing.

Sometimes itinerant preachers would come to a settlement and hold a revival meeting beside a stream where the converts could be baptized in a large group. Farm families often came to the area with their crude tents and fed the horses out of sacks of feed stored in the wagon.

Perhaps the greatest impact on camping as a form of recreation came from the establishment of the Boy Scouts in 1908 in Great Britain and two years later in the United States. The principles of the Boy Scouts that emphasize good physical and mental health, and lots of outdoor activities, were quickly adopted by Americans because the United States and Canada had more forms of outdoor recreation available than almost any other part of the world.

A major breakthrough in clothing and sleeping equipment came when Eddie Bauer invented the quilted goose down design for a coat. Not only is it pounds lighter than woolen and other insulating materials, the quilted-down coat he invented could be compressed into a tiny bundle for storage or carrying. When the coat was taken out and "spanked," the goose down sprang back to its original shape to give its original insulation qualities.

Shortly after this came an event that had an enormous impact on not only outdoor equipment, medicine, and plastics, but also the entire shape of the world. This event was World War II. Wars are filled with contradictions, of course, and one of them is that they kill and maim the young and innocent. They also speed up research and development of new medical and other equipment that improves the lives of those who survive the wars.

For campers, World War II meant that the use of down as insulation was seriously considered for clothing and sleeping bags. Since the silk supply was cut off by the war, industry developed synthetic replacements: nylon, rayon, and similar inexpensive yet lightweight, strong materials. The techniques and processes learned in the manufacture of parachutes and other equipment quickly found peacetime applications.

Some of those are the tents that will sleep six, yet can be carried by a child; parkas that are extremely durable, yet cost and weigh little; pack materials that also are lightweight and durable.

New processes for manufacturing aluminum were developed for airplanes and many other military uses. Those were soon translated into tent poles, pack frames, tent pegs, stove parts, and a multitude of other uses.

Thus, shortly after the war ended in 1945, America was ready for a revolution in outdoor recreation, and industry was frantically searching for peacetime products to manufacture. Before the end of the 1940 s, America was camping in a big way. Those who could not find the proper equipment in sporting goods stores had the option of shopping at the thousands of military surplus stores that sprang up all over the country. Anyone who was alive during that period remembers with great fondness going to the surplus stores with their curious blend of odors of leather, mothballs, oils, preservatives, and new cloth. Those stores offered you parachutes still packed and ready for use, flight suits with down insulation, gas masks, belts, yards of webbing, command tents that would hold a dozen or more campers, miscellaneous spare parts for tanks, barrels of nuts and bolts, special tools that were good for absolutely nothing, steel helmets, crates of C- and K-rations, tail wheels from trainer planes, torpedo-shaped wing fuel tanks, and so forth.

Ingenious outdoorsmen could find uses for many of these products, and the garages and backyards of America were littered with surplus products that the families were going to get around to building one day.

When the war ended, another revolution of sorts began. Most Americans were better off financially than ever before, and Detroit went back into full production of cars and trucks. The great American urge to travel resumed, and many families whose first experiences with camping came during the Great Depression of the 1930s while they traveled America looking for work now began remembering those hard times with nostalgia. They remembered sleeping out beneath the stars in hayfields or along streams. Wouldn't it be fun, they reasoned, to do the same thing again, only this time by choice.

This urge for traveling, and America's great selection of places to see and things to do, almost naturally led to the next step in outdoor recreation—the recreational vehicle. This began as a trailer pulled behind the family automobile, not so much for recreation as for having a temporary home wherever the jobs were available. Americans had always slept in their vehicles, from the covered wagons that went across the plains and mountains to the Pacific Coast to the first trucks and pickups.

But now these vehicles became a luxury item. First the beds of pickups were enclosed with a simple box. Then people added ice chests, portable stoves, and gasoline lanterns and built racks to double as beds and storage areas. Then came the customized campers with natural-gas stoves and heaters, electric lights, holding tanks so indoor plumbing could be installed.

By the early 1950s, recreational

vehicle (RV) was a way of life in America. By the 1960s, franchised campgrounds for RVs were scattered all over America. The public agencies—National Park, National Forests, Bureau of Land Management, state and city agencies—all had to go into crash campground construction programs. America was going on a camping spree.

It has been ever since. Remote areas that had been visited by no more than 100 persons in all history suddenly were being visited by thousands of hikers and backpackers. Old trails that had not been used since the automobile arrived now were becoming gullies from all the footprints. Land that is of no use to farmers has been turned into commercial campgrounds. The banks of artificial lakes all across the continent are now lined with RVs and tents.

Camping equipment manufacturers are always on the lookout for ways to use new technology. As the space age required new materials, some of those found their way into camping circles, such as "space blankets," sturdy but extremely lightweight plastics that are being used on some RVs to help them improve gasoline mileage. Aerodynamics helped modify RV designs so that they created less drag while being driven. New methods of waste disposal, such as using the heat from the exhaust system, have been more environmentally acceptable. New plastics have made all equipment stronger and lighter.

Obviously, camping is not a passing fad of a nation famous for its fads. It is a firmly entrenched way of life for many Americans, and no other form of family recreation comes close to its popularity. And it is still growing.

THE LITTLE CAMPERS

Adults often forget that children approach camping—the whole business of life—in a slightly different way from adults. Adults tend to think of camping as a serious, hard-nosed, man vs. nature activity. They also forget that children not only don't care about the Ten Essentials and calories and carbohydrates, but that most children don't even vaguely want to know.

Lecture them on these things all you want and the best you'll get from them is a polite silence while their agile minds flick from subject to subject. Look closely and you'll see a far-off look in their eyes.

This does not mean that children aren't eager to learn all the things adults know about camping; they would prefer on-the-job training rather than classroom settings.

So traveling with children in the outdoors means you have to meet them more or less on their own terms and not be disappointed when the trip ends without their having memorized the wind-chill chart or the capacity of a gas stove.

As mentioned elsewhere, infants and toddlers will go anywhere their parents take them and are usually manageable as long as they are dry, warm, and well fed. Of course all parents are subjected to the "terrible twos" when the ex-toddlers are amazingly contrary. But since their attention spans are still brief at the best, it is easy to bribe them into your good graces with a new game, a bit of tickling, a promise of a peanut-butter-and-jelly sandwich, or a "Look! There's a pileated woodpecker."

One of the most remarkable characteristics about children is their propensity for becoming totally exhausted almost as quickly as a bolt of lightning. One moment they are running pell-mell through the campground, and the next moment they are sound asleep on the cargo bag in which food is stored. A few minutes later they are awake again and going full speed as though nothing happened. This can be disquieting.

But that is the nature of children. They burn out fast, but their recuperative powers are staggering.

The instructional portion of camping should not be overemphasized. Children love to learn by doing, and learning to do the things adults do is part of the initiation into adulthood. There was a time not long ago (it still exists in some families) when labors were divided strictly according to gender. The menfolk did one kind of work; the womenfolk another. This isn't much of a problem in camping. So many men go off on hunting and fishing trips with their pals that the boundaries are blurred if not erased. Happily it is also relatively easy to get children to do at least a modest amount of work around the camp. Here, much more than at home, it is easier to sell the idea that those who work, eat; those who do not, go without.

Some camping cynics insist that camping and all the equipment it can entail is the ultimate in toys. Since a lot of camping gear isn't used very frequently at home, children tend to think of the stoves and tents and lanterns and sleeping bags as toys. Let them. They'll enjoy them much more if they are in a different category from light switches, electric ranges, refrigerators, and automatic washers.

Before the trip, go over the entire plan with the family. If the first day's drive is a long one, tell them so. But try to plan at least one stop every 100 or 150 miles so they can romp a bit. If you still have two hours' driving before you reach the goal and the youngsters ask if you're almost there, don't lie and say "almost," because they'll get even with you by grouching, whining, or poking each other.

One father likes to tell his children that they still have two hours to drive when they are actually less than two minutes from the campground. Early delivery on a promise is always pleasant.

One nice way to pass the time on a long car ride to camp is to require everyone to help tell a story. One of the parents will start an adventure story, build up to an exciting scene, then turn it over to someone else in the car. Require each member of the group to tell at least two minutes' worth at a time, or longer depending on the ages and inclinations of the group. This seldom works on the way home. Everyone seems to want to sleep all the way home.

When you go on day hikes or longer ones, prepare for all sorts of imagined cases of exhaustion. Children can always manage to have some kind of disaster that requires them to be carried part of the way, usually the last steps back to camp. It is a continual source of amazement to parents how calm children become when carried. After they have won this little victory, it is often easy for the bearer to pretend fatigue, too, and get the little con artist back on the trail.

By the time children are in the post-toddler period, around kindergarten or first grade, they want even more to be a part of the adult proceedings. This is a good time to get them a small pack. It doesn't have to be fancy or large enough for an expedition, but it should be large enough for them to carry their lunch or a sweater or something useful. Let them have a choice in colors or shape.

As soon as they are old enough to be trusted in a store without a leash or being carried, take them along on shopping trips for the camping trip. This not only is instructive; it also makes them feel a part of the whole expedition.

All the playground psychology you've learned at home comes into play when you arrive in a campground with other children. Pecking orders are quickly established among these total strangers. Like strange dogs getting together, children circle each other briefly and seem to know instinctively who is going to be the leader and who is going to be the follower.

One area where a very firm hand is needed is in the matter of campground good manners. Once people—all ages of them it seems—leave their hometowns, they seem to think they should be free to do as they please. Most adults manage to restrain themselves in campgrounds, but many children will take this opportunity to make more noise, break more glassware belonging to other campers, and tease more dogs than they have done in the past year at home. Unless they are exceptional or watched closely, all concepts of territorial boundaries are abandoned and the children will go tramping through other campers' territories. Since no fences exist and many campsites look alike, it is easy for them to confuse a campground with a city park without realizing that adults build invisible privacy areas around themselves wherever they go. This "personal space" is essential for everyone, especially adults, and children have to be taught its existence.

As children grow older and discover such amenities as hours-long telephone conversations and best friends, one way to enhance a camping trip is to permit each child to bring his or her best friend along. By this stage in their development children think parents are pretty boring people and siblings are wretched specimens of humanity. So the alternative to back-seat psychological warfare ("She's looking at me!") and campground brawls ("He put a leaf in my oatmeal!") is the presence of a best friend to give them something to do other than torment each other.

When you do take along a best friend, be certain to discuss your plans, menus, and equipment with the parents. Find out if the friend will eat anything other than graham crackers and pop.

Children need to be part of the whole camping experience, including food preparation and cleanup.

Elsewhere in this book you will find the recommendation that each child be given a blank book, not a diary, to be used as a journal of camping trips. Parents can add to this storehouse of memories by taking photographs of the trips. Children love to look at family albums and will continue looking at them after they are adults. Few things bring more hysterical laughter among children than photographs of themselves when they were just kids. (They start saying that when they're around ten or twelve.)

Some families become addicted to movie film, which is certainly fine, but you can't decide on a moment's notice to go look at the camping photographs. A good argument can be made for taking 35 millimeter slides and showing them on a projector after returning. The best or most memorable photos can be made into color prints for the family album. With the continual improvement in amateur photographic equipment, almost everyone can take at least adequate photographs, and the camping camera does not have to be a sensitive, expensive camera. On the other hand, some campers take along the sturdy underwater cameras that are built for rugged use and can be used no matter the weather conditions.

Like adults, children have selective memories, and their lasting impression of a camping trip might be some embarrassing incident; a horrid little boy who dangled worms in people's faces, or how crabby Daddy was the last morning. But with a good selection of photographs to

jog memories, the unhappy moments soon blend into the whole scene.

Children are basically conservatives, if not reactionaries, at heart. If they had a great time at a campgrund last summer, they not only won't mind returning to it, they're likely to insist that you do. No matter how rosy the picture you paint for them of a new camping area, they approach it with the same apprehension they do moving to a new neighborhood. Parents simply have to live with this conservatism and continually introduce them to new adventures, new geography, and different types of camping trips.

The irony of this conservatism is that while they are afraid of a new piece of geography, they will immediately scamper up a leaning dead tree with total abandon. They help keep camping a lively form of recreation.

Much has been said and written about our obligations to the wilderness on behalf of our children and grandchildren, especially relating to wilderness preservation and expansion of lands set aside for the enjoyment of the people here now and those to come. You can tell children things like this, but since their idea of the future usually doesn't extend beyond tonight's marshmallow roast or tomorrow's sugar-coated cereal with powdered milk, don't expect them to be awed by the responsibility we feel for the outdoors. But they'll understand at about the age they decide parents are decent folk, flaws and all. And they'll appreciate your efforts.

CHAPTER 2

ESPECIALLY FOR FIRST-TIMERS

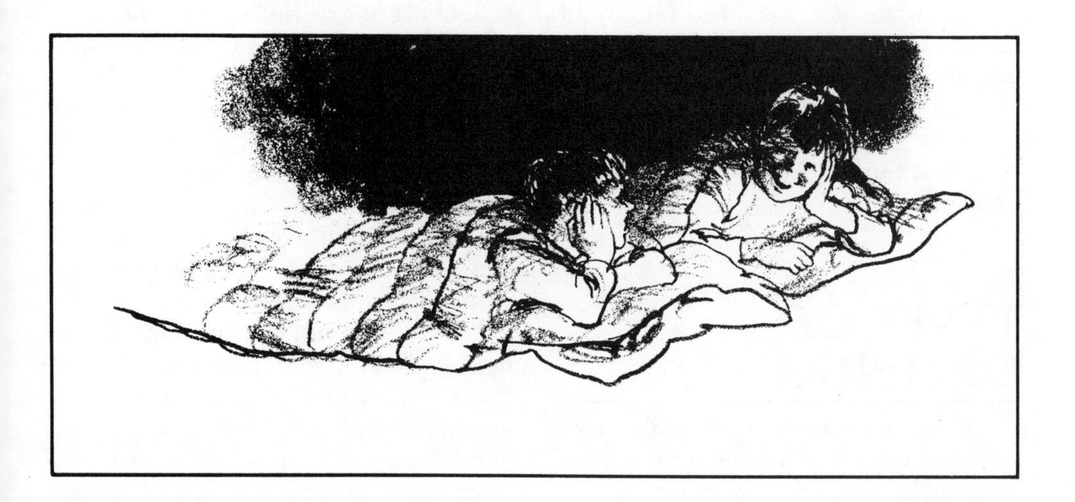

It is essential that a first camping outing for both children and adults be as pleasant as possible. If you plan an introductory outing and the weekend weather report indicates stormy or cold weather, it is best to postpone the trip. Later, as everyone becomes more experienced, camping in inclement weather often becomes part of the excitement of camping; it gives one a sense of accomplishment, and if you are properly outfitted, the discomforts are few.

Many campers, and books on the subject, make this relatively simple form of recreation seem, to the novice, only less complicated than launching a space shuttle. This is due in part to the wide variety of outdoor products on the market, and the many choices we have of places to go camping.

As with all forms of recreation, you can invest enormous sums of money in camping equipment, or you can shop carefully and have the best equipment at a more modest investment. Outdoor recreation has become very specialized, and it is possible, but not necessary, to feel compelled to buy three or four kinds of stoves, tents, and sleeping bags. With only a few exceptions, such as camping at extreme temperatures, you can buy a stove, tent, or sleeping bag that will be sufficient for all your camping needs for years and years.

Consequently, the emphasis in basic equipment is on quality so that most of your investments will be one-time expenditures. Under normal conditions, your camping equipment should last a lifetime with only occasional repairs. It isn't unusual for this basic equipment to last not only for years, but for generations so that it is handed down from parents to children like family heirlooms.

Few purchases of equipment are as versatile as those intended primarily for camping. Stoves and lanterns are a blessing in the case of power failures, as well as backyard cookouts, automobile breakdowns, and the like. Tents and sleeping bags are useful in slumber parties, and many families use the sleeping bags as comforters to keep winter heating bills down. The uses of outdoor equipment are limited only by the owner's imagination.

TYPES OF CAMPING

Camping comes in all manner of sizes and shapes, from overnights in your own backyard to several weeks in the wilderness with your home, kitchen, and bedroom on your back. If you are a first-timer who knows no more about camping than most of us do about the dead civilizations of Sumatra, here is a general breakdown of the forms from which to choose.

Backpacking: As stated earlier, this is the most rigorous form of camping because you carry your tent, food, cooking equipment, clothing, and various tools and medical supplies on your back. It has many advantages, primary among them the ability to go far from the crowds into wilderness areas. In addition to the equipment described throughout this book, you will need to add good boots and a pack to your equipment list.

Bicycle camping has become more and more popular during the past few years as people become more aware of the energy problem and equally interested in physical fitness. For this kind of traveling and camping, lightweight and compact equipment is as essential as it is for backpacking.

A similar form of camping, which isn't quite so concerned with weight but requires compact equipment, is motorcycle camping. Although motorcycles still suffer from the Hell's Angels image, it isn't really fair because the vast majority of motorcycle owners are dedicated family people who are very environmentally aware. The typical motorcycle club outing is an example of quiet, considerate campers who keep the campgrounds cleaner than the average car or RV camper, and you won't hear their motors roaring after dark.

Car camping: Car camping is the most popular form of camping because you carry everything you need in the family car and camp within sight of the car. This gives you more versatility than camping in any other vehicle because you can camp in the most elaborate campground with all the modern conveniences, or find an isolated spot along a country lane or toward the end of a logging road.

The bulk of information in this book is aimed at this type of camping, with some variations. Most car camping is done in designated campgrounds, ranging from national parks and national forests to other governmental or privately owned and supervised campgrounds. Many of these have resident rangers or managers.

Assuming that many readers will want to work their way into the more rugged outdoor sports, such as backpacking, we have placed a great deal of emphasis on lightweight and compact equipment. Even if you don't care to try the Appalachian or Pacific Crest trails, this equipment will serve you well for car camping because it occupies little space and won't weigh down the family car.

Even if you are traveling in a large RV, you will still appreciate sleeping bags that compress easily into stuff bags or may be left flat on the bunks that are folded away during the day. The same storage concerns apply at home, especially when equipment must be stored for months at a time.

Recreational vehicle camping: Abbreviated to RVs by nearly everyone, recreational vehicles can cost as much as a modest family home, or you can spend less than $1,000 on one. Along with the new lightweight and sturdy materials developed since World War II, RVs are an integral part of the camping boom in recent years. There was a time when RV owners and backpackers were no more friendly then Yankees and Rebels in 1865, but the lines between these two forms of camping are becoming more and more blurred, and it isn't unusual to find RV owners who are also backpackers, river runners, and die-hard conservationists. One of the major complaints against RVs is the amount of fuel they consume, but the gasoline mileage for some of the major brands has been gradually improved until now some models get better mileage than many family sedans.

TYPES OF RVS

Motorhomes: They are often called "land yachts" or "land cruisers," because they are at the top of the RV line, both in cost and in accessories. To most RVers, they represent the ultimate vehicle. Virtually every convenience of home can be included in a motorhome—all the kitchen gadgets, easy chairs, sound centers, queen-sized beds—and at night they can be divided into as many as three compartments for sleeping.

They can range up to 35 feet or more, although many are a stubby 18 feet. They are a great answer if you prefer the space of a trailer but don't like to tow vehicles. Also, they are much easier to back up and park than a trailer. They can be driven on all types of roads, and most have enough engine muscle to manage winding mountain roads.

While the majority of motorhomes are built on chassis from the top three or four truck manufacturers, a few are built from the ground up.

There are a few basic similarities, such as in overall construction. Most use a steel cage to frame and strengthen the walls and roof. You have a choice of steel, aluminum, or fiberglass on the outer walls, and you should have some kind of insulation between the outer skin and the inner walls of paneling for temperature control noise-deadening value.

The floor plans are different not only among various manufacturers, but also under the same brand. For sleeping privacy, most are divided by sliding doors into at least two compartments, and some offer three. Usually, but not always, there is a large sleeping area in the rear that doubles during the day as a set of facing gaucho seats with a removable table between. Others sacrifice storage space overhead for two rows of bunks across the rear where four children can sleep in comfort.

Most models have a couch or cafeteria-style table in the front compartment that folds out into a twin bed at night, and another bed can be hung from the top of the driver's compartment and clamped up against the ceiling and out of the way during the day.

There are rear-bath models and side-bath models, each with an equal number of advantages and disadvantages, depending on one's personal preferences.

Some have the kitchen area flush against the wall while more expensive models will have an L-shaped kitchen to separate it and to cut down on traffic through the food preparation area.

About the only place with no variation is in the driver's compartment, where plush, adjustable seats with all the instruments in a convenient location are more or less standard. One popular option is a double-sized chair on the passenger side where an adult and child can sit comfortably. Privacy curtains separate the driver's area, creating an extra sleeping compartment, plus shielding the driver from inside lights when on the road at night.

All motorhomes have at least one full-length closet, and storage bins are found throughout the vehicle, often along the sides on the outside, for skis or items that won't be used often such as spare parts and tools.

A motorhome's size can be a disadvantage when you decide to go exploring after finding a camping site. Many owners get around this by hauling along bicycles or trail bikes or towing a small vehicle behind. The main thing to remember here is that small cars with automatic transmissions should not be towed with the power wheels turning; a small trailer arrangement that holds the power wheels off the pavement is needed. Towing stick-shift cars in neutral causes no problems.

Pickup campers: Pickup campers have been a mainstay of the RV market for years, so much so that it is almost taken for granted that when someone buys a pickup, he or she eventually will buy some kind of "lid" for it, whether it be a complete camper with kitchen, bathroom, and beds, or a simple box that doubles as a cover for hauling building materials and as a basic shelter for weekend trips.

Four-wheel drive vehicles are increasing in popularity to explore the back country, so much so that many four-wheels come equipped with this basic camper already built on them. Also, a number of campers are being built specifically for the small imported pickups as well as the tiny ones made by domestic manufacturers.

Most campers fit right into the bed of a pickup and can be removed quite easily. Many owners have special racks built in their garages so they can back the truck in, remove the camper's bolts or clamps, raise it slightly, and drive it out.

There is a very fine line between a "lid" or a cover as the industry calls it, and the regular camper, and it probably isn't all that important. Many do-it-yourselfers buy a bare shell and add their own equipment as they build in bunks, sinks, stoves, and lighting. Most covers, however, are no higher than the roof of the pickup cab (in order to reduce drag) which makes them too small to stand in, but just right for sitting, eating, and sleeping.

The more elaborate campers, which slide in and are tied down, often project over the cab of the pickup, and this projection holds a double bed. The camper will have a galley-sized kitchen with a two- or three-burner LP gas stove, usually a cafeteria-style table that folds into another bed, plus a toilet and shower. The latter utilities are usually found only in the longer campers, the ones that project backward from the truck bed.

These models are a good compromise between a trailer and a mini-motorhome. They give you everything mini-motorhomes have (except a passage directly from the driving compartment to the camper) plus the option of taking the camper off the pickup and using the truck for other purposes. Some of the largest campers offer an amount of space equal to that of a mini-motorhome and more space than in the increasingly popular van conversions.

Another benefit to a slide-in camper is that, should the occasion call for it, it can be jacked up at a campsite and left while you drive elsewhere.

Since both the increased weight and wind resistance increase your fuel consumption, many manufacturers have found a market for models that expand, or telescope, up and down. They are raised and lowered by either spring-arm lifter systems, hydraulic jacks, or electric motors, then lowered to rooftop level when under way. The advantage of this system is fuel economy; the disadvantage is that you sacrifice some storage and over-the-cab sleeping space.

The average slide-in camper is built on the same principle as the travel trailer-wood stud frames, and skins of either plastic or aluminum. Interior walls are usually wood or plastic paneling, with fiberglass or plastic foam for insulation between the walls.

Before you buy a camper or cover, be sure your pickup is built for the increased load. Dealers have "camper special" pickups for sale that have the stronger suspension, heavy-duty cooling system, bigger wheels and tires, and rack for the auxiliary battery that you need.

Mini-motorhomes: This type of RV is a compromise between the most expensive and the least expensive of the RVs. Until it was developed and made popular, there was quite a gap between the fleet-sized—and expensive—motorhome and the pickup camper. A mini-motorhome costs a bit more than a pickup camper, but less than a motorhome.

The primary difference between a mini and a regular, or standard, motorhome is the kind of chassis each is built on. A mini is built on a van or pickup chassis, and a motorhome is built on a modified truck chassis. This brings the price up considerably before anything is installed.

One major advantage over a pickup camper is that in a mini you can walk directly back from the driving compartment. Another advantage over other RVs is that the front doors are retained.

Minis range from 17 to 25 feet, sometimes more, and they are built lower to the ground than standard motorhomes. They feel much like a car feels from the vantage point of the driver's seat. They offer automatic transmission, power steering, cruise control, power brakes, and virtually every other option available in full-sized motorhomes.

Minis are approximately the same width as conventional motorhomes and travel trailers, about eight feet wide, and the gross vehicle weight (GVW) ranges up to 11,500 pounds. Those with the biggest capacities have dual rear wheels.

More and more manufacturers are using steel framing to enforce walls of fiberglass or aluminum, so construction is similar to that of full-sized motorhomes.

The interiors are often elegant, and always functional. You have a wide variety of floor plans to choose from, and each manufacturer often has five or six basic plans. In all models, the overhang section above the cab is a bed, but after that, you have a choice. Up to eight persons can sleep in a mini-motorhome.

The galley, bathroom, and living areas are as roomy as most motorhomes, and you have a choice of appliance sizes. All have a combination 12-volt to 115-volt electrical system with a heavy-duty alternator and a deep-cycle battery. If you want to install a power plant (generator) on your mini, it is possible to put one on the rear bumper, either as a removable or a permanent fixture.

It has either a demand or a pressurized water system that can be hooked to a city water supply. It has the water, sewage, and wastewater holding tanks, and everything else of this nature available on the big-brother motorhomes.

Minis give you superior gas mileage compared with motorhomes, and a definite break in price. On the other hand, you may find the ride a bit choppy since you're sitting directly over the wheels on a shorter wheelbase. You don't get the storage space of a motorhome, and you may find that luxury items, such as a blender and a microwave oven, are options rather than standard equipment.

Travel trailers: Year after year, this type of RV outsells all others and has formed the basis of the entire RV industry. In fact, the industry started with travel trailers soon after World War II, and their popularity hasn't diminished in the face of competition from other types.

The major reason for their popularity is that they can be towed behind your car or pickup to a campsite, then left while you explore the area or run errands.

You will probably find more variety within this type than in any other you investigate—from the simple, straightforward kitchen-and-bed models to the downright elegant ones that almost resemble a resort condominium inside. Some, such as the fifth-wheel trailers, even have spiral staircases.

Travel trailers come in various lengths, from the stubby twelve-footers that look almost as tall as they are long to the "park" models which go up to forty feet. The latter are not designed for frequent towing, however, as are their shorter relatives.

Each model has its own specifications for a towing vehicle, and you can match a small car to one as easily as a four-wheel-drive vehicle. However, you must make certain your tow vehicle is within the required limits, not only for safety but also to prevent burning out your transmission or engine.

Weight is the major difference between the types. The most expensive tend to be the lightest—those made of aircraft-aluminum skin. They usually hold together better over a longer period of time, while those made of wood-frame construction have a tendency to work screws and glue loose.

Camping trailers: Camping trailers are the most economical type of RVs, in terms of both initial cost and operating expenses. They can be towed behind nearly every type of car or truck because of their light weight and the fact that their low profile while under way creates virtually no wind resistance.

As most models have canvas sides, camping trailers sometimes are referred to as "tent trailers," which isn't a bad definition at all. Their simplicity gives them more of a camping-out feeling than any other type of RV.

You will find that travel trailers offer every luxury of the fleet-sized motorhomes. The only disadvantage, depending on which state you're traveling in, is that passengers often are unable to ride in the trailers while they are under way.

Other than that, the sky and the thickness of your bank account are the limits on amenities. You will get a bathroom, kitchen, sleeping quarters, and both water and waste disposal systems. You will have a refrigerator and LP gas range, cupboards, and generous storage space. Options will include TV set, microwave range, air-conditioning unit and, of course, generator.

This doesn't mean they are simply tents on wheels. Most models have a two- or three-burner gas stove, a sink with a water supply and city water hookup, cafeteria-style table, an icebox or small refrigerator, and lots of storage space. Some of the luxury models even have portable toilets and shower stalls! Some can sleep up to eight persons using fold-out wing beds at each end plus the convertible table and seats. Some come equipped with batteries to power 12-volt lighting, and nearly all have provisions for 110-volt hookups at modern campgrounds.

Nearly all camping trailers will fold down for travel lower than the tow vehicle, so low in fact that you may not need side mirrors for rear visibility.

The trailers are raised and lowered by two basic methods: one system uses a friend's garage. When they are erected and the bed wings folded out (some even fold out on the sides, too), you will see that their size almost doubles. There is plenty of headroom for average-sized people, and a family of four can easily sit inside one while meals are being cooked—or just to get out of heavy weather.

spring-arm mechanism which you guide up and down, and the other is a crank system—the more popular of the two.

Don't be fooled by their apparent small size when you see one being towed down the interstate or stored away in a

Some families just getting involved in the RV way of life buy a camping trailer first, then work their way up through the various types of RVs. Others, however, wouldn't consider owning any other kind of RV and point to the advantages of storage, ease of towing, and a low inital cost which ranges from less than $1,000 up to around $5,000.

TRAILER CLASSES

Manufacturers have placed trailers in four classes by weight. The buyer is able to determine from these classes what kind of tow vehicle is needed for each type of trailer.

Class I—Light duty includes the tent trailers as well as small trailers for snowmobiles and trail bikes. The gross trailer weight is under 2,000 pounds, and the tongue weight is up to 200 pounds.

Class II—Medium duty trailers are usually those with single axles in the small- to medium-size range. Gross trailer weight is 2,000 to 3,500 pounds, and the tongue weight is 10 to 15 percent of the gross trailer weight.

Class III—Heavy duty includes the dual-axle models, which have a gross weight of 3,500 to 6,000 pounds and a tongue weight of 10 to 15 percent of the gross trailer weight.

Class IV—Extra heavy duty includes the fifth-wheel models and those with a gross trailer weight of 6,000 or more pounds and 10 to 15 percent of the trailer weight on the tongue. Nearly all trailers in this class require a pickup for towing, and of course all fifth-wheel trailers must have a pickup for towing.

THE RIGHT HITCH

Bumper-mounted hitches are not recommended for towing any type of trailer. Always use either hitches mounted to the underbody of your tow vehicle or, in the case of heavy loads, hitches mounted directly to the frame.

For trailers weighing more than a ton, load-equalizing hitches should be used. These are designed to distribute the weight between the axles of both the trailer and tow vehicle.

For pulling the light Class I trailers, use a weight-carrying ball hitch that takes the trailer's full tongue weight. These are not to be mounted on the bumper, though, but always on the frame or underbody of the tow vehicle.

Safety chains should always be installed. The chains should cross under the trailer tongue to prevent it from dropping to the road in case of hitch failure. A breakaway switch should be used for a trailer that has brakes. The breakaway switch automatically applies the trailer's brakes if the hitch fails. These switches are available in both electrical and mechanical models.

Towing tips: Keep the trailer's center of gravity low for good stability. It's the same as ballast in a boat.

Stow heavy articles such as canned goods, tools, and books as near to the floor as possible. Stow lightweight items such as clothes and linens on a higher level. Never store dangerous items overhead.

Distribute the weight of the load evenly to keep the trailer from tilting or leaning and to keep from putting too much strain on the hitch and the towing vehicle. Balance the load from side to side, too.

Carry only essential items and keep the load down. Be sure all doors and drawers are secured. Leave nothing loose.

Van conversions: Vans have become the "in" vehicle during the past two or three years, and owners, particularly young people, often turn them into pieces of moving artwork. You see them daily with imaginative paintings on the sides and big stretches of glass, often one-way or privacy glass. Look inside them and you'll see furnishings fit for a king: deep shag carpeting, plush seats, small stoves, refrigerators, TV sets, and stereos.

Of course, not all converted vans are so richly decorated. Most are rather basic with everything other RVs offer—only in less space. However, vans have an advantage over most other RVs: they can be used daily for going to work, running errands, and so forth. Some call them "super stations," because their size isn't a problem when you are looking for a parallel parking space, yet they are large enough for RV vacations.

Since there isn't enough headroom for comfortable conditions while using them as an RV, many owners put a higher roof first on the priority list. There are two choices: a permanent extender can be put on to give you headroom, or you can have an expandable roof that is raised and lowered while you're camping. The permanently expanded roof has the disadvantage of creating more wind drag, thus decreasing gas mileage, but it gives you permanent storage space and doesn't require passengers to crawl about while under way. The expandable roof creates virtually no additional drag and doesn't make the van a broader target for side winds.

Fifth-wheel trailers: The newest member of the RV caravan is the fifth-wheel trailer, first introduced on the market in the early 1970s. It was an instant success, and hardly an overnight campground is without one in residence.

Essentially, fifth-wheelers increase the living quarters by about seven feet without increasing the wheelbase. And they give you a split-level home.

The major differences between them and standard travel trailers are first, the type of hitch required, and second, the extra level above the hitch that is the bedroom. The extra seven feet are above the pickup or four-wheeler that is used as the tow vehicle. Thus a 29-footer has virtually 22 feet of space behind the tow vehicle.

The hitch is designed to be anchored over or just in front of the rear axle of the pickup. It sits in the bed of the pickup and is attached to the frame for absolute security. If you've noticed a semi trailer turning, you'll get an idea of how short a turn your fifth-wheeler can make, because the pickup can pivot at a right angle from the trailer.

Fifth-wheelers also are less prone to swaying and jackknifing because of the stability of the hitch arrangement and in part because of the weight taken by the pickup.

Hitching up is easy. The trailer has a kingpin that hangs down and couples to the hitch in the pickup bed. You have excellent visibility through the pickup window since the hitch is almost at eye level—unlike standard travel trailers that almost always require a second person standing outside giving you backing instructions. And manufacturers, pointing to the safety record of similar hitches on interstate trucks, claim this kind of hitch is the safest.

Fifth-wheelers are usually built of the same type of material as the standard travel trailers. They should be built sturdily, as their increased size makes them more likely to be used as permanent residences by owners. In addition to the living space available, many manufacturers have added such amenities as sliding glass patio doors and glass alcoves that crank outward to form bay windows, with seats, so guests can sit without having to move their feet every time someone walks the length of the trailer.

Parked, and with everything set for a long stay, there is little difference between a fifth-wheeler and the most luxurious permanent mobile home. And you have the advantage of being totally mobile. Crank in the alcove, hitch up to the pickup, and you're off.

Fifth-wheelers will sleep up to eight in comfort. But many of these RVs are owned by couples on the move, who opt for sleeping accommodations for four and use the rest of the space for their other needs, such as a bigger kitchen or a larger living room.

The typical fifth-wheeler has tandem axles for extra support and stability while traveling, your choice of a double or twin beds in the master bedroom upstairs, a bathroom with both tub and shower, a totally equipped kitchen with double sink and options including a blender and a microwave oven, a TV cabinet and antenna option, gauchos that fold out into another double bed, a dressing room just off the master suite, and lots of storage space for both food and clothing.

How much current is drawn? When you consider size in the auxiliary batteries and generators you need to buy for your RV, it is best to first understand the amount of current each electrical fixture in your RV draws. Usually the batteries and generators that come as standard equipment are sufficient for normal use, but many RV owners use more than the normal load. A large group, a pack of small children, or unusual circumstances can cause the system to be overloaded. Your battery might go dead in the middle of the night, cutting off your furnace, or the generator might rebel and refuse to operate until you cut back on the demands made upon it.

Following is a list of wattage draws on the 12-volt system:

■ Dinette light	3
■ Hall light	3
■ Reading light	1½
■ Kitchen light	1½
■ Each compartment light	½
■ Exterior light	1½
■ Step light	½
■ Bathroom light	6
■ Range hood and vent fan	5½
■ Furnace blower	6
■ Water pump	6
■ Monitor panel	2½
■ Vent fans	3

Wattage for 110-volt equipment:

■ Air conditioner	1,100– 2,200
■ Electric heater	1,100– 1,540

CAMPER'S SECRET

If you plan to visit at least half a dozen federal camping areas in one year, particularly national parks, the Golden Eagle Passport is a good buy for you and your family. It costs ten dollars a year and covers entrance fees for everyone in a single vehicle. It's not good for camping fees, however.

The Golden Age Passport picks up where the Golden Eagle leaves off. This is for persons sixty-two and older to replace the Golden Eagle. It has an added advantage as it also gives you a 50 percent discount on overnight camping in all federal camping areas.

Both passes are available at National Park and National Forest Service offices throughout the country.

■ Stereo	220
■ TV	330
■ Microwave oven	1,430

When choosing a generator for your RV, keep these figures in mind to be sure the generator is strong enough for the load. They are rated in the amount of watts produced:

- ■ 2,500—Powers one air conditioner and all lighting
- ■ 4,000—Powers one air conditioner, all lighting and one appliance such as a toaster or electric frying pan, or two air conditioners and nothing else
- ■ 5,000—Powers two air conditioners, all lighting, and one appliance
- ■ 6,500-7,500—Powers two air conditioners, all lighting, and two appliances

Weighing in: It is very important that you keep your RV loaded within the requirements established by the manufacturer. In overloaded RVs fuel consumption increases, and structural damage can occur. The RV becomes a safety hazard as well.

Of course the simplest way to determine weight is to go to a trucking company and have your RV weighed. But if that is inconvenient, you can figure your weight, using the following list of sample weights as a guide:

■ Tableware for eight	5 pounds
■ Electric coffee pot	4
■ Portable toilet	18
■ Set of tools	10
■ 12-inch TV	20
■ Toaster	4
■ Sleeping bag (not down)	5
■ Standard tent	50-60
■ Portable lantern	6
■ Air mattress	6

Most states do not require you to stop in at highway weigh stations.

Sample weights: Here are some sample weights of RV fixtures (and people) to help you estimate the *gross vehicle weight*, which is the RV's weight plus that of all it is carrying. Your manufacturer will provide you with the optimum GVW figure for your particular RV.

Four passengers (two adults, two children)—500 pounds
Clothing for four—100 pounds
Food for normal circumstances—200 pounds
Auxiliary (20-gallon) water tank (filled to capacity)—200 pounds
Extra (20-gallon) fuel tank (filled to capacity)—160 pounds
Auxiliary battery—49 to 140 pounds, depending on amperage
Auxiliary deep-cycle battery—56 pounds
Towing hitch—40 to 125 pounds, depending on class of trailer
Trail bike—150 to 250 pounds
Bunk ladder—6 pounds

The energy problem: You've undoubtedly heard the complaints about how RVs are fuel hogs, and there is no escaping the fact that your car uses more fuel towing a trailer, and a motorhome uses more fuel than does a car.

What you've heard isn't the whole story. The Recreational Vehicle Industry Association (RVIA) has shown that your family on vacation in an RV actually consumes less energy than it would at home. In the first place, your family won't be using the electrical energy at home, nor will you be utilizing the water and sewage system. By comparing your gasoline consumption during the same length of time driving to and from work, running errands, and so forth, you will be surprised how little difference there is between what you consume on vacation and what you consume under everyday conditions.

Another factor to consider is that many family vacations in RVs are taken reasonably close to home, and the RV is often used in lieu of a motel or resort room. Many vacations consist of driving to one spot and staying there for a week or more, then returning home.

Forget about the energy problem a moment and concentrate on costs in general. Statistics from the RV industry show that the typical family of four will save more than $1,000 a year on travel by using an RV. Lodging alone is a big factor in saving: most commercial campgrounds charge no more than $7.50 a night. Compare this with $30 a night for motel charges during the tourist season —often more—and your vacation is already considerably less expensive.

Most families using an RV on vacation prepare for the majority of their meals in the RV, treating themselves to a breakfast and an occasional dinner out. It is simply the difference between eating at home and eating in a restaurant. Many families enjoy getting up early and preparing a pot of coffee and some orange juice, driving an hour or two, then having brunch in a restaurant. Such treats are part of the joy of traveling; you can sip your coffee and juice while on the road.

With all vacation costs taken into consideration, it is a rare family that can't take a trip in an RV and save money, no matter how far they travel. All reputable commercial campgrounds have free recreational facilities such as swimming pools, horseshoe pits, tennis courts, and playground equipment. Often the campgrounds are near skiing, either downhill or cross-country, for winter guests.

Other RVers find places away from established campgrounds to park and unload their boats and water skis or fishing equipment. They can have a family outing on a lake or river that ordinarily would be possible only if they owned a summer cabin.

SANITARY SYSTEMS

There are five major types:

Permanent freshwater flush toilet—This is the most common and the easiest to install and use. It is virtually odor free and uses no electricity. On the minus side, it requires a holding tank with chemicals to be added, an extra supply of water for flushing, and can freeze in skiing weather.

Electric recirculating—This operates on a small amount of recirculated water, needs emptying only every three to five days, and doesn't require a special holding tank. However, it requires 12-volt power, is one of the most expensive types, requires special chemicals after each dumping, and often gives off a strong chemical odor.

Incineration—This is often an option for the more expensive RVs. It doesn't require a holding tank, uses no water for flushing, uses no chemicals, won't freeze in the winter, and can handle sanitary napkins and disposable diapers. But if the battery is low it won't function properly, and it needs at least fifteen minutes between uses to operate properly—a problem with large groups and at certain times of the day when use is heavy.

Waste destruction system—In this type, waste is burned during travel or when the main engine is running while standing. When used constantly and properly, the holding tank won't have to be emptied. And since the waste is burned, there is no danger of bacteria causing contamination. The disadvantages include the need for a holding tank, the need of liquids to hide odors, and the fact that if the system cools down while in use, it won't work.

Portable toilet—This is the most basic type of toilet. It is the least expensive and can be used not only in RVs but in cottages, at home, or in the sickroom. Many people feel this toilet far too primitive, however, since the waste must be carried to a dumping station and the whole system cleaned before reuse.

PLAYING IT SAFE

Since an RV is a combination home and vehicle, the potential for accidents and injuries definitely exists. But by following basic safety rules, you can make your vacations and weekend trips as much fun as you hoped them to be.

The first safety tip comes from the RV industry, and that is to be certain the RV you buy or rent has the Recreational Vehicle Industry Association (RVIA) standards seal. This seal informs you that the RV has been built according to the basic standards of the industry as a whole.

While safe operation of vehicles rests with the driver, there are basic rules for manufacturers to follow which are set down by the American National Standards Institute. The standard for RVs, called Standard A119.2, contains more than 500 requirements for plumbing, heating, and electrical systems.

Here is a brief rundown of the standards:

Fire and life safety—The vehicle must come under the minimum flame-spread ratings for interior walls and ceilings. A vehicle with fuel-burning appliances or an internal combustion engine must have a fire extinguisher with the proper charge. Also regulated is the location of the gasoline filler spout so it is away from potential ignition sources such as the exhaust pipe.

Plumbing systems—This standard prevents the use of substandard materials that would cause contaminated water. The drainage system could be hazardous if not installed properly, and the specifications prevent sewer gases from accumulating. Standard-sized pipes to insure adequate flow of water are required, along with sanitizing instructions, an adequate drainage system, adequate venting, and proper drain outlets. A final testing of the complete system is required.

LP gas systems—This standard involves safety-relief valves for excessive pressure, controls the location of the LP gas container, provides specifications for all pipe sizes, covered tubing, and appliances, and tests for leakage. Sufficient clearance is required to prevent ignition of adjacent surfaces.

Electrical systems—Essentially, the same requirements for an RV that apply to a home are used.

In addition to the RVIA standards, there are the normal common sense safety precautions one should take while traveling in an RV. These apply especially to the self-contained mini-motorhomes, fleet-sized vehicles, and van conversions.

Since most RVs aren't designed for

drag racing, you must always allow extra space between yourself and other vehicles. If you are following one, remember that your extra weight will lessen the vehicle's stopping power. It is not what you are accustomed to in the family car.

The same applies to passing. It will take you longer to get around that car ahead. You are driving a vehicle of greater length, so allow plenty of space before you pull in ahead. Learn to "read" your right-side rear-view mirror at a glance. And always watch that potential blind spot right beside you. Unless you have a convex mirror, or a small stick-on convex one attached to the bigger mirror, a car can pull up beside you and you'll never see it.

The first time you drive or pull an RV, you will understandably be a bit nervous. Test it on familiar roads before taking off on the interstate and by all means practice backing up. In spite of what you may think, you will be very lucky indeed to find only campgrounds with pull through campsites. Sooner or later you will find yourself in a spot where you have to back up the RV, and nothing is more satisfying than being able to do so without bumping into a picnic table, knocking over the water pipe and starting a gusher, or having a crowd gather to watch the fun.

WHICH IS BEST FOR YOU?

Generally the type of camping you set out to do will depend on the equipment you own. But if you are thinking of renting or buying special camping vehicles, consider the points listed.

Recreational Vehicles

Advantages

All the comforts of home
Don't have to pack and unpack
Can lock up when going on a hike
No worries about space
Can cook anything you want

Disadvantages

Most expensive form of camping
Isolated from the outdoors
Limited to larger campgrounds
Don't learn about the outdoors
Can't pack and go in a few minutes

Car Camping

Advantages

Great mobility in choosing site
Gives family true outdoor experience
Teaches children self-sufficiency
Easier to find isolated camping areas

Disadvantages

Requires more advance planning
Requires more compact equipment
Requires more outdoor knowledge
Car is crowded en route

While we're on the subject, learn to back up using only the mirrors. It is easier once you learn, and the mark of a professional.

If your RV or tow vehicle is equipped with a speed control device, by all means use it. Most RV owners plan to drive a bit below the 55-mph speed limit (some states require it) for safety and consideration. And keep to the right. You present an imposing target if you're driving in the fast lane and those following you have a restricted view ahead.

If you are driving on a two-lane mountainous road with frequent turnouts, keep an eye on the traffic piling up behind you and turn out to let it pass whenever possible. Many states require you to use a turnout if five or more vehicles are following you.

You should always compensate for sudden wind currents, such as those encountered when passing a large truck or emerging from a tunnel or a viaduct. It is best to have a firm grip on the wheel in these cases, and increase your speed slightly rather than following your first instinct of slowing. Treat wind gusts much the same as driving on ice—keep power on.

A WORD ABOUT PETS

Few subjects get campers' backs up more quickly than pets, yet the world seems divided almost evenly between those who wouldn't leave home without the family pet and those who wouldn't think of taking a pet camping. It is clearly a situation to which there is no solution that will satisfy every camper. The National Park Service has for a number of years banned pets from its campgrounds and trails, but they are welcome at nearly all other campgrounds.

The main issue, obviously, is the same with all outdoor etiquette; your neighbors should hardly be aware you are present, which is impossible if your dog is a barker or loves to cruise the garbage cans. The general rule of thumb is that if you've had complaints about your dog at home, then leave it there when going camping.

On the other hand, there is no more comfortable companion to have on a camping trip than a well-trained and lively dog. Parents rest easier when their youngsters leave their sight with the dog accompanying them. Dogs can keep many of the camping pests, such as food-stealing ground squirrels, away from camp and serve as the first line of defense against other creatures. Just hope, of course, that your dog knows enough about skunks and porcupines to bark at them from a safe distance.

Many outdoors people travel constantly with their dogs and make or buy packs for the animals so they can at least carry their own food on outings. A child who decides all adults and siblings are wretched people will work his or her way back into the flow of things if a dog is around for affection and as a companion with whom to file complaints.

The basic rules of most neighborhoods should apply to camping with a dog: it should be confined to the family's camping area, on a leash, and should not be permitted to roam through the entire campground, setting pet haters' teeth on edge.

Cats present another kind of problem for camping trips. Although they seldom cause many problems in the campground, they do have a tendency to wander off and get lost. City cats are notorious for having absolutely no sense of direction. Sometimes, out of fear of a dog or from sheer exuberance, they decide to climb a tree. Then they get stranded where nobody can retrieve them. Their independent nature and lack of innate wilderness know-how can make them more of a problem for owners than their dogs.

Since it is easier to leave cats for several days alone in a house with an ample food supply and a litter box handy, it is best to leave them home. If they could talk, they'd probably thank you for it.

CAMPER'S SECRET

Noise is one of the worst problems in busy campgrounds, and most supervised ones have hours posted when all radios, phonographs, and tape decks must be turned off. It is more of a problem in the less developed campgrounds, such as many in national forests. RVs with electrical generators that operate the air-conditioning system and other 110-volt appliances are one of the worst offenders. You should plan on turning it off no later than when darkness falls, and preferably more than an hour earlier since most campers associate the early evening hours with silence. If you're a tent camper near an RV with a generator, don't hesitate to ask the owners if they'd mind shutting it off. Most will comply.

PREPARING FOR THE NEW EXPERIENCE

It is very important that you and your family think of camping as perhaps the most natural form of recreation available. It is not a venture into the great unknown; on the contrary, it is simply doing what people have always done and always will. Millions throughout the world live all their lives doing what we call camping, many by choice. Camping is as natural as a backyard barbecue; only the location is different.

Granted, some people approach camping with vague fears, especially children who are reared in urban areas and know nothing more about the outdoor experience than what they learn in the local playgrounds surrounded by streets and buildings. In these cases, and with some apprehensive adults, you may have to ease into the sport gradually.

One method of overcoming this is to find a nearby campground and make one or more trips there for picnics and day hikes. Let the family become acquainted with the area and see how much campers are enjoying themselves.

Encourage them to talk to the campers and ask questions about their equipment, where they've camped, and what they do while camping. Stay until after dark so they can see the campers settling down for dinner and bed. This will remove much of the mystery of camping, and in many cases will make the trip home a bother because they will wish they were still there for the night. On these occasions it is easier to sell first-timers on the idea that food and drink taste better when brewed over a campfire or campstove, and that you sleep better outdoors.

In other words, you often have to sell camping to first-timers because it is in our nature to approach new experiences with misgivings. Children tend to be very negative about a new experience, whether it's an unfamiliar dish, a first meeting with a cousin, or trying a new form of recreation.

When planning a camping expedition, involve **everyone** totally in the trip preparations. Take the whole family to the outdoor store to look at equipment. Let them read this book and others on the subject. Involve everyone in the menu planning (but with the understanding that parents have the last word on nutrition, of course). If it isn't a total family effort, someone is likely to feel he or she is being dragged along.

Be democratic, too, once on the road. While parents may want to find a campsite tucked away from the crowds, children usually like to be around other children. Thus, parents may have to forsake total peace and quiet and isolation in favor of happy camping children. Many of the larger state parks and national forest campgrounds have common playground areas where children can strike up new friendships while playing team sports. If you choose your campground carefully, you will be able to combine your desire for privacy with the children's desire for action, since most large parks offer large campsites. Check first with the ranger or friends and see if campsites are crammed together or if they have a shield of plants or other obstructions between them.

For specific information about choosing a site, see PART IV, RESOURCES.

OUTDOOR GOOD MANNERS

It is difficult to improve upon the Boy Scouts of America's Outdoor Code, which follows:

"As an American I will do my best to:

"Be clean in my outdoor habits. I will treat the outdoors as a heritage to be improved for our greater enjoyment. I will keep my trash and garbage out of America's waters, fields, woods, and roadways.

"Be careful with fire. I will prevent wild fire. I will build my fire in a safe place, and be sure it is dead out before I leave.

"Be considerate in the outdoors. I will treat public and private property with respect. I will remember that use of the outdoors is a privilege I can lose by abuse.

"Be conservation-minded. I will learn how to practice good conservation of soil, waters, forests, minerals, grasslands, and wildlife; and I will urge others to do the same. I will use sportsmanlike methods in all my outdoor activities."

Children learn outdoor good manners from the example of their parents. If cutting dead trees is banned in your camping area, don't make an "exception" because you love big fires. Keep your campsite clean and in better condition than you found it.

A few specific items might be added to this admirable code of outdoor ethics, including:

Radios: They're handy for keeping up with the news of the world (unless you're easily depressed) and local weather reports. But it is virtually impossible to hear the wind in the trees and the birds singing if you or one of your neighbors has a radio going full blast. One small transistor with a fresh battery used for information rather than entertainment is all you really need on a camping trip.

Lanterns: Part of the pleasure of camping is sitting on a log or stump, or the ground, watching day end and night begin, watching the stars appear one by one and listening to the night sounds that are quite different from those during daylight hours. Then when it is totally dark and your eyes have adjusted, you can go to sleep in the soft darkness. Thus, lanterns have a way of interfering with your enjoyment and should be used only when necessary. Too many of the gasoline lanterns make a constant hissing noise that carries long distances and interferes with your neighbors' enjoyment of the evening.

Power toys: This includes scooters, motorbikes, motorcycles, chainsaws, generators, and the like. Campers appear to be divided into two distinct camps: those who love power toys and those who hate them. One child on a motor scooter can make life miserable for an entire campground, and if you are a silence lover, you may have no other recourse than leaving in search of a campground where these toys aren't popular or where they are not permitted. Many campgrounds have a 10 P.M. curfew on noisemakers, and you should honor that rule. Better yet, set your own curfew before darkness falls.

Vandals: Always report vandalism to the campground ranger or manager. Strangely, some parents strongly resent other campers implying that their child (or spouse or even themselves) is committing an act of vandalism, so it is best not to take matters into your own hands and try to deal with it directly. Vandalism is one of the major problems in the outdoors, as is its cousin, littering.

PART II
GETTING READY

CHAPTER 3

CAMPING EQUIPMENT— THE ESSENTIALS

It wasn't too many years ago that a camping trip took on the characteristics of a trek across the Oregon Trail in a covered wagon. The pioneers in family camping as a form of recreation had to make do with many standard household items that were manufactured with no concern for weight or bulk. Cooking equipment included cast-iron skillets, steel pots and pans, and second-best tableware. Conspicuously absent were devices such as nesting pots and pans or tableware that collapsed into small components for ease of carrying. Stoves small enough to fit into a coat pocket were rare.

Shelter was often a heavy cotton or canvas tent that was too heavy to carry more than a few feet at a time and was subject to mildew and leaks late at night. Clothing was bulky leather coats or heavy woolens, which got increasingly heavy the longer the rain fell. Rain gear was both stiff and heavy.

With today's ultralight insulated clothing and sleeping bags, and the new fabrics for garments and tents, these items occupy little space and weigh only a fraction of their earlier counterparts. Now it is possible for a family of four to go on an extended camping trip and carry everything they need for a week in a medium-sized car. Dehydrated and freeze-dried food has been improved to the point that a dinner for four takes no more space than a pair of rolled jeans and weighs less.

In order to keep your investments in equipment to a minimum, it is best to purchase the ultralight and most com-

THEN

NOW

pact equipment available. This will enable you to use the same equipment on all camping trips, whether you're in an RV or car camping, backpacking or bike touring.

Less is best. As an illustration of how revolutionary the development of down-filled products was, Eddie Bauer recently compared the weight of the outdoor gear he used prior to his invention of quilted down garments and sleeping bags in 1935.

THEN	NOW
One pair of Hudson's Bay blankets, 72" × 90", pinned into a bag 36" × 84"; comfortable to 10° F. without campfire; subzero with fire: 12 lbs.	Skyliner goose down-filled bag, 72" x 32". Comfortable to -30° F. without fire: 7 lbs., 12 oz.
Alpaca lined hooded parka good to 10° over wool shirt and wool underwear: 4 lbs.	Eddie Bauer Expedition Parka worn over medium-weight wool shirt and undershirt; comfortable to -40°: 2 lbs., 13 oz.
Malone mackinaw pants worn over wool underpants. Comfortable to 10° without fire: 4 lbs., 12 oz.	Cotton field pants worn over down underpants plus duafold two-layer underpants. Comfortable to -20°: 3 lbs.
Total: 20 lbs., 12 oz.	Total: 13 lbs., 12 oz.

Not shown in this example is the equally major savings in bulk. All the down-filled items can be compressed into small stuff bags, saving at least one-fourth of the space taken by the heavier woolen items at the left.

Most investments in camping equipment are one-time expenditures if you shop carefully. With proper care, tents and cooking equipment will last for decades. Shoes, boots, socks, and some other articles of clothing will eventually wear out, of course, but footwear can be repaired repeatedly, and insulated and good woolen clothing will last for years and years. Thus it is not uncommon to see dedicated campers wearing articles of clothing that have been washed and mended so many times that the clothing looks as though it were borrowed from a stage hobo.

Campers become loyal to their equipment, at times superstitiously so, and they do not worry about things looking new. Often a reverse snobbery results, and those campers who would never think of wearing a shirt or blouse with frayed cuffs and collars at home think nothing of wearing a tattered parka with ripstop nylon patches decorating it. These clothes become the on-trail equivalent of the now-classic tweed and corduroy coats with elbow or shoulder leather patches.

Owing to a combination of space-age technology and intense competition among manufacturers, plus rigid requirements by suppliers who deal directly with purchasers, outdoor gear is among the most carefully designed and

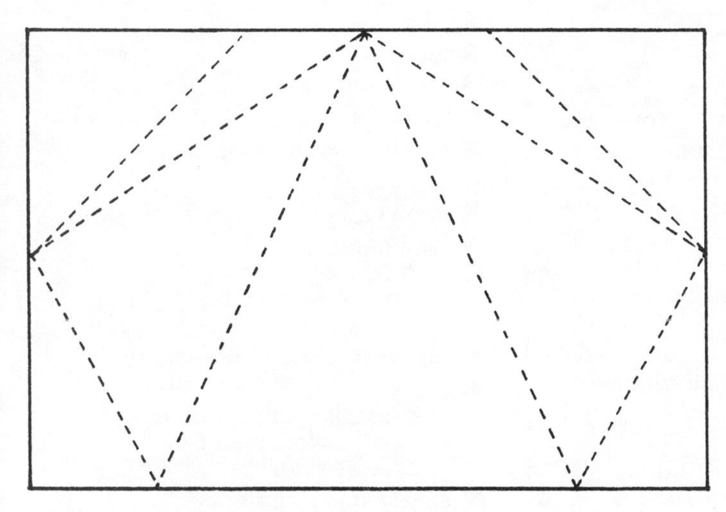

Eddie Bauer designed a simple tent made of a single piece of waterproof "balloon cloth" that could be used in a variety of shapes—half-pyramid, A-shape and lean-to. Such tents are still popular with many campers.

manufactured products on the market. It is built to withstand the widest extremes in temperature and moisture in the world, and the roughest use. Thus, when you buy a tent or a stove or a backpack from a reputable dealer, you can expect it to have a longer life than almost any other product you buy.

Basically, your needs for camping are centered on the essentials for sustaining life anywhere: food, clothing, and shelter. You can make these needs as basic or as complex as you wish, and with all the technology involved in manufacturing and processing these essentials, the choices become wider, making it difficult to draw up a list that will be satisfactory for every camper.

Following is a list of basic equipment needed for virtually every camp, whether it is a drive-in campground or one in the back country. The list is also suitable for any number of campers in a group, or even a solo camper.

A family of four or six can carry all the camping equipment and food in the family car, provided that equipment is made of ultralight, compact modern materials such as goose down and nylon.

CAMPER'S SECRET

Many professional guides wouldn't think of leaving on a trip without a pair of pliers hanging from their belt in a leather sheath. Pliers are indispensable for many camping chores. You can use them as pot holders, to move the grill around, to remove stubborn bottle caps, as tongs, and for a host of other cooking tasks. They can be used for repairing clothing and tents. They can double as a light hammer—for almost any task that calls for a wrench or hammer. The longer you carry a pair, the more uses you will find for this simple tool.

EQUIPMENT CHECK LIST

The following check list of camping equipment is broken down into the needs of each individual camper and the equipment needed by the group as a whole. Some overlap is inevitable.

For Each Individual

- sleeping bag
- insulated ground pad
- plate
- knife, fork, spoon
- cup
- water bottle
- watch
- suntan or sun-screen lotion
- toothbrush, toothpaste
- comb
- soap
- mirror
- towel and washcloth
- insect repellent
- hooded parka or rainshell
- cap or hat
- sweater (light wool or goose down)
- long pants
  - cotton for warm weather
  - wool or wool blend for cool weather
- rain jacket and pants (or poncho)
- shirts
  - cotton for warm weather
  - wool or wool blend for cool weather
- sturdy shoes and boots
  - boots for hiking
  - lightweight shoes for camp
- socks (lightweight and wool)

- underclothing
- gloves or mittens

For the Group

- tent(s)
- stove and fuel
- nesting cook set
- pot holder
- tableware
- can opener
- cook spoon
- food and seasonings
- salt, pepper, sugar
- beverage mixes
- spatula
- grate
- water bucket
- dish scrubber
- tarp or dining fly
- toilet paper
- first-aid kit
- flashlight with spare batteries and bulb
- camp lantern
- reading material, games, etc.
- backpack
- repair kit (needle, thread, etc.)
- rope (for guy lines, clothesline, etc.)
- small hand-operated winch
- lashing cord for packing
- automobile repair kit
- emergency provisions (see "The Ten Essentials" below)

Optional

- camera and film
- a few coins for phone calls
- protective lip cream or stick
- pajamas
- inflatable pillow

The Ten Essentials

- extra food
- extra clothing
- matches in waterproof container
- candles or chemical fire-starter
- first-aid kit
- maps in waterproof case
- compass
- flashlight with spare batteries and bulb
- pocketknife with sharpening stone
- sunglasses

The importance of these Ten Essentials cannot be overemphasized. You will find yourself needing them while enjoying the outdoors in all kinds of weather conditions, from the best to the worst. So make them as much a part of your camping experience as coffee or hot cocoa, whether you are camping in a crowd at a private or public campground or miles away from the nearest human being.

Extra food: You should always keep an extra day's supply of food for each person in your party that is not to be touched except in emergency situations. Some campers have been known to select something particularly untasty to be certain their will power isn't put to an impossible test. Obviously, candy bars, trail gorp, or other tempting items should not be in the emergency provisions, but lightweight items such as freeze-dried or dehydrated food should be in the package.

Extra clothing: You should always have dry and warm clothing to wear in case a sudden storm blows in or you take

CAMPER'S SECRET

Most campers always take more gear than they will need. One way to eliminate this unnecessary gear is to sort through all the equipment after the trip and see what you didn't need (the Ten Essentials excepted).

an unplanned dunking while fording a stream. Always keep warm, dry clothing in your pack, even if it means wearing dirty clothing an extra day or two. Be sure it is wrapped in waterproof plastic inside the pack.

Matches in waterproof container: Small waterproof containers for matches should be kept filled and tucked away and never used except in emergencies. The best match containers are made of steel with a rough striking surface on the sides or inside the lid. Again, do not use except in emergencies.

Candles or chemical fire-starter: Fire-starting items are almost as important as extra food and clothing. A few stubs of household candles scattered throughout the camping equipment, or one or two full-sized candles, are always welcome for starting fires with wet wood. Better yet, buy some chemical fire-starting pellets or strips of chemical fire-starter, and never use them for ordinary fire-starting. Neither type takes much room in a pack.

First-aid kit: This should include adhesive bandage strips for small cuts, gauze pads of various sizes, a full roll of adhesive tape, salt tablets, an analgesic, a needle for treating blisters, and a first-aid manual. It is also recommended that you carry a sheet of moleskin for covering blisters, a single-edge razor blade, water purification tablets, first-aid cream, a mild antiseptic such as Bactine, antihistamine tablets, and a snakebite kit. (See CHAPTER 10 for a complete first-aid check list.) Of course, you should carry any prescription drugs you must take regularly or in case of emergencies.

Maps: These include road maps and topographical maps of the offroad areas you are visiting. Most recreational stores stock the "topo" maps or their equivalent, such as national forest or national park maps. Store them in waterproof cases, such as clear plastic cases with zippers or snaps. Also, be certain you know how to read them. Instructions are in CHAPTER 9.

Compass: As with maps, a compass is of little value unless you know how to use it in your area and with the maps you buy. Learning this can be a form of camping activity for the whole family. (See CHAPTER 9.)

Flashlight with spare batteries and bulb: Few emergency items are more likely to be rendered useless by campers than flashlights. Some campers find it impossible to face bedtime without reading a chapter or two of a book. Always keep one set of batteries and a bulb sealed and tucked away.

Knife and sharpening tool: Some campers get carried away with knives and wear a bayonet-style belt knife or an enormous Swiss army knife with a score or more blades. The best belt knives have short blades, and the best pocketknives are based on the old Boy Scout knives with a big blade, a can opener, an awl, and a screwdriver. Some of them have a blade that doubles as a can opener and screwdriver. There's nothing wrong with the biggest Swiss army knives, but for emergency purposes, the smaller pocket knives will suffice. A small sharpening stone should be carried.

Sunglasses: The only camping site that might not require sunglasses as an emergency provision is a deep, dark forest. However, they should always be carried. In the case of camping in open country, in the desert, and in snow, they are absolutely essential. If you wear prescription glasses or contact lenses, be certain your sunglasses are prescription also. Keep with your glasses an elastic holder to keep them from falling off.

EDDIE BAUER'S CHECK LIST

Now compare this list with one compiled several years ago by Eddie Bauer when he was outfitting hunters, fishermen, mountaineering expeditions, polar explorers, and many others.

"Because thousands of young outdoor people have asked what I have personally found necessary on my treks into remote areas, I have prepared a check list of those items. Some were homemade, but few people will go to the trouble of making their own today. During my years of outfitting I have made it my business to field-test all the new products, including many that I have created, to meet the needs of our constantly changing public taste, environment, economy, and other factors.

"It's a special joy to help those who choose to rough it, to leave the beaten trails, to hold costs down, to live off the land as their forebears did.

"Essential to me during my long life of outdoor outfitting and outdoor living are certain items I depended on in the backwoods of California, the Rockies, the Pacific Northwest, Canada, and Alaska. These items need not be costly, just simply adequate whether going solo or with a companion. As a check list, it has been useful to me, my friends, and customers. It is wise to anticipate one's needs for an outing, and this list may prove useful in selecting certain items that are often overlooked or forgotten.

"I suggest you start a list of your own. My list started in 1920, and it is still useful. The total weight needed went down sharply after 1935 when I came on the market with northern goose down.

"Before I invented the quilted down jackets and sleeping bags, I carried up to 120 pounds on a packboard. Everything was heavy. The woolen underwear was heavy, the woolen pants and mackinaw coats were heavy. Sometimes we wore sheepskin shearling-lined coats, and they were heavy too. They would take on water, and it was all you could do just to carry the clothing you were wearing.

"In addition to the heavy clothing I wore, I would take along a waterproof groundcloth and horse-blanket safety pins to pin the blankets together. Now that 120-pound load I carried wouldn't weigh any more than 70 or 75 pounds with the dehydrated foods, down clothing and sleeping bags, and ultralight fabrics.

"**1.** A lightweight .30-caliber rifle with auxiliary chamber for reduced loads, plus ammunition. It always went with me on treks when there might be a need for food or the chance meeting with potentially dangerous creatures.

"**2.** A short-jointed 4-oz. flyrod, reel, line, and terminal tackle. In a survival kit, rod and reel are omitted.

"**3.** A Hudson's Bay ax with leather head sheath, 20- or 24-inch handle.

"**4.** A belt knife in leather sheath with a 5-inch blade, plus a small sharpening stone to keep the edge keen.

"**5.** A Silva Polaris compass and topographical map of the travel area.

"**6.** A small waterproof matchbox for the old-fashioned wood matches. I always carry them on my person for emergencies.

"**7.** A small folding, or otherwise compact, wood saw. There are several small styles to choose from.

"**8.** My 1920 and 1940 homemade wooden-frame two-and-one-half-pound packboards. These have served me and friends all through the years carrying loads upwards to 150 and 200 pounds. Today aluminum packs do equally well or better. Many are on the market.

"**9.** A Norwegian Bergen metal-framed rucksack is ideal for short trips, and convenient with its four side and top-cover compartments. Your sleeping bag rides outside.

"**10.** My shelter tent which I made in 1915 of unbleached muslin sheeting, seven by twelve feet, and waterproofed with paraffin melted in benzine. It sets up as a half-pyramid for two, as an A-tent for several, or may be used full-sized as a lean-to shelter. The first one was replaced after twenty years with one made of balloon cloth, and I have used that one all through the years.

"**11.** Bedding: I have managed, under all conditions in temperatures from moderate to minus 40 degrees Fahrenheit, with a variety of bedding, which I will describe by the years used.

a. 1914 to 1935: Temperature fall to 20 degrees without fire for warmth, and down to minus 10 degrees with a campfire reflecting warmth toward my homemade tent: **Two all-wool trapper blankets.** Probably 90 percent of all outdoors people used blankets, although a few did invest in sleeping bags insulated with Kapok wool batting or down.

b. 1935 to 1980: I have used one or both of my trapper blankets periodically but usually a sleeping bag of my own manufacture. I have three of these accumulated through the years for these uses:

Backpacking: Comfort range from moderate to minus 40 degrees. A goose down mummy bag.

Car or packhorse: Moderate to minus 50 degrees. A heavy-duty 90-by-90-inch goose down rectangular bag.

Car or boat: Mild to 20 degrees. A rectangular sleeping bag insulated with three pounds of DuPont Dacron, plus a washable, removable flannel liner. I have used this bag so much during the past twenty years that the nylon outer fabric has worn through from my whiskers.

"**12.** Miscellaneous: These are periodically useful articles that have been in my inventory as long as sixty-five years, including:

a. An army-type trench shovel, lightweight and about two feet long. The blade is about 6½ inches wide by 7½ or 8 inches long. It is very useful around camp.

b. An ultralight air mattress, 44 x 28 inches. It is excellent for backpacking, and useful both as a bed and flotation device. When weight is less important, one can select larger, more comfortable mattresses.

c. Waterproof poncho (eight feet, seven inches long by 47 inches wide). It weighs 23 ounces and slips over the head to shed rain over pack and person. It has many uses around camp, boating, and horseback.

d. Campfire grid, 20 to 24 inches, with folding legs, plus a lightweight fabric bag or cover.

e. Plastic water bag and water flask. The bag holds up to three gallons, and rolls to pocket size. Flask is pint-sized and I carry it on my person.

f. Vacuum bottle. I carry two pint-sized.

g. Flashlights. I have several to choose from and select the ones that may be useful for specific outings—two-cell pen type, two-cell medium sized, and a large two-celled.

h. Plumber candles. One or more can be useful at times.

i. First-aid kit. No larger than necessary for personal needs. Mine includes

safety pins, needles, buttons, and thread.
j. Cooking kit and tableware. Aluminum nesting pots, with or without lids, 10-quart and 2-quart to use on the grid, over hot coals, or to hang over the fire. Take only the sizes needed.

Nesting tin cups, stainless-steel forks and spoons, one or two stainless-steel plates with straight sides, one sheet steel fry pan. Again, take only those needed for the trip.

Useful at times: Paper plates, plastic plates, paper cups, plastic cups and bowls, plastic spoons and forks when going light.

Knife: I use my belt knife.
k. A few nails, assorted sizes, common 6-, 8-, and 10-penny. You'll find them useful.
l. An 8-inch file, flat mill bastard type.
m. A 100-pound pulley. Ultralight, compact, and useful in moving or hanging heavy objects.
n. Braided cord and wire. Both are useful in camp and on treks. I take drapery cord, cotton or nylon, and baling or galvanized telephone wire, as much as I think will be useful.
o. A strong "S" link, closed on one end, to use with wire for cooking to hang a pot or kettle from a tripod of poles over an open fire.
p. Slingshot and rabbit snares, to take small game when necessary. I have always made my own, but now slingshots with an effective range of 225 yards, and weighing seven ounces, are available.

"And finally, always have in mind the possibility of getting wet to the skin by slipping and falling into a stream or in a downpour of rain. It is wise to keep food and matches in waterproof bags or containers. Also keep in mind the windchill factor and carry a windproof parka or substitute garment of ultralight backpacking weight."

CAMPER'S SECRET

Consider going with more experienced campers, preferably good friends or relatives, before striking out on your own. Camping isn't as complex as it sounds, but after camping with experienced people, you will have a better idea of what to expect when you go it alone.

SURVIVAL KITS

Most recreation stores stock survival kits that contain important items. Some are in cans no larger than pipe-tobacco cans; others are larger and contain more sophisticated items. A typical kit will have the following items:

- tube tent
- whistle
- candle
- matches
- duct tape
- aluminum foil
- candy bars for energy
- bouillon cubes
- dextrose cubes
- herb tea bags
- signal mirror
- aid supplies
- fire-starter
- nylon cord
- razor blade
- waterproof survival instructions

Another popular item in many such kits is the "space blanket" which is extremely light and compact, yet reflects up to 80 percent of your body heat when you wrap yourself in it.

As with all basic equipment lists, neither the Ten Essentials nor survival kits are carved in stone, and each camper will have his or her own items to add. Boaters have their own particular needs, as do private pilots. The Coast Guard, Federal Aviation Administration, and search-and-rescue associations should be consulted for their lists.

As an aside, the Tacoma (Washington) Mountain Search and Rescue Council spent months designing a survival kit that would fit inside a tobacco can. The kit has an eight-foot tube tent, a whistle, a signal mirror, matches, a candle, wire, bouillon, tea bags, sugar, salt, and instructions for using the can as a cooking container.

After long discussions, the council members decided to include the tea bags—not for their food value, but simply to give the stranded or lost camper something to do. Decisions made early in an emergency situation can easily be the most important ones, and the council reasoned that if campers had something to do for the first few minutes—namely, to build a fire and wait for the water in the tobacco can to heat enough to brew a cup of tea—they would calm down and be better able to handle the emergency.

Additional emergency equipment for automobiles and RVs is listed below, along with other reminders.

- Check insurance premium dates, pay ahead for period of trip.
- If traveling with pet, make sure you have pet vaccination records.
- Check vaccination dates; get shots if expiration occurs during trip.
- Check credit card expiration dates; renew if needed.
- List credit card numbers on sheet stored away from cards.
- Lock all doors, including garage, and all windows.
- Turn radio on; tune to twenty-four-hour station and turn up volume.
- Make sure you have driver's license, vehicle registration, spare keys.
- Store valuables in a safe place.

A SPECIAL NOTE

Most adults will deny it, but nearly all of us have the equivalent of a security blanket without which we feel badly equipped for a trip. It may be totally useless, but we like to carry it. Children are more open about their attachment to favorite belongings, such as a particularly disreputable stuffed animal with all the fur worn off, a favorite book, or a toothbrush that looks as though it was beaten with a hammer. The point is that whatever makes you feel comfortable, and doesn't take up a lot of space, should be taken along.

So many personal items are taken for granted at home that it is wise to make a list of what is absolutely necessary, such as a pair of reading glasses or a magnifying glass for threading needles or reading minute instructions on food packages. You may also need a paperback book, a sketch pad, or similar items for passing away those hot, quiet afternoons when you decide that relaxation is a major part of your vacation. Remember, going on camping trips does not mean you must rough it completely and shed yourself of everything you enjoy.

ROPES

Ropes (or lines, depending on your choice) are one of the most indispensable parts of your equipment list. You will find more uses for rope than you can imagine while sitting in the family room planning a camping trip. You'll need it for stringing tarps, for repairing tents or reinforcing them, to lash gear onto the family car, as a clothesline, for hanging food from a tree branch, for crossing swift streams, to replace broken boot or shoe laces, or for learning to tie knots on slow afternoons.

For the heavier work, such as hanging the camp larder from a tree branch or even to pull your car from the mud on a small winch, a sixty-foot length of 5/16-inch nylon climbing rope is usually sufficient. This takes little space and weighs less than two pounds, yet has a 2,800-pound capacity.

For lighter work, repairs and so forth, pick up a 100-foot length of the so-called clothesline rope, the small nylon rope you find in virtually every hardware store and some grocery stores. You will find uses for this on every outing.

Remember that when you cut off a length of nylon rope, you should hold both severed ends to an open flame to melt the individual strands into a mass so the rope won't unravel.

Manila and other natural-fiber ropes should be tied off on the ends with heavy thread.

LIGHTING

Even the pioneers of America had artificial lighting for their prairie schooners and cabins, but the state of the art has improved considerably. Now we have everything from the old dependable "gas" lanterns to battery-powered fluorescent lights that can be recharged after each camping trip.

For many campers, the old stand-by is the white gas lantern with the glass globe, the delicate mantles, and mild hissing noise that is as much a part of camp at night as shooting stars.

Relatively new on the market is a much smaller lantern that fits neatly atop a butane cartridge and starts with the flick of a starting switch, something like a cigarette lighter. These are becoming more and more popular because they are lightweight and occupy little space. Also, since the butane cartridges are disposable, you don't have to worry about spilling the fluid when filling it.

Also relatively new is the battery-operated fluorescent light, perhaps the brightest of the camping lanterns. It can either be operated from the rechargeable battery or plugged into a 110-volt outlet (which also is used to recharge the battery).

Flashlights: A bewildering array of flashlights are on the market, from the tiny models that hang from your key-chain to the 12-volt models that are almost strong enough to double as a search light. Some are disposable; when the battery goes out, the whole thing goes in the garbage.

CAMPER'S SECRET

When shopping for a camping flashlight, consider buying one that has a clip inside for a spare bulb. These cost very little more, and you'll never have to dig through the packs for a spare bulb that was probably left at home anyway.

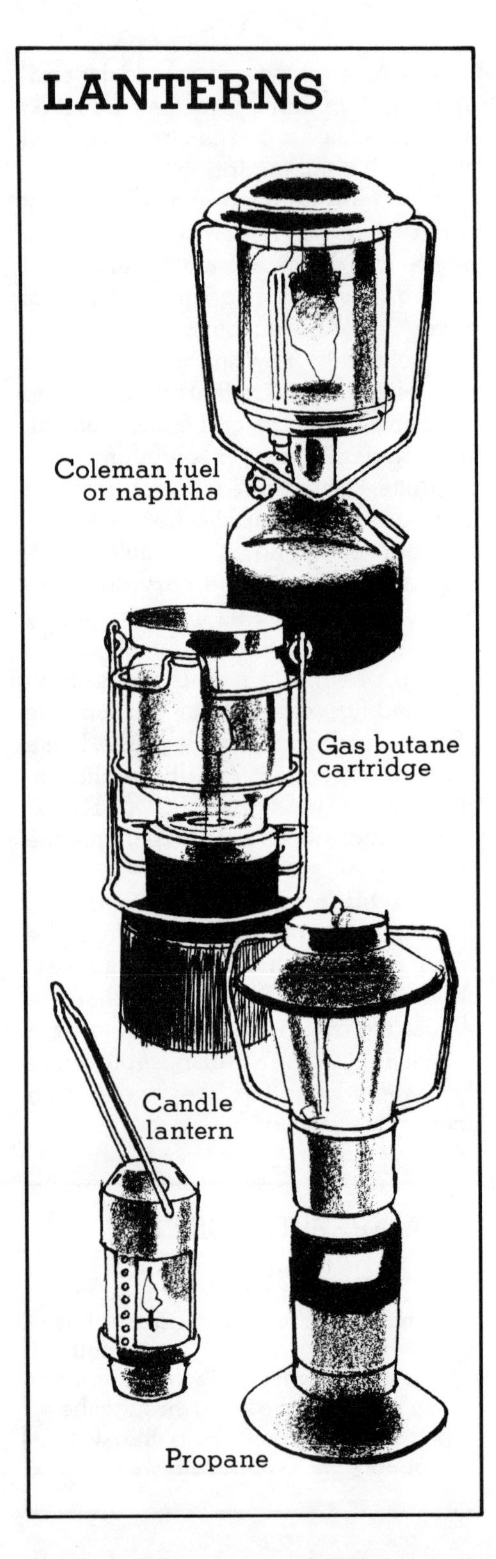

The best for most purposes are still those that have the main elements of lens, bulb, and batteries, all of which can be replaced. The important thing, of course, is reliability. When you wake in the night and go stumbling off toward the latrine, you do not want a flashlight that abandons you halfway there.

The best models are waterproof, and many of the plastic ones will float should you drop them into a lake or stream.

The most practical models have a clip inside to hold a spare bulb, the one part many people frequently forget to carry.

Kids (and some adults) are addicted to playing with flashlights, even at high noon, so that by the second night of camping, you often have at the best a weak yellow light. Whether this is the case with your group or not, always keep a spare set of batteries tucked away out of sight. And hide the flashlight at dawn if you have to.

A number of small candle lanterns are available, which are simply metal cases for candles with transparent lenses. These are obviously much safer to use than plain candles and have the advantage of being windproof.

Another convenient light is the adaptation of miners' lanterns that you slip over your head. There are battery-powered and are much more convenient than hand-held flashlights because they free your hands and point in the directions your head is turned. These also are ideal for the camper who can't face a night in bed without reading a chapter or two of a novel.

Candles: Also available are small candle lanterns, a small cylinder in which a candle is placed. These can provide enough light in a tent so you can tell the difference between a sleeping bag and a parka, but few people can depend on

them for nighttime reading. Their main value seems to be psychological—a small ray of light in the blackness.

PACKING

Few things are more frustrating to the beginning camper than trying to cram everything you think you need on a trip into the family car that you normally think is crowded when the whole family goes to visit a relative on Sunday afternoon. The key phrase here is **everything you think you need,** because no matter how long you've camped, you will want to bring more than is actually needed.

This is one of the best arguments for buying the newer ultralight and compact camping equipment and freeze-dried and dehydrated food now available. If you own a medium-sized car—or even a sub-compact with average trunk space—and you avoid packing breakables, you can put a medium-sized ice chest, camping stove, cooking utensils, food, two tents, clothing, six sleeping bags, and a few other items in the trunk. Other smaller items can be stuffed in nooks and crannies beneath the seats, at your feet, and along the rear window.

If at all possible, avoid using cartop carriers for anything other than bicycles, skis, and so forth, because the enclosed cartop boxes are brutal on gasoline mileage. However, if you're forced to cram six people, all their equipment for a week, plus a pet or two into a four-person vehicle, then you have no choice but to use a car top carrier.

The best means of carrying your own gear is in a backpack with all its compartments for ease of separating items by their use: sleeping bag pad and tent attached to the bottom, toilet articles in one side pocket, eating utensils in another, camping tools in still another, and so forth.

The difficulty of using several backpacks with frame intact is that they are often difficult to arrange in the car trunk for a party of more than three. If you know you won't be going on anything more strenuous than a day hike, you can remove the pack from its frame.

If you don't own backpacks and don't plan to use them, the next best alternative is a duffel bag for each person in the group. The longer the zipper in a duffel bag, the better. One of camping's little frustrations is rummaging around in a duffel bag in search of something on the very bottom when the zipper is at one end. The best choices, like packs, have more than one compartment so you can divide the gear into easy-to-find places.

Obviously you won't want to take anything along that will break easily, no glass objects, no delicate wooden models, and as few items with sharp edges as possible.

When everything is stacked beside the car ready for stowing, run an inventory to be sure that en route snacks, rain gear, reading material, and car games are not in the group that will be stowed in the trunk. And have a plastic bag or grocery bag in the car with you for the amazing amount of debris that accumulates between your garage and the campground.

Stow the unyielding objects, such as coolers, campstoves, etc., first and then stuff the softer items around them, such as sleeping bags and clothing. Keep the cooler accessible since you may be stopping for a snack or soft drinks en route. Also keep the car's tool kit in a convenient place. It is better to stow things beneath the passengers' feet than on the rear window shelf because it is both illegal and dangerous to obstruct the driver's view.

CAMPER'S SECRET

Use every available inch of everything for packing. If you're carrying a two-burner stove, you can pack most of your cooking implements and perhaps your dishwashing soap or other items in it. Do not put food in a campstove, since it may well have a residue of fuel or soot.

TRIP CHECK LIST

Vehicle Preparation

- Get engine tune-up.
- Check battery level and battery connections.
- Check tire pressure and tread; check spare.
- Check transmission fluid.
- Check radiator level.
- Check all lights—interior, brake, backup lights, etc.
- Check trailer hitch for tightness, connection points, and hitch ball.
- Clean inside of windows.
- Check tool kit for all essentials.

RV Preparation

- Check trailer tires for pressure and tread.
- Check all trailer lights and turn signals, and interior lights.
- Check trailer battery level and terminal corrosion.

- Check all air vents for blockage (furnace, refrigerator).
- Fill water tanks; test water pump.
- Check gas-light mantles.
- Add chemical to holding tanks.
- Make sure power cord, water hose, jacks, and disposal hose are aboard.
- Close windows and roof vents tight.
- Check and top off propane tanks.

Travel Information

- Prepare detailed route plan.
- Identify potential scenic areas, recreation stops, other "don't miss" places.
- Estimate travel time and time of arrival.
- Make and/or confirm reservations.
- Advise campground of any delays.

Recreation Gear

- fishing tackle, rods, and reels
- camera, film, accessory items
- binoculars
- tennis rackets
- balls (volleyball, softball, football)
- Frisbee
- swimsuits and swimming gear
- games, cards, toys, car toys
- life jackets

Kitchen Gear

- can and bottle opener
- heating kettles or assortment of pans and kettles
- stirring and serving spoons and forks
- eating utensils
- pancake turner
- wiener or marshmallow cooking forks
- hot pads
- plastic tablecloth
- serving platter
- coffee pot
- pitcher
- sealable plastic containers

Camping Gear

- tent, poles, stakes
- ground cloth
- sleeping bags
- cots, air mattresses
- lantern
- stove
- catalytic heater
- cooler
- charcoal grill
- folding chairs
- folding table
- ax, saw
- water jug
- flashlight
- stove stand
- lantern stand

Securing Your Home

- Notify police of exact dates of your trip.
- Stop mail delivery at post office or have mail picked up by a neighbor.
- Stop deliveries (paper, milk, etc.).
- Arrange for someone to mow grass or shovel walks while you are away.
- Arrange for a friend or neighbor to turn on different lights every few days.
- Arrange to have plants watered.
- Empty refrigerator of perishable foods.
- Unplug all unnecessary electrical appliances.
- Turn off gas to kitchen range.

CHAPTER 4

TENTS AND SHELTERS

Tents are your home away from home, and you should exercise the same care in buying one you would when buying a home or vacation home. Under normal circumstances you won't spend many waking hours in a tent, but should your camping trip coincide with the local rainy season, you will become intimately familiar with the interior of the tent. So it is important that you choose one in which you will be comfortable.

The first generation of tents tended to be rather dim and confining—like a house with all the shutters drawn. In spite of their having windows, most tended to be dark and dismal even on a bright day. But the newer fabrics and better designs have turned today's tents into light and airy dwellings. The walls of solid canvas have been replaced in some designs with a fine mesh that lets both light and air inside—and yet keeps bugs out. At night the flaps can be unrolled to cover the mesh, giving privacy and comfort.

Gone, too, are the days when all tents were so heavy it took at least two people to carry one, when leaks were taken for granted and mildew a fact of life. Today there is an almost bewildering variety of tents of all fabrics, designs, and weights. Although the large wall tent is still popular with campers who stay within sight of their vehicles or who use pack animals, there are roomy tents available so light that a six-year-old can carry one.

Most tents on the market today are made of lightweight, sheer nylon. The basic tent is intentionally not waterproof on the top and sides, whereas the floor and perhaps a few inches up the walls are coated with waterproof material to keep groundwater out. The walls and top are made of breathable nylon material so that moisture will pass through rather than condense on the walls and ceiling. Strung over the entire tent is a rain fly, usually nylon coated with waterproofing. The fly is supported a few inches away from the tent to allow air flow that carries away the moisture that comes from bodies and breathing inside. The rain fly also protects the tent against damage from ultraviolet rays.

Nearly all tents are equipped with a fine-mesh screen over all doors and windows to keep out insects. A common description of this mesh is "noseeum-proof" because the irritating little bugs called no-seeums, plus gnats, chiggers, mosquitoes, flies, and bees, are the bane of older designs without this durable netting.

In addition to this netting, which is usually zippered at the entrances and either permanent or closed with drawstrings at the windows and vents, tents also have flaps that cover the entrances and windows for privacy and comfort.

Because they are porous, very few tents, including the canvas models, are designed to keep you warm. But they do protect you from rain, heavy winds, snow, and sun. In extremely cold situations, such as climbing expeditions or winter steelheading and cross-country skiing, you can use tent heaters to keep much of the cold out, but carefully observe all safety precautions if you use them. For the most part, you must depend on your clothing and sleeping bags for warmth.

Some of the classic canvas wall tents have zippered and fireproofed openings in the roof or high on the walls for stovepipes so you can use a woodstove and have the coziness of a mountain cabin.

Since it is impossible to find a tent that is good for all kinds of camping, before purchasing one you should decide how you plan to use it. A good rule of thumb regarding the heavier wall tents is that if you never plan to camp more than a hundred yards from your car or boat, or if you will use it on horsepacking trips, you will have no problems. You will have enough space to set up cots, if you prefer them, and you can set up a stove as well.

However, most families expect to do some backpacking, and this automatically rules out the tents weighing more than 12 pounds. You are a candidate for a lightweight tent weighing not more than ten pounds.

Rather than follow the manufacturer's recommendations on how many persons a tent will sleep, you might want to compute your tent needs on a square-foot basis. You can assume that each person is six feet tall and needs a three-foot-wide space for sleeping. Add some room for personal equipment, such as packs and clothing, and you get 27 square feet per person in each tent. For extended periods of camping, such as a week or longer, increase the square footage requirement to approximately 30 feet.

CAMPER'S SECRET

You can rent tents, sleeping bags, and some other equipment before deciding if you want to invest in camping gear. Always test it before leaving the rental store, and be sure you know how to use the equipment.

These figures are only estimates, of course, and you may decide you need less, especially if you have young children or if you are a firm believer in traveling light and plan to crowd three adults in a single tent.

With all these factors in mind, it is well to remember that in spite of all the things you should consider, most campers purchase a backpacking tent weighing less than ten pounds and make it work for all kinds of camping. A backpacking tent weighing six or seven pounds can easily be used for car-camping trips, river trips, extended backpacking expeditions, bike touring, and so forth.

Another possibility is a large tent for adults and a small "pup" type for children.

When shopping for a tent, it is wise to talk to people who already own tents, and, if possible, to rent or borrow a model similar to one you want to buy. In some cases, your individual needs will dictate the design and shape you choose. For example, you may want to connect tents so that children won't be totally separated from their parents, yet can have their privacy, too. In this case, a properly placed tarp or rainfly can be used. This

CAMPER'S SECRET

When you buy your tent, always try to buy a swatch of matching material for repairs and carry it with you at all times. Watch places of extreme wear, such as sleeves for the poles and around the entrance, and when signs of wear appear, make the repairs before the material wears through or tears.

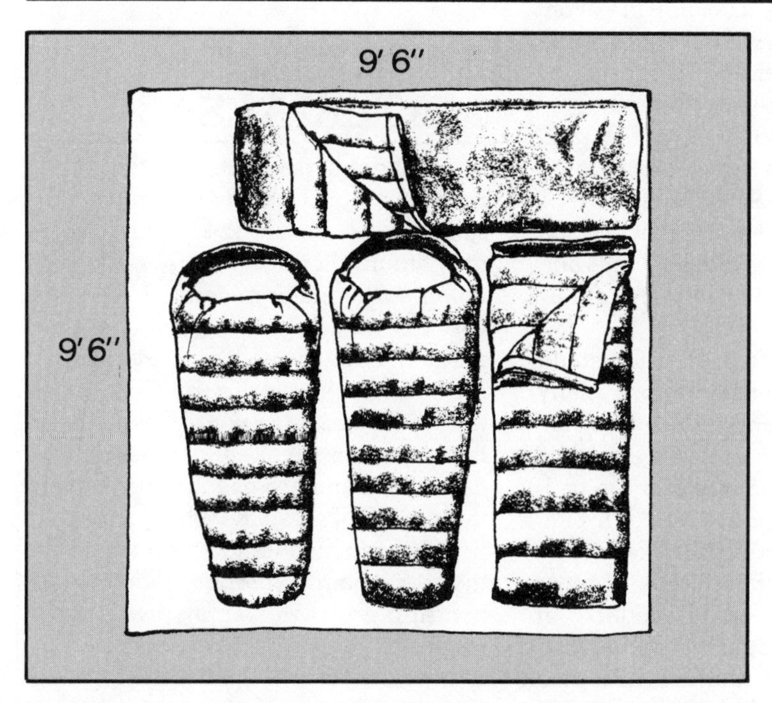

Rather than follow the manufacturers recommendations on how many persons a tent will sleep, you might want to compute your tent needs on a square-foot basis.

system is useful in poor weather, since you can visit back and forth without getting wet or cold.

Many backpacking models come with two carrying bags so that the load can be distributed nearly evenly between those using it: the tent in one bag, and the poles, stakes, and rain fly in the other

Many families who camp in RVs take along a small tent that will sleep two or three children and pitch it outside the RV so the children can sample sleeping out. Also, children like to get away from adults (and vice versa) occasionally. Such tents are convenient for the hunters and fishers of the family who get up before dawn, then relax with a nap at midday before going out again in the evening.

You will find tents available in a wide variety of colors, and it doesn't seem to make much difference whether you select a vivid or a subdued color scheme. Some veteran campers say it helps to select a bright color in the event you're forced to spend a lot of time inside to wait out bad weather.

The colors range from International orange to purple to blue, green, and white. Some tents have as many as three separate colors, almost like a circus tent. Frequently the floor will be one color, usually a darker tone, and the walls and ceiling a brighter color with the rain fly still another color.

Since tents come in an astonishing variety of sizes as well as shapes, many families purchase tents for use as weekend or vacation homes. These are large enough for a small family and are usually erected on a special wooden platform built for that purpose. The classic American Indian tepee is a popular design for this use and is available in many specialized outdoor stores. Patterns for do-it-yourselfers are also available.

Another type of tent that has been popular in Europe, where recreational opportunities are more limited than in North America, are the very elaborate dome-style tents with private sleeping quarters in the wings and with communal spaces in the middle. As with the tepees, these are erected only when in use and stored at home the rest of the year.

There is enough demand for the standard wall tent that measures at least nine by twelve feet to keep several manufactures in business. These are getting lighter and lighter in weight as cotton duck is replaced by synthetic fabrics. They offer more headroom than the smaller designs that can be used for backpacking and are popular with families who tow a small trailer with camping gear behind their family car.

BASIC TENT MATERIALS

Although research into tent materials and construction continues at a steady pace, during the past few years most manufacturers have settled on nylon as the major material. A very few use Dacron, and some have experimented with various other materials that are lightweight and durable. Cotton, cotton duck, and canvas are used only in the larger tents that are not meant to be carried far and are used for base-camp tents that will sleep the whole family.

Ripstop nylon and nylon taffeta are the most common fabrics found in tents, ranging from the small two-person mountain tents to the larger three- and four-person models that usually weigh less than 12 pounds, rain fly included.

They are of almost equal weight and durability.

Taffeta is a tightly woven nylon that is soft to the touch and permits air and moisture to pass through rather than collecting into condensation.

Ripstop nylon is also tightly woven but has heavier threads approximately three-sixteenths inch apart to prevent tears from spreading. Nylon will eventually break down from ultraviolet light unless the tent's rain fly is kept over it for protection.

Dacron is used by a few manufacturers but it is more expensive than nylon and more difficult to work with. Its major advantage is that it is virtually impervious to ultraviolet light damage from the sun.

BASIC TENT DESIGNS

In spite of the bewildering variety of designs available today, the basic tent shapes are still as follows:

- A-frame, or pup tent
- pyramid
- dome
- wall tent
- "bakers" (lean-to)

A-frame: The first major design was the A-frame or pup tent favored by soldiers and Boy Scouts. It could be erected with a line strung between two trees and the tent hanging over it, or with a single pole at each end. Then each corner was stretched taut and held down by a stake in the ground. Later versions had aluminum or fiberglass poles forming an A-frame at each end so that no poles were inside the tent.

Pup tents are still very much in demand, and the design is constantly undergoing modifications. One disadvantage of the earlier designs was the need of several lines to hold the tent in place, which invariably became tripwires for anyone walking nearby. People who owned the earlier models soon learned to expect frequent shouts of rage from their children as the family clown pulled up the stakes holding down the lines, collapsing the tent onto the inhabitants.

Today many tents of the classic A-frame design have no exterior lines. Most avoid this by having stakes all around the edge of the floor and a tension bar across the top holding the A-frames at each end taut and erect. Various devices have been created to keep the rain fly a few inches away from the tent wall so that the tent will "breathe" and avoid condensation on the walls.

Pyramid: Like the A-frame, this is a basic—and classic—design that has been used for centuries by Mongols, Lapps, and North American Indians. So basic is the design that its modifications during this century have been relatively minor.

Early canvas tents of this design, however, often depended on a single center pole that had a disquieting tendency to collapse on the occupants, or always to be in the way. The next step, after aluminum and fiberglass poles became common, was to insert three or four lightweight poles in sleeves sewn into the cover, thus removing poles from inside the tent.

This taller design is particularly popular with campers who find it difficult to pull on their trousers without standing up. Many models have tabs sewn into the walls about three feet up so that the walls can be stretched out, giving occupants a bit of additional sleep-

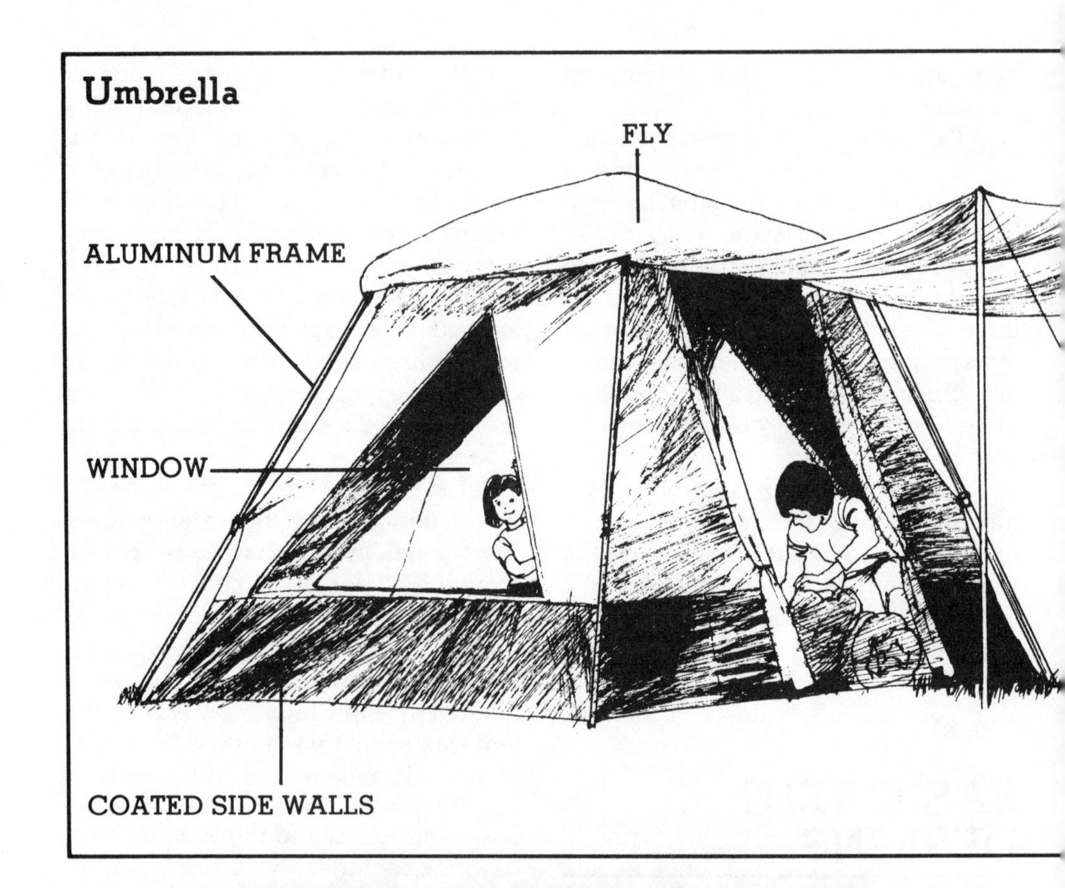
Umbrella
FLY
ALUMINUM FRAME
WINDOW
COATED SIDE WALLS

A-frame

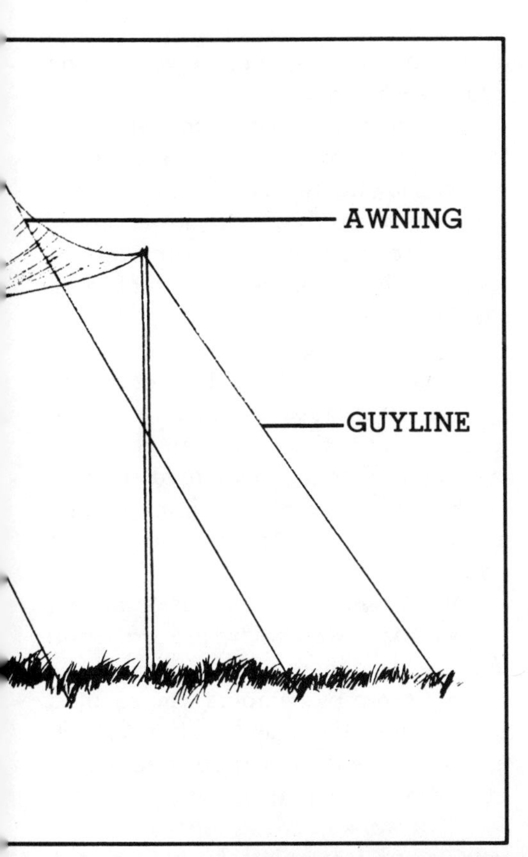

TENTS

Some things to watch for in tent construction include:

1. Reinforced pole pockets, preferably webbing, so the poles won't rip away from the tent.
2. Catenary cut on ridgelines so tent can be kept taut.
3. Noseeum-proof netting over entrances and vents.
4. Nylon coil zippers.
5. "Tub" floors that are waterproof part way up walls.

CAMPER'S SECRET

Never go camping with tents you haven't put up at least once at home. Always pretest them so you can cut down on the time it takes to get them erected, and to inventory them to see which pieces are missing.

Dome

ing space without having the walls against their sleeping bags or faces.

Some pyramid-style tents are variations of the design, such as a half-pyramid with the front open to catch heat from campfires reflected inside. This design permits you to erect two half-pyramids facing each other with a rain fly or tarp stretched over the openings to give you a weather-free opening. An example of this versatile tent is Eddie Bauer's first tent, shown on page 43, one he designed for himself and later sold through his stores.

Dome: This design is the newest on the market, and there are perhaps hundreds of variations from which to choose. Although many people buy a dome simply because it is more attractive than other designs and reminiscent of Buckminster Fuller's designs or Canadian Eskimo igloos, this tent has almost an equal number of advantages and disadvantages.

First the advantages: It is self-supporting, meaning that no stakes or lines are required under normal situations. This makes it simple to clean; just empty it, pick it up, and shake out the dirt and sand. It is more aerodynamic than most tents because it offers no flat surface for the wind to strike and whip the tent.

One disadvantage is that the dome can be hard to set up, especially in the dark or with the wind blowing, when it threatens to become a kite while you are inserting the poles through the sleeves that encircle them. The dome also tends to weigh more because more poles are required.

Although no stakes or lines are required to keep dome tents erect, you should never leave them unoccupied without staking them down. They are so light that even a slight breeze can send an empty dome tent tumbleweeding off into the wilderness.

The most satisfactory models seem to be the modified dome tents, those which use only two poles and have flat sides. Because they have fewer poles, modified dome tents weigh substantially less and they are easier to erect in all situations.

However, you should remember that no single tent is all things to all campers. The disadvantages of the pure dome tents do not bother most campers who own them. They become quite adept at erecting them even in a midnight gale and would not consider owning another style.

Wall tent: This is a classic design that was used well before the turn of the century by miners and hunters and is still preferred by campers who set up a base camp and never move from it, such as hunting parties or groups spending an entire vacation at one site.

The walls of these tents come up four or five feet before forming a roof. Most are made of cotton or canvas, although some of the new lightweight materials are finding their way into this design. A system of poles and stakes keeps the walls and top taut. Some models require guy lines, but most do not. Some also have awnings attached to form a porch.

A few camping outfitters build half cabins with floors and use the wall tents as a covering and roof, taking the tents down at the end of the season. Wall tents are actually a temporary cabin, and you can set up housekeeping for a week or a month, complete with ice chest, cookstove, bunks, and lanterns.

Some available models of wall tents are designed especially for owners of vans. These have a vestibule that connects the tent with the van's door and are held to the van with padded magnets or other devices.

Baker tents: These wall tents have a roof that slopes backward from the front with a flap that can be used as an awning or porch during the day and dropped as a covering flap at night. They are one of the simplest to erect, though not recommended for camping where bugs are a problem, unless you equip the tent with netting or add the equivalent of a screened porch with insect netting.

CAMPER'S SECRET

If your tent needs guy lines to support it, one way to help avoid the problem of people tripping over the lines day and night is to hang pieces of white plastic or a similar material to the lines. These "flags" will show well on moonlit nights as well.

BUYING A TENT

It isn't a law of the outdoors, but in most cases the less an item weighs, the more it costs. Although weight isn't so important for regular family or RV camping near your source of transportation, you will be wise to keep it in mind when buying tents and sleeping bags. Will these items ever be used for backpacking or bicycle touring? Will they be used by adults only, or will teenagers, who may be more careless than adults, be using them for group outings, slumber parties, and the like?

If you are planning to make these items last for years and years, you should consider buying the top-of-the-line models.

Choosing the material, design, color scheme, and other factors of a tent is at first an overwhelming task to the first-time buyer. But a few outings in different styles of tents will help you narrow the choice, so the best way to determine your needs is either to rent a variety of tents or to go camping with friends and listen to the pros and cons of each style.

In addition, be sure to consider the following before purchasing your tent:

Size: How many people will sleep in it? Will it be too large for some campsites? How often will you move it on each outing?

Weight: Will you use it only for car camping, or will you be using it for backpacking, canoe trips, bike touring, and other forms of travel when weight is critical?

Design: Do you require lots of head room? Do you want a self-supporting design or one with guy lines? Will you be camping in the winter and require a vestibule, cook hole, and tunnel entrance?

TIPS ON TENT CARE

Storage: Before storing your tent after use, be certain it is totally dry and clean. Erect it in your backyard, carport, or family room and sweep out all dirt, sand, and other foreign matter you may have left in when breaking camp. Turn stuff or storage bag inside out to clean and dry. Wipe all poles and stakes. Do not leave the tent in sunlight longer than necessary or ultraviolet rays will cause the fabric to deteriorate. When it is out to dry, erect the rain fly also since it is less susceptible to ultraviolet rays. Do not

Always have the salesclerk show you how to erect your tent, then practice erecting it in your backyard or family room.

leave the tent in a car trunk or other hot place since high temperatures can damage the fabric, too. Store in a cool, dry place in its storage bag.

Leaks: Buy seam sealer, available at outfitters, and use as directed, even on new tents, because the majority of leaks occur in the seams. Apply sealer as needed throughout the tent's life.

Cleaning: Use either a very mild liquid detergent or a baking soda and water solution. Be certain no residue is left on the fabric. In all cases, follow the manufacturer's or dealer's recommendations.

Staking: More and more tents are self-supporting and have no lines to stake down. However, always take aluminum or hard plastic stakes and use them as anchors to avoid watching the tent roll away like a tumbleweed in a strong wind, sleeping bag and all.

Poles: Keep the poles and fiberglass wands wiped clean and free of corrosion. If a burr develops on a metal pole or metal connector for fiberglass wands, buff it down with fine sandpaper or a nail file. Otherwise, a hole in the fabric will develop very quickly.

Shock cords: If your tent is not equipped with shock cords around the base and the rain fly, it is wise to install them because the tent fabric will expand and contract with heat and humidity changes. These cords, made of heavy-duty elastic, are available in various degrees of stiffness and can be cut to the desired length.

Zippers: Keep the zippers coated with a light silicone lubricant to prevent freezing or snagging. Stick or liquid lubricant can be purchased at most camping equipment stores.

Familiarity: Always understand clearly how to erect the tent you buy or rent, and have the clerk show you how before leaving the store. Practice putting it up at least once before going on a trip to be sure all the parts fit. A prime source of frustration is trying to erect an unfamiliar tent in darkness, wind, or rain.

Chemical problems: Be wary of using chemicals such as hair spray inside your nylon tent. Hair spray can damage the fibers.

Fire safety: Most tents manufactured since the mid-1970s are treated with chemicals to render them fire-resistant. This does not mean they won't burn if exposed to an open flame. But it does mean that they won't turn into an instant torch as some earlier models did. Always check the specifications of a tent before purchase to be certain it is treated with a fire retardant. It is required by law in most states, and nearly all manufacturers comply with these laws in all their tents.

Obviously you should not cook in your tent unless absolutely necessary, such as when you're winter camping and caught in a blizzard. Some tents designed for winter use have a zipout cook hole in the floor so that spilled fuel and food will go into the snow and the tipped-over stove will fall harmlessly.

CAMPER'S SECRET

If your tarp doesn't have metal grommets, you can make temporary ones by wadding a small handful of moist dirt and wrapping a corner of the tarp around it. Tie the lump with a line, and you have a grommet. Of course, it is best always to have a few ball-and-wire grommets with you.

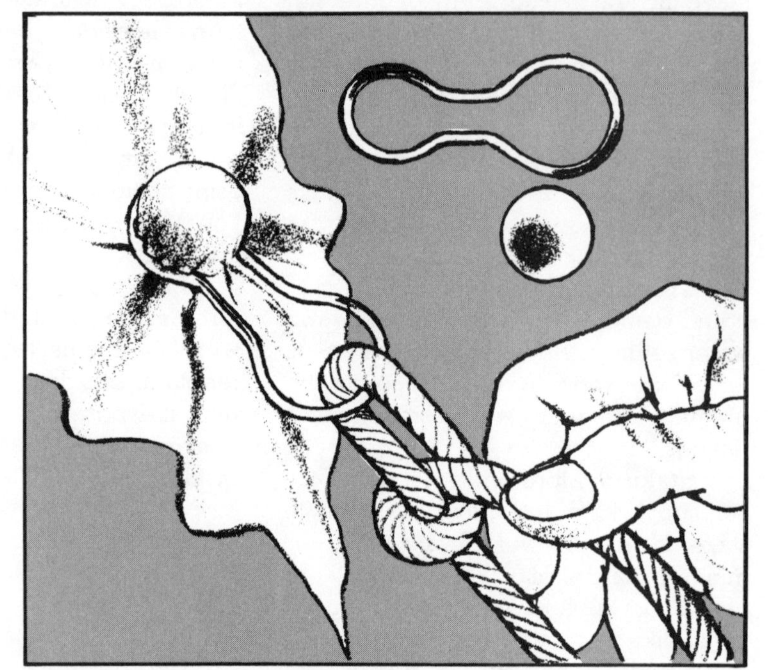

Ball-and-wire grommets are simple to use. You should carry one or two extra with you in case you want to use a poncho for a temporary shelter or in case a grommet in your tent tears.

OTHER SHELTERS

Many experienced campers, particularly those who go on long backpacking trips in warm weather or on river trips, carry only the minimum in shelter: a simple waterproof nylon tarp or plastic sheet about nine by twelve feet. They use this for making either a simple lean-to at night or an open-ended pup tent.

Another choice is the tube tent, which is little more than a plastic sack a person can crawl into, usually open at both ends so it can be suspended above you with a piece of rope stretched between two trees or bushes. Some campers buy an inexpensive plastic tube tent to stake out a campsite while they drive their car or RV elsewhere during the day.

For most car campers, these shelters are barely adequate and are carried for emergency use or as an adjunct to regular tents.

CAMPER'S SECRET

Nearly all tents and lines have a tendency alternately to shrink or to sag. One way to help adjust this condition is to attach a loop of shock cord at the end of the lines where they connect to the poles or stakes. Shock cord is a sturdy elastic that can be purchased in small loops or by the foot. Many tent poles that are hollow and must be fitted together can be kept together by running shock cord through them with tension on them when the poles are together. When it's time to fit the pole sections together, give a snap of the wrist and they pop into place.

CHAPTER 5

BEDDING

Until Eddie Bauer invented quilted goose down-filled garments and sleeping bags in the mid-1930s, campers had to rely on sleeping gear that was only a little lighter and less bulky then tents. They carried blankets or bags stuffed with various battings that were much like homemade quilts.

The modern sleeping bag made of ultralight nylon filled with goose down didn't come into its own until World War II, when Eddie Bauer manufactured them for Army Air Corps crewmen and mountain troops.

The first generation of sleeping bags without goose down simply did not keep campers warm. They often had to take turns getting up at night to stoke the fire in freezing weather. Sleeping in hunting cabins was a bit more comfortable, but in subzero weather the woodstove often had to be kept glowing through the night.

With today's bags, you can sleep in comfort even on the world's tallest mountains without having enormous layers of wool weighing you against the ground. The difference is the down bags and insulating pads beneath you to prevent the conduction of ground cold upward.

Campers also have another new generation of synthetic insulation from which to choose. This insulation, which comes in a variety of kinds and is sold under brand names such as Polarguard, Hollofil, and Thinsulate, is gaining popularity for use in both sleeping bags and garments. Although it is not an across-the-board substitute for down, it is being improved constantly and has a significant niche in the outdoor market.

These synthetics are made of polyester and are manufactured in sheets or batts, then cut to size for each garment or sleeping bag. These batts create millions of tiny air pockets to provide loft, as in goose down, where the dead air creates an insulation barrier.

The advantage of these products over goose down, besides their lower cost, is a resistance to absorbing moisture so that they retain their loft even when wet. If a synthetic-filled item becomes soaked, it can be simply wrung out and the fill springs back to its original shape. Conversely, once saturated, goose down does not regain its loft until it is totally dry, and wringing it out compresses the down further.

The major disadvantages of synthetic insulation are that it requires more weight and bulk to provide warmth, and its life span is much shorter than down's. Some synthetic insulations will not last more than three or four years of hard use. Goose down will last several generations with the proper care.

The rule of price and weight is a major factor here: the least weight and the best insulation cost more.

SOME TERMS YOU SHOULD KNOW

Down: Down is the underplumage of waterfowl closest to the underbody. It does not have a shaft like a feather. It has a quill point that a network of arms cluster around, radiating out like an octopus's arms. The best down comes from birds that live in Arctic conditions, where the harsh cold requires millions of the tiny clusters of down for the bird to survive. The best down of them all, and the most difficult to obtain, is the eiderdown from the Arctic eider duck. This down must be picked from eider nests by hand, and the supply is obviously very limited. The difference between this and regular goose down is that eiderdown has tiny Velcro-like hooks on it which keeps the down clustered together. Throw a handful of eiderdown into the air and it will cling together in a ball; goose down separates.

However, goose down is the most superior insulation available. Duck down is the second best, and the synthetics come next. A recent study by the U.S. Army's research laboratory in Natick, Massachusetts, states: "In terms of sheer warmth per pound, down is still the best. It's also unmatched in its ability to be compacted without losing resiliency, so you have to say we have no synthetic fibers that will outperform Mother Nature in every respect."

Thinsulate: This is the trade name for one of the newest synthetics on the market. It consists of uniform fibers that have a small crimp in each to hold themselves apart to create the dead-air spaces needed for insulation. Its manufacturer says it requires less loft than down, but also requires more weight.

Loft: Loft is the thickness of a bag or article of clothing when it is fluffed out and dry. The thicker the loft, the better the insulation value. Thus, when you remove

CAMPER'S SECRET

Try to buy different colored or shaped duffel bags and sleeping bags for each member of the family. It will make identification much easier.

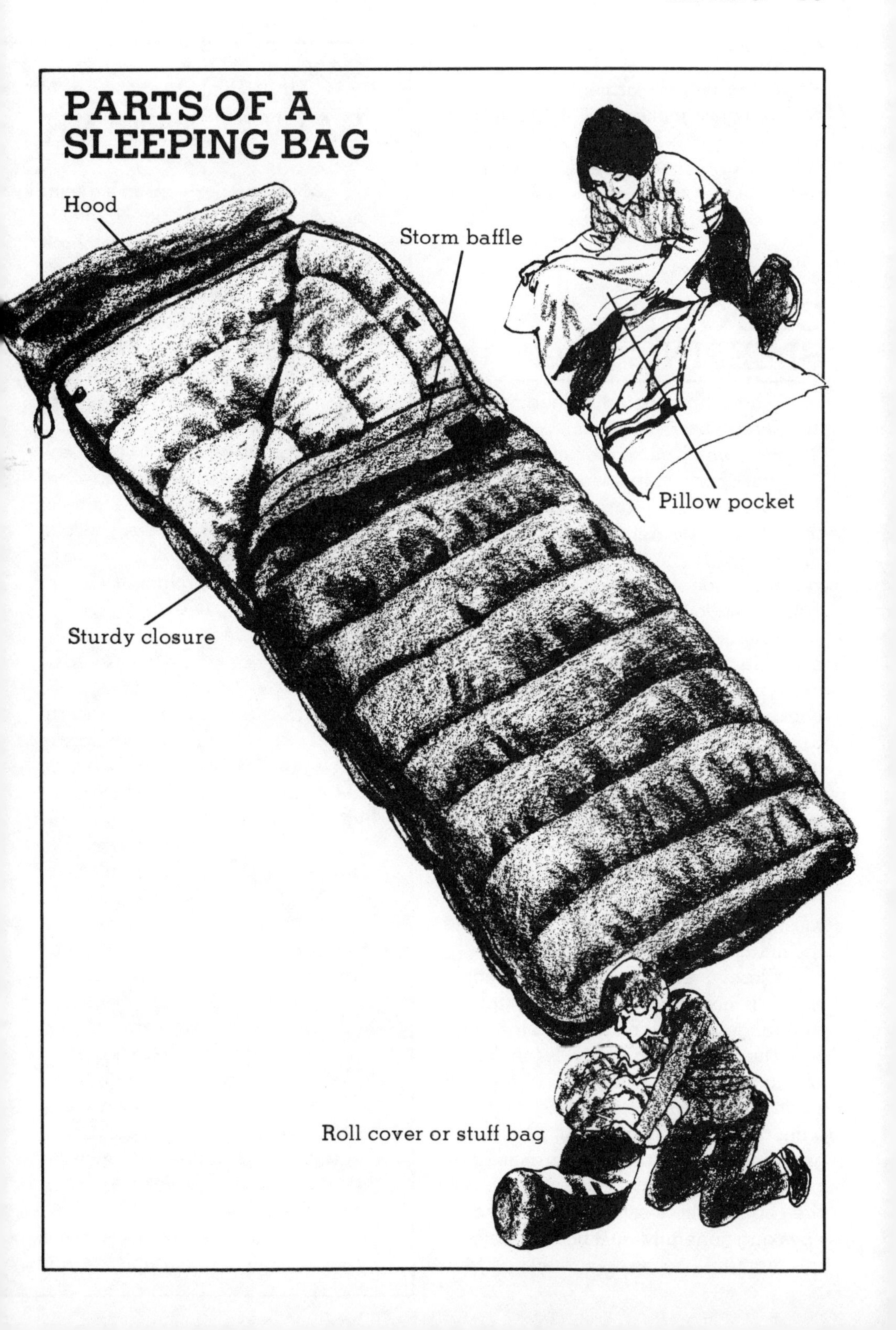
PARTS OF A SLEEPING BAG
Hood
Storm baffle
Pillow pocket
Sturdy closure
Roll cover or stuff bag

a sleeping bag or garment from its stuff bag, you must shake and fluff it to restore the loft, to make all the millions of individual clusters of goose down or synthetic fibers spring back to their normal shape.

CHOOSING A SLEEPING BAG

When shopping for a sleeping bag, there are a number of factors to consider.

First, what will be the nature of your activities? For backpacking, hiking, and bike touring, in the warmer months, you should consider a lightweight bag that you can carry on your pack and that is versatile enough to stretch your season from spring to fall.

For fishing, hunting, boating, and camping, look for a bag that is durable, full-sized, and more comfortable than a backpacking bag since you most likely won't be carrying it for miles and miles.

For canoeing and kayaking, consider a synthetic bag that is both lightweight and compact.

For mountaineering, expeditions, ski touring, and snowshoeing, you will want a bag that is still lightweight, but also the most thermally efficient and durable on the market.

Closely related to the information above is consideration of the temperature range you will be exposed to.

The next consideration is the shape of bag best for your uses. There are two basic shapes— rectangular and mummy, with modifications of each. Generally speaking, the shape you purchase depends on the use to which you will put the bag and the weather conditons. Still speaking generally, on a basis of weight, the mummy bags are warmer because

HOW TO SALVAGE DOWN

Millions of down-insulated items are discarded through charitable organizations such as Goodwill and St. Vincent de Paul thrift shops. Look for the brand names prior to 1968 such as Bauer, Temco, or Comfy. Other products with excellent down are World War II military items such as flying suits and sleeping bags.

Good down will outwear several generations of fabrics, and clothing pattern are readily available in fabric shops. Sometimes you can use the discarded clothing or bags as a pattern if it is your size. Or you can design your own pattern to serve your needs.

Once you find a discarded item insulated with good down, place it in a laundry basin in very warm water with a mild, nonchlorine detergent to both wet and cleanse the down. Launder it the same as other clothing.

Then drain and rinse it, and hand-squeeze it to damp only. This will allow you to remove the down by hand so it can be placed in a suitable "downproof" muslin or ticking bag.

Use these bags for down storage only while you are making your new clothing or bedding. When an item is ready for inserting the down, the wetting procedure is reversed. Wet the down, squeeze out the surplus water, then open the storage bag and remove the down as needed.

By wetting the down, you make it easily manageable and keep it from flying around the house. Water doesn't hurt down; it comes from waterfowl.

It is best to "spank" the down to one end of the container before wetting it. This will keep it gathered into one mass rather than clinging to the entire storage bag.

Fabrics for down products must be nearly "zero" in porosity, or you will be troubled with down seeping through and clinging to other fabrics. The outer fabrics should be water-repellent and durable, such as nylon or blends of long-staple cotton and nylon or polyester, or very tightly woven 100 percent cotton.

The inner fabrics that retain the down need not be water-repellent but should be long-lasting. I prefer ripstop nylon or equally downproof satins and taffetas.

Cotton ticking or unbleached muslins are popular down retainers for pillows and cushions. Slipcovers can be made for both decoration and ease in cleaning.

Once you have chosen your garment or bedding material and prepared it to receive down, measure the amount of down you want in each part of the item and insert it. If it is to be quilted later, simply insert the down and close up the compartment.

Use long stitches; ten to an inch is sufficient.

Dry out the down by tumbling on low heat, or lying flat in the sun, spanking it occasionally to break up the lumps. Then you can spank the down evenly throughout the compartment and make your cross-stitches. Don't worry about catching some down in the stitches; it can't be avoided.

If you are an experienced clothing maker, you can try a more complex item that uses tube construction, such as a cold-weather sleeping bag. I've found the best way to insert the down all the way in these tubes is to get a piece of plastic pipe that fits into the tube and then make a plunger to fit inside the pipe. This can be made with a dowel.

Work the damp down into the pipe, then shove it into the tubes with the plunger. You can work it down by shaking the bag. Don't worry about lumps; they will disappear when the down dries and regains its loft.

How to measure down: As an experiment, I recently ran a test on a pillow filled with northern goose down to find a simple way to measure and weigh down. I soaked the pillow in rather hot water, using a mild detergent, then shook the wet down away from one end. After squeezing away the excess water, I removed enough down to fill one cup and presed it firmly in the cup.

When the cup of high-grade down dried, it weighed 1.74 ounces (49.4 grams). Using this as a base, it would take roughly nine cups of wet down to make one pound of dry down.

Keep in mind, though, that this was unused down. Older down, down that is mixed with feathers, or duck down may require from 50 to 100 percent more to provide the insulation of northern goose down. Unless you know the quality of the salvaged down, it is always best to give yourself some leeway by increasing the down.

It isn't scientific, but one method of testing down for its quality is a simple fatigue test. If you sit on a down pillow or cushion for a half-hour, high-quality down springs back rapidly to its full loft. You can do the same thing with a jacket or sleeping bag by watching a half-hour television show while sitting on it.

Eddie Bauer

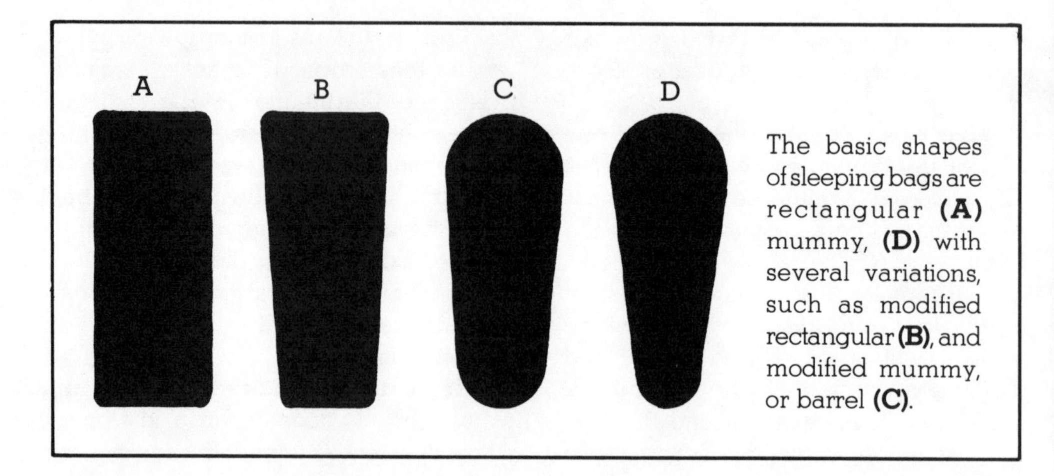

The basic shapes of sleeping bags are rectangular **(A)** mummy, **(D)** with several variations, such as modified rectangular **(B)**, and modified mummy, or barrel **(C)**.

they permit less convection of cold air. They are also more compact and easier to carry. Most mummy bags have an insulated hood with drawstrings around the shoulder area to keep them snug and prevent warm air from escaping. Hoods on the best models are so large that you can cover your entire head, leaving only a small breathing hole. Some hoods in the more expensive bags have a pocket on the ground side in which you can stuff a jacket, sweater, or inflatable pillow for more comfort.

The rectangular bags are usually heavier than the mummy shapes because they use more material. They come in weights suitable to a range of conditions, from warm summer nights to Arctic winter. They have an advantage in that they can be completely unzipped and turned into comforters. Or two can be zipped together and turned into a double bag. They are also ideal for backyard slumber parties and for use in RVs and summer-cabin vacations.

The next consideration is materials and construction. We discussed goose down and synthetics earlier. Since down is much lighter than its competition in synthetic materials, it offers the advantage of lightness.

An average goose-down bag will keep you comfortable in a wide range of temperatures, from ten degrees below zero to fifty degrees above. If the bag becomes too warm in the summer, it can be partially unzipped. For increased warmth, you can wear wool clothing or goose down while inside the bag.

Sleeping bag covers are made of a variety of materials. Down-filled bags must be made of strong and tightly woven fabrics to avoid leaking down through the weave. Hi Count Taffeta or ripstop nylon is used in many goose-down bags because it is virtually leakproof. Another popular material for the fabric that touches your skin is Tri Blend, a blend of polyester, cotton, and nylon that gives the comfort of cotton next to your skin while resisting shrinkage. A third material is Ramar, a rugged poplin blend of 60 percent cotton and 40 percent nylon that is extremely resistant to tears and abrasion and combines the strength of nylon with the comfort of cotton.

The methods of construction are important, too, because they help determine not only the warmth of the bag, but also longevity. The major methods for goose-down bags are as follows:

Sewn-through: The interior and

exterior fabrics are sewn through to form tubes, or channels, to contain the down or other insulation. This technique is common in lightweight bags for use in mild temperatures. It is of limited value for cold-weather camping because the stitching leaves uninsulated areas that allow heat to escape.

Slant-box: Slanted interior walls are created to form baffles that overlap all seams and down channels. The best bags use a stretch fabric in the sidewalls to accommodate the stretching and flexing of the body. This method eliminates cold spots and helps distribute the down evenly throughout the bag.

The most common construction methods for synthetic insulations are the following:

Double offset quilt: The seams of two quilted layers of insulation are offset to maintain even lofting and prevent cold spots. This method is used in bags designed for moderate weather.

Sandwich method: An inner layer of insulation is sandwiched between two quilted layers of insulation and attached to the bag at the edges to prevent shifting. This is a common method used for cold-weather bags.

Other factors to consider are those not immediately obvious, such as the following:

Proper stitching: Eight to ten stitches per inch is best. Fewer stitches may snag or loosen; more may cut or weaken the fabric.

Differential cut (in all bags): The outer shell of mummy bags is tailored larger than the inner shell to permit full lofting and allow you to flex your knees and elbows without creating thin cold spots.

Side block baffle (in goose down bags): A baffle is inserted opposite the zippered side to keep down from shifting from the top to the bottom.

Baffled foot section (in goose-down modified mummy bags): A baffle is placed across the end of the bag to maintain even loft and warmth.

Differential loft (in modified mummy

METHODS OF DOWN CONSTRUCTION

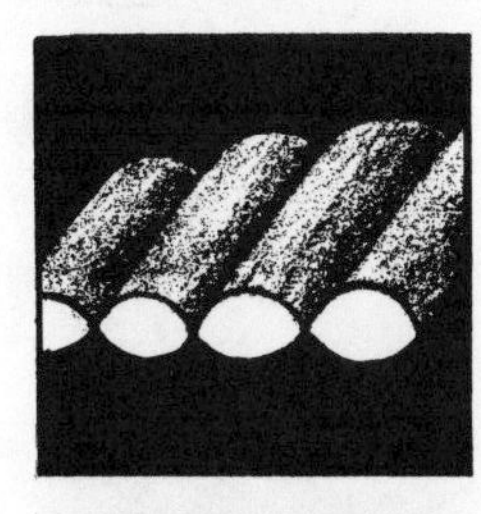

Sewn-through

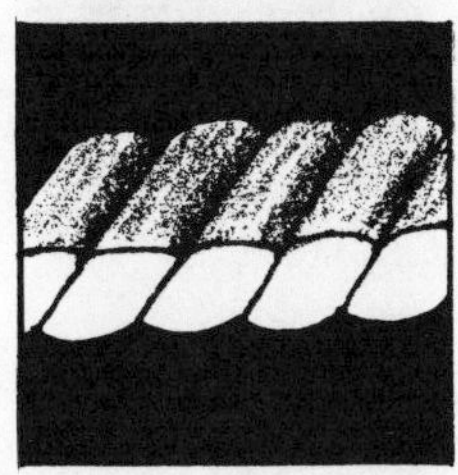

Slant-wall

Offset quilt

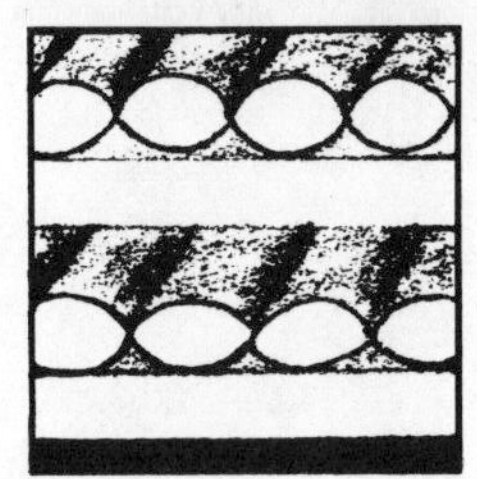

Double offset

bags): Sixty percent of the insulation is on top, where the body weight does not compress it, for better heat retention.

Visible features to watch for include the following:

Sturdy closures: Look for durable zippers, snaps, Velcro fasteners, and draw cords that operate smoothly. Do not settle for second best since these closures are subject to extreme stress. Two-way zippers are preferable because they allow easy and variable ventilation control.

CARE AND STORAGE OF SLEEPING BAGS

When you return from a camping trip, a sleeping bag should be laid out flat to air out and to allow moisture to escape. Then it should be stored loosely to allow the seams and insulation to relax. Do not store your sleeping bag in its stuff bag for prolonged periods because it may cause the insulation to break down and places undue stress on the construction. A good rule of thumb is to use the stuff bags only when transporting the bag.

Goose-down bags can be dry-cleaned or hand-washed. If you have yours dry-cleaned, choose a shop experienced in the care of goose-down products. Check with your outdoor equipment store for recommendations—all Eddie Bauer stores keep a list of shops with experience in cleaning down. A shop that does not have this experience can damage the down.

If you decide to hand-wash, use a detergent designed especially for down, available at most outdoor equipment stores. Wash gently and rinse thoroughly, then tumble dry at the cool setting. It takes a long time to dry a bag or coat thoroughly in an electric dryer.

Synthetic insulation can be machine-washed with a mild detergent. Tumble dry at a cool to warm (not hot) setting.

Do not dry-clean synthetic insulation. The dry-cleaning chemicals will damage the insulation.

PADS

Whenever down or synthetic insulation is pressed flat against the ground, it loses nearly all its insulation value. Down is more susceptible to this than synthetic insulation. Before the development of closed-cell plastic foam pads, campers had to rely on evergreen boughs, leaves, extra clothing, or more covers. The foam pads are seldom more than an inch thick and are cut to the same length as your sleeping bag. The most popular material is Ensolite, and the closed-cell construction forms a barrier between your body and the ground. It insulates so well that it enables you to sleep in comfort on a glacier or lake ice, provided you don't toss and turn and roll off the pad.

The closed-cell foam pads are lightweight and easily roll into a tight package, and they can be rolled separately or wrapped around your sleeping bag, giving it some extra protection if it is being carried through brush or across boulders. The pads can also be used for sitting on beaches and can be stacked in one end of the tent during the day.

Another alternative is the nylon-covered pad made of polyurethane foam. The washable cover protects the pad from snagging and tearing, and weighs only two pounds for one and a half inches of insulation and padding.

Pads come in a variety of thicknesses ranging from one-half inch to two inches. If you plan to combine backpacking with family camping, you will probably want a thinner pad to hold down both weight and bulk. A three-eighths-inch pad is about the minimum that will give you both comfort and protection from the cold.

AIR MATTRESSES

Most campers can be divided into two opposing attitudes toward air mattresses: those who love them and those who detest them. The detesters are frequently campers who tried air mattresses once or twice several years ago when they were as susceptible to tiny holes as a pin cushion. Or, if they didn't deflate halfway through the night, the mattresses created cold spots on the sleeper's back because the mattress did not offer insulation against the ground, or cold air was permitted to move around inside it.

Other campers consider an air mattress almost as important as a comfortable sleeping bag. These mattresses are especially useful in converting first-time campers to your form of recreation: a first-timer who doesn't sleep well is not going to be eager to go out again.

One alternative is to combine the air mattress with a closed-cell foam pad, provided you can get them to stay together through the night. They are also useful when combined with a camping cot, especially those that are already insulated.

A boon to campers' bedrooms is the relatively new self-inflating mattress. These are made of a plastic foam and are covered with a waterproof nylon. They are inflated simply by removing the cap and letting air rush in when the foam springs back to its normal shape. Since the covers are waterproof and airtight, the air inside is heated by your body and does not escape. The mattress rolls into a tight bundle for carrying and storing. A half-length pad weighs one and a half pounds, and a full-length (seventy-two inches) pad weighs two pounds.

COTS

An alternative to sleeping on the ground with the foam pads beneath you—admittedly, they are rather firm—are the insulated camping cots recently made available. Until these cots were developed, few campers used cots during cold weather because they were so cold. But now it is possible to sleep off the ground and still be warm.

Camping cots are popular in fixed campsites, such as hunting and fishing, and when space and weight are no particular problem. They are also useful back home for guests or for drowsing in the backyard. New compact models on the market fold into a small bag not much larger than a sleeping bag.

HAMMOCKS

These sling beds, invented by the Mayan Indians of the Yucatán Peninsula, are sometimes used by campers for fun or as a pleasant accessory. They are lightweight and stow away in a pack or cargo bag. They are easily strung between two trees neatly, and are a very nice place to spend those hot, lazy afternoons while you wait for the sun to drop behind the trees and the fish to start rising again. Few people use them as a bed, although the Mayas have for centuries.

CAMPER'S SECRET

Small pumps, either lightweight foot pumps or the old-fashioned bicycle tire pumps, are still the best for inflating air mattresses because you invariably get some moisture in the mattress when you inflate it with lung power. When you inflate a mattress with a pump, you won't hyperventilate and reel around the campground seeing stars at midday. A sprinkle of talcum powder or cornstarch helps keep the insides dry so that the compartments won't adhere to each other.

CHAPTER 6

THE CAMPING WARDROBE

Preparing for bad weather on a hot day is something like going to the supermarket just after a big meal—not a very appetizing task. When you leave for a camping trip with the sun beating on your back as you load the car, it is difficult to think in terms of warm jackets and sweaters and rain gear. The children will grumble that they don't want to carry those hot old sweaters and they simply hate those stuffy old raincoats Mother always makes them take along.

While there are many campers who seem to survive well in jeans, T-shirts, and tennis shoes, it is courting extreme discomfort, if not danger, to go unprepared for the worst. It is far better to follow the old adage: always expect the worst and you'll never be disappointed.

In warm, clear weather, jeans and T-shirts are perfectly acceptable for campground wear. But come evening or early morning, a goose-down vest or a heavy shirt or wool sweater will be welcome, if not necessary. A sudden shower can drop the temperature immediately, and if you are camping in the high country, that temperature drop can be sufficient to cause primary hypothermia, which is marked by uncontrollable shivering. (See CHAPTER 10 for more information.) Obviously for cold-weather camping your clothing considerations will be different. (See CHAPTER 11.)

Dressing for camping is essentially common sense. Yet, because camping seems removed from normal circumstances, many people are baffled by it and tend to forget that it isn't like home, where you can step inside the house

The layering system of dressing is essential for outdoor comfort and safety. Start with wool socks and warm underwear; add a pullover sweater over your shirt, a parka and hat, and a poncho or rainsuit for rainy weather.

when the rain comes or turn on the furnace when the chill arrives with darkness.

In general, the best insulation for clothing is natural fibers. Cotton is best for hot weather because it gives protection from the sun while allowing air to flow through the fabric to your skin. Wool shirts and pants are the best for warmth and also provide the most protection from moisture in case of rain or snow. Wool can absorb more moisture than any other fabric and still offer warmth. Goose down is the best insulation of all for jackets and sleeping bags.

However, synthetic fabrics have their own values and uses that make them indispensable for camping clothing. The best rain gear is made of nylon coated with a waterproofing agent because raincoats, rain pants, and ponchos made of this material are very lightweight and fold or roll into very small packages when not in use. Coverings for sleeping bags are made of various petrochemicals, as are tents and shell parkas that slip over sweaters and coats as windbreakers.

Many sleeping bags and lightweight parkas and coats are insulated with synthetic materials. These are recommended for beginning campers, particularly those with young children, because they cost less and are simple to keep clean. Also, the bit of extra bulk that synthetics produce is of less consequence than on backpacking or bicycling trips where weight and bulk are primary considerations.

A word of warning to fastidious parents: you may as well forget about keeping the children looking as though they are ready to sing in a school concert. Camping is a vacation from clean clothes and bodies. If your child is a magnet for dirt like Pigpen in the Peanuts comic strip—and most children are—you had best forget about your rules of cleanliness except at mealtime and bedtime. Getting grubby is part of the fun of a vacation for children, and getting to wear the same grubby jeans two days in a row is positively wonderful to active youngsters.

Beginning from the ground up, these are the basic articles of clothing suggested for a weekend camping trip. An additional change of outerwear and two more changes of underwear will usually get you through a week with perhaps one laundry session.

FOOTWEAR

The old joke that the big toe is the thermostat of the body has an element of truth. If your feet are cold or causing pain, you are far from a happy camper. Footwear for outdoor use has become as specialized as all other types of clothing, and it is possible to have a closet virtually filled with shoes and boots for every conceivable use.

For the purposes of this chapter, we will concentrate on footwear suitable for summer camping. CHAPTER 11 deals with footwear and other clothing needed for stretching the camping season throughout the year.

A cardinal rule of camping is never to wear a pair of shoes or boots that haven't been worn for at least several hours before the trip. Only by wearing them will you break them in and find potential blister spots or other problems.

Most campers can get along very well with no more than two pairs of shoes or boots, usually one pair that is suitable for day hikes and normal camping activities, and a second pair for wet-weather use. Since the majority of camping areas are near streams, lakes, or an ocean, a

lightweight pair of waterproof boots will permit you to wade streams or explore the shores without getting wet.

In recent years the National Park Service has begun outfitting some rangers with lightweight walking or running shoes to protect the trails and meadows from the deep indentations made by the lug-soled boots commonly known as waffle-stompers. The old environmental slogan "Take only photographs, leave only footprints" is being rewritten to avoid even footprints if possible, especially in areas of heavy use.

Blends of wool and nylon, or cotton or pure wool socks are preferred by most campers. All are ultimately more comfortable against the skin than synthetics. The first choice is usually blends of wool or cotton with nylon or polyester, which are not only cheaper but also help the socks retain their shape through repeated washings.

Moccasins are very popular among campers who stay at one fixed camp. They are a boon for getting in and out of tents (where you should never wear shoes to avoid damaging the fabric) because a tap of the toe against the heel of the other foot removes them easily.

For those cool evenings and mornings, particularly in the high country where the temperature might drop to freezing at night, goose down-filled booties are excellent for around the camp. Many models have leather or synthetic soles with down insulation from the top of the foot to the ankle. These are easy to keep clean and can be worn in the sleeping bag for a bit of extra comfort on cold nights.

An alternative is slipper socks, with the same sole but heavy wool sock material over the ankle.

In most situations, however, a pair of comfortable shoes good for walking along

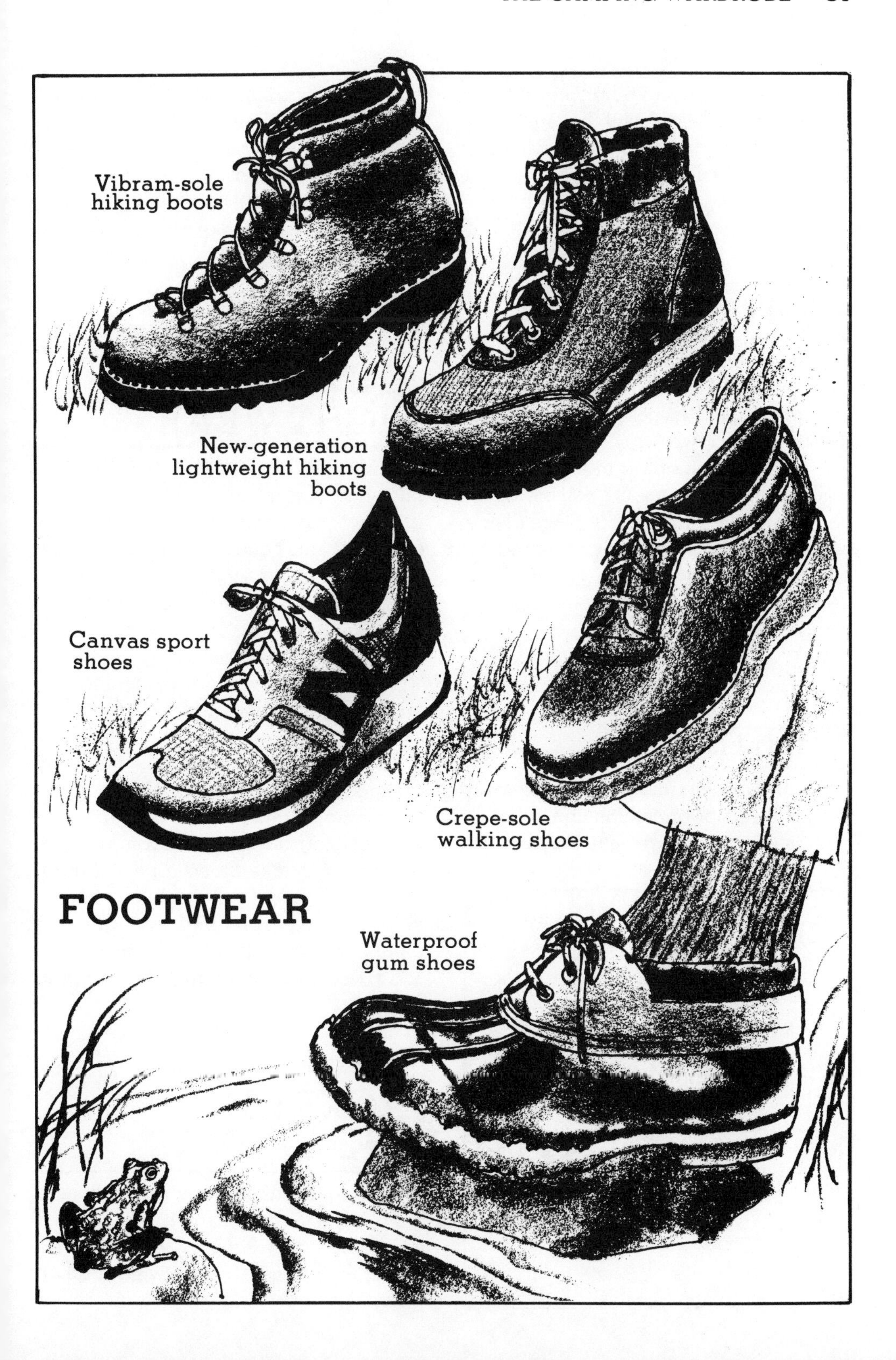
Vibram-sole
hiking boots
New-generation
lightweight hiking
boots
Canvas sport
shoes
Crepe-sole
walking shoes
FOOTWEAR
Waterproof
gum shoes

established trails, plus a lightweight pair of waterproof boots, will be sufficient. Extra warmth can be gained by slipping on a pair of wool socks over your regular socks.

PANTS

Jeans and shorts are suitable for most outings. Try to keep one dry pair always on hand in case of unexpected dunkings or showers.

If you are camping in areas where chilly nights are common, a pair of woolen pants can be taken along in place of one of your pairs of jeans. They are good for evening wear and in case of rain since they continue to provide some warmth even when wet.

SHIRTS

Sturdy washable shirts, preferably long-sleeved for protection against night chill, sun, and insects, are suitable. As with pants, one shirt should be woolen to ward off the chill and damp weather. T-shirts are fine for the heat of the day, but they should be considered a foundation from which you can build layers of clothing, from shirts to jackets or rainwear.

CAMPER'S SECRET

When moving from campground to campground, take advantage of the car's motion by letting it help with the laundry. Place the dirty clothes in a sealed container with detergent and water and wedge it between other containers. The car's motion will act as an agitator. When you arrive at the new camp, just rinse your wash and hang it up to dry.

UNDERGARMENTS AND NIGHT WEAR

Although some campers have been known to buy military-style khaki undergarments, not because they are superior garments but because they don't show dirt as readily as white ones, this is still playing sleight-of-hand tricks with yourself. Comfort is the primary consideration. Be sure they don't bind or chafe.

The only other suggestion is that you take along either wool-blend long johns for colder weather or goose down-insulated sleeping garments for the really cold nights and mornings.

Some campers prefer sleeping in their clothing, but this isn't recommended. It is better to take along a pair of lightweight pajamas. If you must sleep in your clothes, make sure they are dry and clean, or at least free of caked dirt and sand to avoid damaging the sleeping bag.

JACKETS AND PARKAS

Under normal summer conditions, a lightweight jacket is usually sufficient for car camping. A school jacket or a ski jacket is fine, as is a lightweight down jacket.

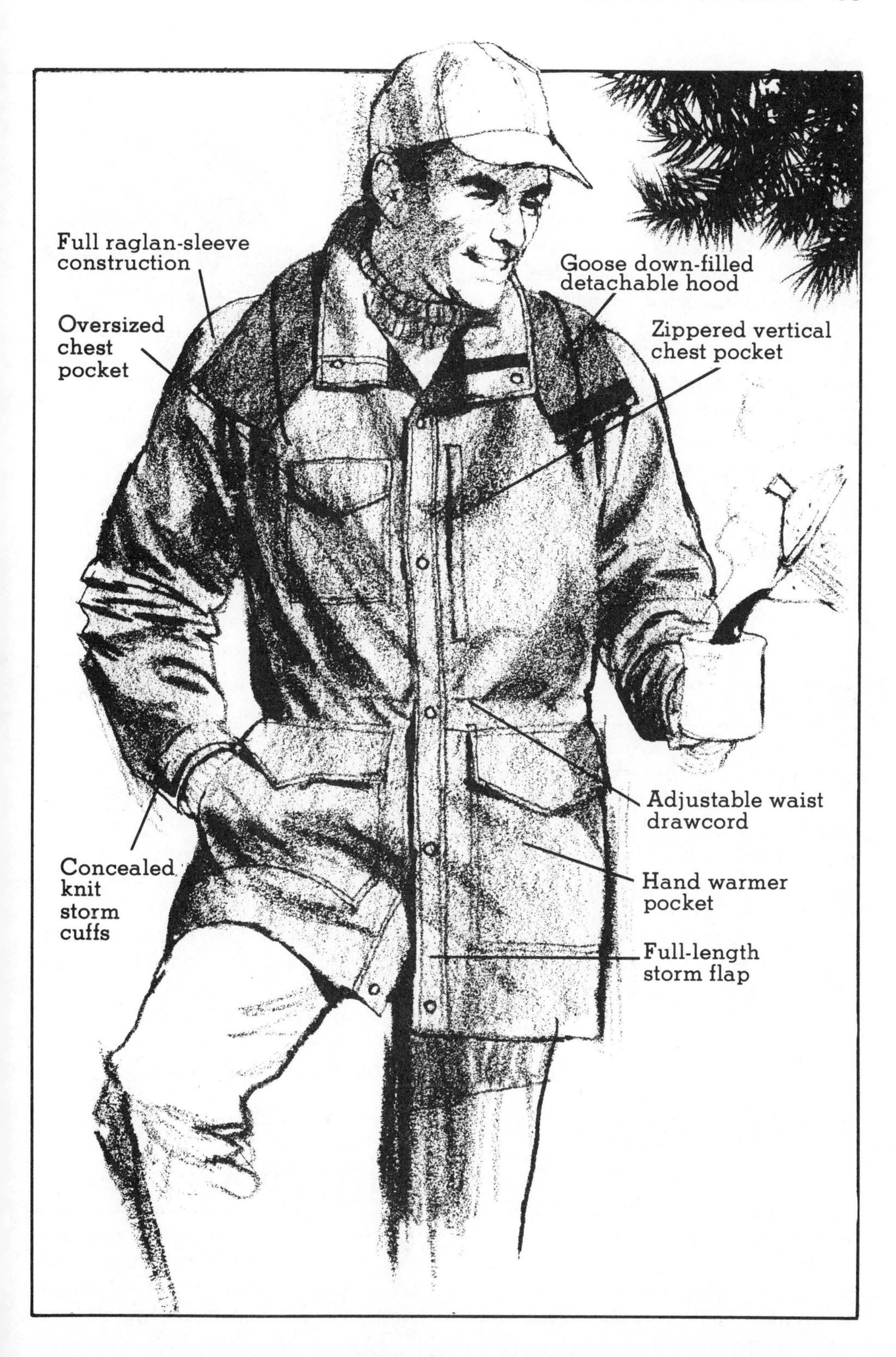
Full raglan-sleeve construction
Oversized chest pocket
Goose down-filled detachable hood
Zippered vertical chest pocket
Adjustable waist drawcord
Concealed knit storm cuffs
Hand warmer pocket
Full-length storm flap

An alternative is to dress strictly on the layer system and, instead of carrying a bulky jacket, use a nylon shell that is called either a wind parka or simply a shell. This is extremely lightweight and will fold into its own pocket to make a wallet-sized package. This, plus a warm shirt and a wool sweater, offers both warmth and protection from light rain.

Parkas come in a variety of styles and are popular casual clothing in many parts of the country. Some are lined with a satinlike nylon or, for colder conditions, with a flannel lining.

Among the many advantages of parkas over jackets are that parkas drop below the hips and keep your waist and kidney area warm, and that they are usually constructed with a hood.

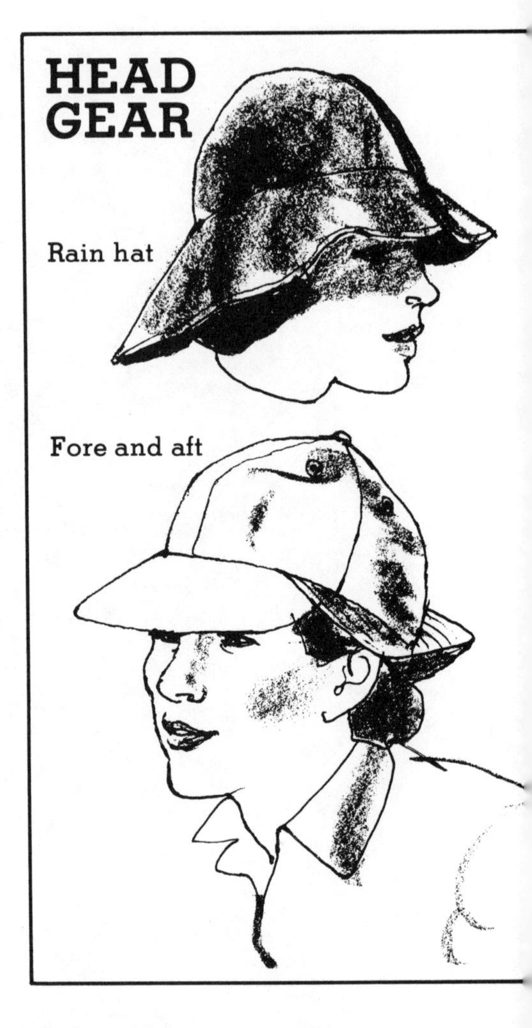

GLOVES

Each member of your group should have a pair of gloves or mittens, even if you're camping in August. Not only do they keep your hands warm on chilly nights, but they are also useful in handling hot pots and sooty grills and pots, and they protect your hands while performing campground chores. In extremely cold weather, mittens are far superior to gloves, especially those with removable wool liners.

HEAD GEAR

You should have some kind of head gear along even though you may never wear a cap or hat in town. More heat is lost through your scalp than any other part of the body, and few things feel more welcome in chilly or wet weather than a warm wool watch cap or ski cap. Although nearly all parkas have hoods, few of those designed for summer use have much insulation. Thus, a watch cap or similar cap beneath the hood makes a big difference.

RAIN GEAR

Rain gear comes in a variety of coated fabrics and designs, and each has its own avid supporters. Until only recently most campers said that none of it was

really good; some kinds just weren't as bad as the rest. This was because rain coats and pants were almost invariably stiff and heavy. Since the fabric kept moisture out, it also kept moisture from the body in. Before long, a camper was as damp from perspiration as he or she would have been from standing outside in the rain.

The perfect rain gear still hasn't been invented, but a new fabric on the market is getting closer and closer. The first generation of rain gear was the famed "oilskins" worn by seamen. These were much too stiff and heavy for campers, however, so when nylon came onto the market around World War II, manufacturers seized upon this lightweight fabric as an alternative. By impregnating nylon with waterproof solutions, they gained a very lightweight fabric that was also much more pliable.

These fabrics still had their disadvantages, and it wasn't until the 1970s that a third generation of waterproof fabric was developed. Known in the trade as PTFE fabrics, their trade names include Goretex® and Klimate®.

This fabric was a miracle for outdoors lovers. Although it is porous and

CAMPER'S SECRET

A poncho is one of your most versatile articles of clothing. It can be used as an emergency rain fly, as a sunshade, as a windbreak, as a ground cover for your tent or a seating area for a picnic, as a privacy curtain in big tents or around the latrine. Most come with grommets on the corners and sometimes scattered along the edges so they can be laced together. If they don't have grommets when you buy them, grommet kits are inexpensive and easy to use. Another use for a poncho is as a temporary tent to mark a campsite in a busy campground. Stretch out the poncho like a pup tent, then dash back to the ranger station to pay for the campsite. And two of them strung over a line between two trees makes a good fair-weather tent or a place for you to stretch out away from the crowd for an afternoon nap.

permits perspiration vapor to flow through the fabric and disappear, it is totally waterproof and will not permit moisture to come through to your clothing and body. This microporous film is sandwiched between two layers of outer and inner fabric, such as cotton or nylon (usually the latter).

The first batches of the material had problems. It had to be kept absolutely clean to function; even a fingerprint could plug up the micro opening. And it had to be cleaned frequently with denatured alcohol to keep it efficient. But new laminates have been developed that can be washed using warm water and mild detergents.

Since the introduction of PTFE fabrics on the market, field testing has gone on constantly and modifications continue to be made. But no other product on the market approaches it for practicality.

Following is a rundown of basic rain gear designs.

Coat: Many campers still prefer a simple waterproof coat that can be matched with pants. The coat can be zippered, buttoned, snapped, or pulled over the head. Most designs have a hood attached, and most have some method of keeping the cuffs snug, using Velcro strips, elastic, or snaps.

Ponchos: These are popular with campers everywhere, as they can be slipped on over a backpack and they allow lots of room inside for movement and ventilation to evaporate perspiration. They can also double as a ground cloth for tents, as a tarp for a tent vestibule, or as a shelter over the cooking area. In a sense, ponchos are little more than a tarp with a head hole and hood (which can be tied off to form a waterproof tarp) and snaps along each side.

Their disadvantages are obvious. Ponchos tend to flap in the breeze, even when tied securely to the body with a belt or piece of rope. They catch on underbrush, make a lot of noise, and can be awkward around the campfire or stove.

Cagoule: These are pullover parkas that reach down to or below the knees. Cagoules are almost portable tents because most are cut spaciously enough for you to sit down in them, pull them over your knees and feet, and let the rain fall. The bottom can be wrapped upward and tied around the waist for more freedom while walking. Because they become personal tents, so to speak, they are popular with mountaineers and other people in harsh climates. They are an extension of the parka design used by Eskimos for centuries.

CAMPER'S SECRET

Since up to half of your body heat is lost from the head and neck, where the blood flows closest to the surface, a wool turtleneck sweater and a wool cap will conserve most of that heat.

PART III
AT THE CAMPSITE

CHAPTER 7

SETTING UP CAMP

FINDING THE CAMPSITE

If you plan to stay at one campsite longer than an overnight stop, you should try to shop for it as carefully as you would a hotel or resort. This will, in effect, be your neighborhood for the duration of your visit, and you will want it to be as pleasant as possible.

First, do some research before leaving on the trip. Word of mouth can be trusted only so far, and brochures usually accentuate the positive and ignore the negative aspects. Government campgrounds are usually described reasonably accurately, but these information sheets or little tent symbols on maps won't tell you if you will be camped next to a swampy area, if this particular campground is a popular hangout for teenagers with their hot cars, or if the water tastes as though it had been drained from the radiator of an elderly tractor.

These are the exceptions to the rule, of course, but such things do happen if you're not careful in your campsite selection. There are too many campgrounds available throughout America to have to put up with nuisances on a regular basis.

Federal and state campgrounds are usually the best maintained, especially the more elaborate campgrounds in national forests or those in large parks. Some of the nicest and most peaceful ones are tucked away several miles from the nearest busy highway or town. The

farther you go from the rest of the population, the more privacy you usually can expect.

If you are going out for the first time, or going into an area new to you, it is best to load up on brochures and maps well in advance of the trip. Write the federal and state agencies listed in PART IV (RESOURCES) and get as much information as possible. Check with your local outdoor store for suggestions and other publications on camping in the area you will visit.

Most dedicated campers soon accumulate a bulging file of maps, books, brochures, and other material on campgrounds. You can store this literature in a filing system of your own design. An inexpensive cloth or plastic case will do, as will a briefcase that has seen better days. Take along the literature you'll need for the trip at hand and leave the rest at home.

If you are traveling in an RV, set aside an area near the driver's compartment for maps, guidebooks, and other literature and leave it in the vehicle all the time, extracting only those items needed for each particular trip.

Nearly every major camping area of the country has a selection of booklets and books on the natural history and hiking trails of the area. The National Parks Association connected with the major parks in the system is a good source of guides to the parks and the surrounding area.

A smart rule to follow while searching for the elusive perfect campground is to give yourself plenty of time to explore. If you are leaving on a Friday for a weekend's camping and don't know

the area well, find an alternate campground for Friday night on the way, then leave early the following morning to search for a spot to stay Saturday afternoon through Sunday afternoon.

Once you've settled on the campground, you need to stake out the specific site. Ask several experienced campers to describe the perfect campsite, and you'll get a wide variety of responses. It may be easy to visualize the perfect one, but you'll probably never find it. Instead, after trying several, you will soon develop your own list of favorite places and keep returning to them from time to time throughout your camping career. Some campers become attached to a single site and visit it exclusively for their only camping trip of each year.

The basic considerations in choos-

CAMPER'S SECRET

When camping in a desert, never set up camp in a wash, those dry stream beds that look as though water hasn't flowed through them in the past century. Flash floods are common in the desert, and a rainstorm several miles away from you that leaves your area dry can send a flash flood down the wash in which you're camped with no warning. Many campers have been drowned in this manner. Keep to the high ground, always.

ing a good noncommercial campsite include scenery, prevailing weather, privacy, nearby activities, distance from home, and other similar factors. In general, a good campsite should have these characteristics:

- a good source of pure water nearby
- a site high enough to avoid being inundated by a nearby stream
- a site that is level and not subject to flooding in case of rain
- soil that is hard-packed where tents and kitchen will be placed so that tent and awning stakes will hold
- a good source of firewood
- a frequent breeze to help keep the bug population down and to blow away campfire smoke
- sunlight in the morning and evening but shade in the hot midday
- a suitable spot for a latrine, if no public bathrooms are available
- some security against fires
- nearby trees for stringing clotheslines and hanging food to keep away from wildlife

Hang food out of reach of bears and other animals.

If you are a beginning camper, you might consider making your own check list of campsite requirements, but after a few outings you will instinctively look for the characteristics you require.

Some campers start with a list of requirements, then throw it away after a few trips because they find themselves returning again and again to places that offer few of these amenities; they simply like those more obscure sites because they are close to a waterfall or are high on a shelf of land above a canyon, or for other personal reasons. Sometimes you will find a campsite that is tucked back in a dense grove only a few feet from a well-used trail, giving you the privacy of a patio just off a freeway exit. Campers have been known to become very secretive about their favorite haunts and to share their locations only with family members or close friends—something like a condominium partnership.

Setting up a campsite is largely a matter of personal preference, much like decorating a home. Yet there are basic components of a campsite just as there are of homes: living area, bedrooms, kitchen, and bathrooms.

In most campsites where campfires are permitted, the fire ring is usually in the middle of the site. The natural terrain will dictate, to a great extent, where the other components are placed. You will want your tent or tents erected on level spots to avoid sleeping on a hillside, and preferably at least fifteen feet from the campfire to avoid the danger of sparks landing on the tent. The kitchen should be convenient to the campfire, yet a few steps away so that people can ring the fire for warmth or run from the smoke without blundering into pots and pans. The bathroom will be as far away from the campsite as possible and downwind from the prevailing wind.

In more organized campgrounds, such as most government and privately owned sites, these factors are already taken care of for you. All you need do is park in the assigned space, pitch the tents, and unload the kitchen onto the picnic table.

When you pitch the tents, it is best to have them facing away from the prevailing wind. This helps keep the breezes from whipping the tent door during the night and will prevent rain from being blown in. This is a consideration only, and not critical with most modern tents, which have rain flies that completely cover the entrance. Also, when more than one tent is being used by a family, it gives a sense of security to have tents facing each other.

It is a rare camper who cannot set up a camp without rearranging the "furniture" according to personal tastes. Many campsites have stumps or logs or big blocks of wood suitable for sitting or use as a kitchen table. Part of the fun of setting up a campsite is organizing your temporary home.

A good schedule for setting up camp begins with reaching the campsite well before dark to save much frustrating fumbling in the dark, and to give yourself a wider selection.

On arrival at the site, most campers first set up the tents. Then the sleeping bags are stretched out or hung from nearby trees or over bushes to air out before dark. Then the kitchen is laid out and a campfire started. Keeping in mind that many government campgrounds, particularly national parks and national

BEACH CAMPING

Camping along the seashore and some stretches of the Great Lakes presents different problems for campers to resolve. The major one, of course, is the presence of sand. The other, on seashores, is the tides.

You will need something to replace your tent stakes because they will not hold in sand. The "deadman" weights made of stones or pieces of boards will help, as will snow anchors made for winter camping. Both of these anchors are tied to the tent lines, then buried and stamped down firmly.

Since you will be standing, walking, and sitting in sand, it is going to be everywhere, especially in the bottom of the tent. If you have large pieces of tarp or unused ponchos, these should be stretched beneath the tents and out a few feet to form a patio of sorts. Sand should be swept out of the tents constantly, and everyone should sit down in the entrance of the tent and remove shoes and socks and shake them out before entering the tent. Sand is an excellent corrosive and will wear holes in the floor of a tent rapidly. At the best, it will weaken the floor more in a weekend's use than a month's normal use.

Another potential problem on seashores is the tides. You will see the tideline along the beach, and no matter what the tide table says, you should never take a chance on sleeping below the last high-tide line. Few things are more frightening and dangerous than waking up with saltwater lapping in your tent.

wilderness areas, do not permit campfires except in the heavily organized campgrounds, you should be aware of such restrictions before leaving home. In any event, you should always carry the campstove and ample fuel for the trip. It is best to call ahead and see if firewood is available at the site.

If you are in a more primitive camping area with no firewood available for purchase at the ranger's office, wood gathering is one of the first chores after a site has been selected. Government recreation specialists speak of the "human browse line," which refers to dead branches on trees that people pick off for firewood. You will note in heavily used camping areas that dead branches have been broken off up to a height of six or seven feet. This does not damage the trees, but live trees and saplings should not be cut unless prior permission is obtained from whoever owns or manages the land.

If you find poles lying around a camping area, these should not be burned or damaged, because someone no doubt left them there for tent poles or similar uses. Even though you may not need them, the next camper might be using a tent that requires such poles.

One of the most important rules of outdoor etiquette you should always practice is to leave a supply of wood and kindling behind when you leave a campground. This rule harkens back to the pioneer days when users of wilderness cabins always left a wood supply in case the next occupant might arrive suffering from hypothermia, frostbite, or worse, and an immediate fire was a life-or-death matter. Granted, this seldom happens in lowland camping during the summer months, but good manners require that we leave the campsite in better condition than we found it.

CAMPFIRE BUILDING

In areas where campfires are permitted, everyone in camp should be taught how to build wood fires quickly and simply. As each member of the family becomes more and more adept at getting them started, the conditions should be made more and more difficult so that fire-building becomes easy for everyone.

You will find that children enjoy this combination of learning and play; in fact, fire-starting holds what seems to be a primeval attraction to children. The problem is keeping their enthusiasm under control. It is sometimes difficult to convince children of the importance of fire safety in the outdoors when they

FIRE-BUILDING

Tepee fire

Log-cabin fire

Star fire

have used almost a box of matches trying to get dry wood to burn with no success. Just as it is easier to grow weeds than vegetables, it seems very difficult to start a campfire but simple to start a forest fire.

The one essential to starting a campfire is good tinder, something that will burn easily and provide enough flame and heat to ignite the other components of the fire.

Some of the best tinder comes from birch bark, dry moss, dry pine needles, crushed cedar bark, and very small dry twigs. An alternative when you can't find small tinder is to make dozens of notches in a piece of dry wood, leaving the chips and slivers still attached. This will give you both tinder and fuel in one piece.

The Boy Scout test of starting a fire with one match is excellent discipline for everyone, beginner or seasoned camper, because it forces you to construct the fire properly the first time.

Tepee fires: This is the classic structure for campfires. The tinder goes on first, with space between the twigs or needles or whatever the base is. Then slightly larger sticks are stacked loosely around the center, with a space on the windward side for air, either wind or your breath, to enter.

A common mistake made when building this fire—and all others—is to put too much fuel on too early rather than let it get a good start and gradually add the larger fuel.

When the basic tepee has burned, it will leave a good bed of coals for the remainder of the wood, which can be laid flat.

Log-cabin fire: This is a good fire to use for quick warming rather than simply cooking. As with the tepee design, this one starts with the tinder; then add larger sticks stacked log-cabin or rail-fence style all around it, and finally cap it off with a loosely built lid of firewood. This fire will heat much faster and create a bigger blaze than most other types.

Star fire: This is a flat version of the tepee fire and gives you more control over the fire and the cooking heat it produces. As the sticks burn, you shove them into the fire in amounts sufficient to keep the fire going. One chore for a member of the party is to keep a good supply of tinder and kindling available

CAMPER'S SECRET

Unless you are camping right beside a stream or lake, always keep a bucket of water handy in case of fire. Plastic jugs or buckets that collapse are excellent for this. You might carry one that is marked "Fire."

CAMPFIRE SAFETY

More and more camping areas are either banning campfires entirely or restricting them to designated fireplaces or concrete-ringed firepits. It is illegal in many camping areas to cut down any timber, dead or alive, and consequently more and more campers are having to bring their own wood or charcoal. It is a disappointing situation for most campers, who find it difficult to imagine camping without the smell of woodsmoke and the fascination of staring into a campfire at night.

However, fires are still permitted in many areas and you should never lose your fire building skill or fail to pass it along to your children, if for no other reason than for use in emergencies. Following are some dos and don'ts of campfire building.

- Don't build new fire rings when one is already there.
- Don't build a fire on the edge of a water source. Keep it as much as 200 feet from the water to prevent pollution.
- Don't build a fire next to a tree, near exposed roots, near vegetation, or on anything other than bare ground.
- Don't build a fire with a clean rock as a reflector.

- Don't burn any garbage other than paper and organic trash or other materials that will burn. By putting paper, metal, glass, and other material in the fire, you're only starting a garbage pit.
- Don't leave the campfire unattended.
- Burn only dead wood that is lying on the ground.
- Keep your fire small.
- Use wood that doesn't create sparks, if possible. Green wood, some driftwood, and woods with lots of pitch such as pine, alder, spruce, and hemlock are notorious for sparks.
- When you're done with the fire, literally drown it, then keep poking around the deluged area for bits of burning material beneath the mud. Smooth over when completed.
- If you must build a fire in an area with lots of humus, carefully cut out chunks of the sod and stack them neatly to one side. Dig down to bare dirt and ring with rocks. Then build the fire. When you're through with the fire, drench the entire area with water, and when you're certain it is completely out, remove the rocks and fill the hole with enough dirt so that you can replace the sod to its former depth. If you're neat, nobody will suspect a fire has been there and you've treated the forest properly.

at all times. It is a good idea to store enough for a fire in a container in the corner of your tent or some other dry place.

Trapper fires: After the fire is built, you need something to put the skillets, pots, and pans on. This fire has logs or rocks on each side of it to serve as racks and to concentrate the heat. It should be laid out with the opening in the direction the wind blows. It can also serve the purpose of drying out logs before they're put on the fire.

Reflectors: To send the heat toward your tent or to keep it away from equipment stacked nearby, you can build a reflector of a variety of materials. One of the best materials for the reflector is aluminum foil, but rocks or stacks of wood will also work.

Trench fire: This is a version of the trapper fire, except that it is in a hole in the ground with the sides serving as a rack for your utensils. These are good fires for sandy beaches and you can line the holes with rocks. After you have made certain the fire is completely out, it can be buried with no trace.

As everyone's skill in fire-building increases, the task can be made more and more difficult—setting time limits, using wet wood, or limiting the campers to one piece of wood for building a fire. These tests of skill not only help fill a long afternoon or evening, but also teach survival skills that will make campers more secure in the outdoors, and better camping companions, too.

Preparing the site: In most camping areas you will be using the same firepit others have used before you. If an established fire ring is available, always use it instead of creating another one since fires damage the soil for a few inches down. Also, if the fire ring is changed, before long the entire campsite will have an ash and charcoal floor.

Be sure all vegetable matter—roots, branches, leaves—is removed from the ring. Fires may be difficult to start, but once they get started on an underground root or wood covered with soil, they will travel along it.

Do not build fires against boulders. Although boulders make excellent reflectors, the burn marks on them will last for years, if not decades.

If you are camping on a sand or gravel bar along a stream or lake, try to burn everything in the fire so that all you have left are ashes, which can be buried and will soon wash away.

Putting out the fire: When you are packing to leave the campsite, permit the fire to burn down before you leave. Completely drench the fire area for several feet around it. Keep pouring water into the soil until the entire area is soaked, and keep testing the surrounding soil for hot stones or underground pieces of wood still smoldering. It is far better to leave a sooty puddle behind you than a smoldering stick or root.

Leave enough tinder and wood behind for the next party to get a fire started. It is not only the courteous thing to do, but also a safety measure in case the next group has someone suffering

CAMPER'S SECRET

You can make your own fire-starter briquets by using an empty carton. Fill the depressions with a mixture of melted paraffin, sawdust, or wood chips, then cut them apart into individual briquets, wrap in plastic, and carry for use as a fire-starter.

from hypothermia or, less drastic, they arrive late and need a fire for both heat and light while setting up camp. In any event, it is simply basic outdoor good manners.

In the colder regions of North America, Canada and Alaska especially, it has always been a tradition to leave unoccupied cabins with at least as much firewood as when you arrived, plus good tinder or kindling and a good supply of matches and food. Many lives have been saved through this courtesy, and it should extend to all camping areas.

LATRINES

Next to sleeping outside, outdoor bathroom facilities are a major hurdle for many first-time campers to overcome, particularly those who have no experience with life beyond the city limits. This problem is similar to that of the farmer whose wife repeatedly asked him to build a patio so they could eat outside in good weather. "I spent half my life eating in the house and going outside to the bathroom," he replied. "Now you want me to eat outside and go in the house to the bathroom!"

The camp's bathroom facilities are not an insurmountable problem, of course, and after the initial camping trip, most people are so enthusiastic over the pleasures of camping ("The coffee **does** taste better") that the lack of modern bathroom facilities becomes only a minor inconvenience.

If your campsite is not within an established camping area with running water for showers, bathroom, and laundry, there are a variety of choices for a latrine.

Some campers prefer carrying portable toilets along that can be set up behind a screen of trees or brush, or behind a tarp or poncho. Some models come equipped with chemicals and disposable bags, and some simply have a plastic bag that is disposed of after each use. Most of these require a bit of pioneer carpentry to build a base.

Other models are self-contained and designed for use on boats as well as for camping. The base is built in and can be placed on the ground almost anywhere.

This isn't a factor at organized campgrounds where an outhouse is always part of the equipment, and in the more elaborate campground flush toilets are used. However, a word of warning: unless the outhouses are cleaned frequently, using them can be unpleasant. You may want to carry your own portable toilet if you plan on camping at several places you haven't visited before. Again, call ahead to the managing agency for information on this and other areas of concern. We have come so far since the pioneering days that the vast majority of Americans have never used anything other than indoor plumbing, and the thought of not having it available may come as a shock. Thus, it is important to introduce children to camping early. In the case of adults, it might be best to stay at organized campgrounds first, then gradually ease them into the wilderness in stages.

In more primitive situations, however, most seasoned campers eventually come to depend on a latrine that is dug in the form of a trench and filled in as used. Other choices are what some call "cat toilets," meaning you dig a hole as needed and fill it in after use.

All bathroom facilities should be downwind from the camping area, and should never be within two hundred feet of a stream or lake. Human waste should

LATRINES

Portable toilet

Trench latrine

be buried at least six inches or even a foot deep. A good safety precaution is to leave a mound of dirt over the covered hole so other campers won't accidentally dig in the same spot.

It is best to scout out bathroom areas in different directions from camp, one for males and the other for females.

Be wary of building them near places frequented by insects, and be particularly watchful for hornets that may be nesting nearby.

GARBAGE

The whole subject of garbage can be stated simply: pack it in and pack it out. Unless you are at an established campground with frequent pickup of garbage cans in the area, do not scatter leftover food around the area on the assumption that it is biodegradable. Even though it may be, it won't rot and nourish the soil for some time, and it will turn wildlife nearby into scavengers dependent on campers for food. It will attract some of the less desirable elements of the animal kingdom, such as bears, porcupines, and skunks, who will lurk around the campground as a threat to your food supply. Besides, garbage stinks. You may not throw much away, but the campers before you and after you might feel the same way, and before

"Cat" toilet

CAMPER'SECRET

Although most campers prefer keeping the toilet paper in the main camping area—if nothing else, its absence serves as a signal to others that the latrine is spoken for—some store it on site in a tin can suspended upside down from a tree branch or stuck in the ground nearby.

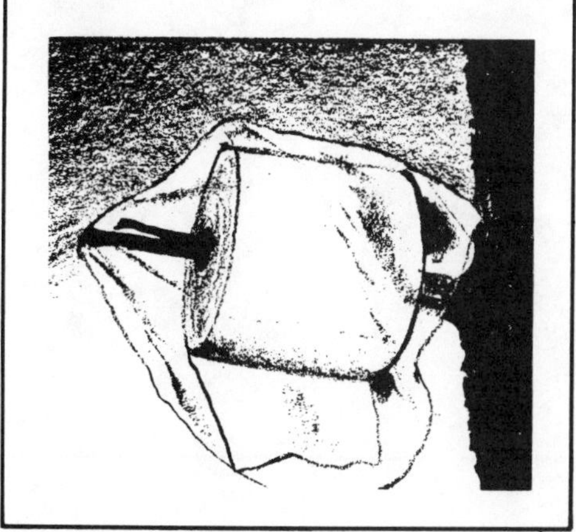

a camping season ends, that campsite can become ringed with garbage.

Some campers dig grease pits to dispose of bacon fat and other types of grease. This should be avoided, too. Grease will burn in the campfire. If you are camping in an area that prohibits campfires, store the grease in a plastic container or a tin can and pack it out with you.

Most camping food comes in plastic or foil packs that can be wadded into small balls that weigh virtually nothing. Take along a nylon sack with a drawstring and stuff all wrappers and containers into the sack, carry it back to the nearest garbage drop, then wash the sack when you return home.

DIVIDING THE LABOR

An odd thing happens to children on camping trips. The lazy one whose idea of a strenuous evening is getting up occasionally to change television channels will often be the most willing worker on a camping trip. While going outside to get a load of firewood for the family fireplace at home is beneath this child's dignity, out in the woods it is often a grand adventure that he or she is unwilling to share with other siblings.

The mechanics of erecting a tent fascinate most children, and learning how to operate a white gas campstove is akin to learning to drive an automobile. Nearly all camping equipment has a toy-like mystique for children, and organiz-

ing work parties and dividing the camp labors usually is not the problem encountered when assigning home chores.

Parents of today who unthinkingly assign camp chores according to gender are going to receive one of the short, pithy lectures on equal rights between the sexes. Girls aren't automatically the cooks and dishwashers, nor are the boys wood gatherers and splitters. Your past experience with the camping companions will dictate how to assign tasks.

Some families work best on the volunteer basis. Something needs to be done, and everyone pitches in to help. While one is mixing the pancake batter, another is mixing breakfast juice and another is replenishing the firewood or getting the campstove going.

In most cases, though, it works best to organize a rotation system before the trip begins and give the children a voice in the proceedings. Who will do the dishes the first day? Who will do the laundry, or will each person be responsible for his or her own laundry? If you are eating freeze-dried or dehydrated food for most of the trip, anyone can prepare it. But who knows how to prepare a mushroom omelet?

Nearly every family has a trader in its midst who can, through a series of complicated maneuvers, spend most of the trip standing around watching while others do all the work. While nothing is wrong with parents letting the children trade jobs occasionally, they should step in whenever they see one child on the verge of taking advantage of another. A feud between children can cast a pall over a camping trip and should be avoided.

Experience at home and on pre-

vious family trips will indicate whether to depend on the volunteer system or to assign tasks. Often it is a mistake to force someone to cook who doesn't know the difference between a fire with moderate heat and one with extreme heat. Each person should, however, know all camp chores, and if you deem it necessary to assign unwanted chores, especially the cooking, let the others decide (within reasonable limits) what they will cook when their turn comes.

Some families prefer making the cook responsible for everything, start to finish, including washing the pots. The wood gatherer knows in advance how much to bring, such as enough for the rest of the evening and kindling and wood to start tomorrow morning's fire. When dividing the camp chores into units, be as specific as possible on what is expected.

Each person should be responsible for his or her own camping equipment; sleeping bags, clothing, eating utensils, and toilet articles. Those using each tent are responsible for erecting it (with perhaps a bit of help from adults), keeping it clean, and stowing it away when the trip ends. This kind of responsibility should begin on the first outing, no matter how young the child is. Most children secretly enjoy responsibility that they can handle. It gives them a sense of importance, and something to do while getting accustomed to unfamiliar surroundings.

In all cases, safety should be the first consideration. Teach children how to handle potentially dangerous equipment, such as campstoves and gasoline lanterns, and how to build campfires. But always supervise them, and do not let them perform these tasks unless one of the parents is nearby.

Only through experience will you learn to organize and run a campsite to suit your own needs and tastes, and until you become seasoned in this form of recreation you most likely will take many more things than you will use or need during the trip.

Unless you are an organizational genius, and your children are compulsively neat, you had best assume that the first outing or two is going to be chaotic at times. Someone will misplace a drinking cup or leave it beside a stream on a day hike. The reader of the family will fall asleep while reading a Nancy Drew mystery by flashlight, and you won't have an extra set of batteries. A field mouse or porcupine will gnaw holes in all of the hot chocolate mix that somebody forgot to put with the other food suspended from a line high in a tree. The family glutton will eat three days' supply of peanut butter at one sitting when nobody else is watching.

Nearly every trip, no matter how many you've been on, has its little catastrophes. But the more you camp, the fewer and more minor they become. In most cases, they become part of a family's legacy of "remember that time" stories that are told for generations.

CAMPER'S SECRET

Beware the birds. In many areas you'll have to watch out for certain birds that can be absolute pests around campgrounds. Crows, or ravens, can be among the worst because they are as intelligent as they are mean. They delight in stealing your food, and can swoop down and fly away with a piece of bread or almost any food they can grasp. They also love shiny objects and will steal them for the fun of it. One camper told of a flock of crows that tore a styrofoam ice chest to shreds because they know that was where the food was kept. Magpies, "camp robbers" and a few other species of fearless birds can be interesting to watch but infuriating come mealtime.

CHAPTER 8

THE OUTDOOR KITCHEN

A happy crew is a well-fed crew. A pot of coffee on the back of the grill . . . hot water for tea or hot chocolate . . . bacon or sausage frying in the morning . . . drink mixes that quench your noon thirst while giving a slight puckering sensation . . . a dutch oven filled with vegetables and meat that is ready to eat just before sundown. . .

A family that knows it will have tasty food on a camping trip will look forward to each outing, and you will hear much less muttering about missing favorite television shows. Poor planning is the only reason modern campers come home grumbling about the food, and often the best compliment paid to the campfire cook is silence on the subject.

Thanks to refinements in freeze-dried and dehydrated food technology, grocery shelves are laden with your choice of food that is as tasty when cooked as it is light in weight. Consequently, those "golden" days of camping are gone when novices wondered if they were expected to supplement their diets by grazing in nearby pastures.

It isn't necessary to pack up your microwave oven and food processor to have well-balanced and tasty meals. An easy way to test that statement is to go through the shelves at your grocery store and see how many items you normally buy that are prepared simply by dropping them into boiling water. You'll find that garden-fresh vegetables are about the only kind of food that doesn't fit into this ultralight—and long-lasting—category.

The major challenge to campfire

cooks, then, is to make life as simple for themselves through the shelves possible. Some campers may genuinely enjoy cooking as a hobby, but not at the expense of enjoying those activities only camping offers.

Family camping trips should be vacations for everyone in the family, especially the cook. By the same token, a group that has delicious and nutritious meals is going to have more tolerance for bad weather, too much wind, or too much heat than a group that sacrifices good food in favor of other considerations. Nor is a camping trip an appropriate time to continue the family struggle to make the youngsters eat their broccoli or spinach; vacations are a time to avoid these gastronomic debates.

CAMPER'S SECRET

Take along familiar food the first time. If you're going on a weekend trip, prepare at least one good meal—perhaps even catering to the children's preferences so they won't have a memory of cold mush for breakfast.

Everyone should be involved in the food preparation and cleanup afterward. You will know from past experience if your family can assign a different cook each day, or if it works best to give each member a specific job to perform during the trip. Obviously, each member should be responsible for his or her own tableware, and if one has a history of being careless about matters of hygiene, a parent should hold frequent inspections of cups, plates, and eating utensils to cut the risk of food poisoning due to laziness.

The best rule to follow in food preparation is to keep it as simple as possible. Grocery shelves are laden with excellent foods that require only the addition of water to reconstitute into delicious courses. Food manufacturers have become very adept at producing these for the so-called salt and pepper cooks, a term that aptly describes what much camp cooking is.

COOKING IN THE OUTDOORS

Although campfire cooking is the traditional way, more and more campers are depending almost entirely on stoves because they are more dependable, you have more control over them, and many areas do not permit campfires, anyway. Your camping stove can range from the old standard Coleman two-burner white gas stove to the more recent pack stoves that weigh only ounces. Butane and propane stoves are very popular, too, and come in one- and two-burner models. None of these is excessive in weight, and all are relatively easy to keep clean.

The major problems with cooking over open fires are that you can't control the heat as simply and quickly as with a stove, and you always have the problem of soot on the bottom of your pots and pans. This can be partly alleviated by coating the pots and pans with a thick application of soap or wrapping them in aluminum foil, but stoves remain much simpler to use.

Also, the governmental agencies that administer many of the camping areas are stressing stoves more and more because they are more environmentally responsible ways of cooking.

Stoves: Camping stoves come in three major fuel categories: white gas, propane and butane, and kerosene. Each obviously has its advantages and disadvantages, and each has its own band of loyal users.

White gas stoves are probably the most common in America because the fuel is readily available everywhere in the United States, and the fuel can be either standard white gas or commercially marketed naphtha such as Coleman and Blazo fuels.

These stoves come in two major types: those that must be preheated by pouring a bit of alcohol or gas on the burner, and those with pressure pumps that force fuel into the burners. Under normal summer conditions, these stoves present no problem, but in cold weather the preheating can be a trying and time-consuming chore. This preheating is accomplished by pouring a small amount of fuel or alcohol onto the burner and igniting it with a match. After a minute or so the burner will be hot enough to turn the liquid fuel into a vapor.

The butane and propane stoves have gained popularity for warm-weather camping because no priming is necessary and the fuel comes in cartridges, making it unnecessary to transfer fuel from container to stove. These stoves burn hot and quiet and are by far the simplest to operate.

Their disadvantages are the bulk of the cartridges, which must be carried out with you and discarded. And butane, the most popular, will not vaporize at freezing level. If you are going to use it in cold weather, you must literally sleep with it during the night to keep it warm. Propane is not so severely affected by cold weather.

A disposable butane cartridge will provide up to three hours of burning. Propane is a much higher-pressure fuel, and the containers must be heavier to withstand the pressure. Both are sus-

PURIFYING WATER

Unless your camp is served with treated municipal water, never take chances with its purity. There are several methods of purifying water and you can select the one that is simplest for your type of camping.

Boiling: You should boil it for at least five minutes, longer if it is convenient. Let the water cool and the sediment settle to the bottom, then pour off the pure water and discard the sediment. To remove the bland taste caused by the boiling, aerate by pouring back and forth between two containers.

Purification chemicals: Several chemicals are available to kill bacteria in the water by letting it dissolve and sit for the specified length of time. New products are being introduced on the market all the time, including the so-called water stick, which is a cigar-shaped drinking straw filled with charcoal and chemicals that will purify up to 300 gallons of water. With it you can suck water directly out of a stream, knowing it is pure when it reaches your lips.

CAMPER'S SECRET

A new cookie sheet can be used as a good fire reflector to bake biscuits or pies.

ceptible to overpressurizing when left in the hot sun, and safety valves sometimes open to relieve the pressure.

An advantage for car camping and similar activities is that many accessories have been developed for butane and propane, such as small lanterns hardly larger than a flashlight that fit atop butane cartridges, stoves that nest for carrying into a shape not much larger than saucer, and small heaters.

White gas stoves, as noted earlier, are the most common and versatile of the choices. White gas is used on the one- and two-burner campground stoves and the tiny stoves that nest into cooking pots. Since it is volatile and produces carbon monoxide, white gas must be handled carefully, especially in enclosed spaces such as tents. You should make it a rule never to use any stove inside a tent unless the tent is equipped with a zipout cook hole so the stove is on the ground or snow. Tents are treated with fire-retardant chemicals, but most are built of petrolem-based products and **will** burn.

Another disadvantage is the necessity of priming the stove before it will burn. Some models have pumps to create pressure in the fuel tank. Others, which are popular with backpackers and climbers, are the models that must be preheated by pouring a small amount of fuel or alcohol onto the burner. Most

CAMPER'S SECRET

Fill bottle caps with melted paraffin and place a short length of cotton string inside for a wick. These can be used either as fire-starters or for about fifteen minutes as a candle.

STOVES

White gas Coleman stove

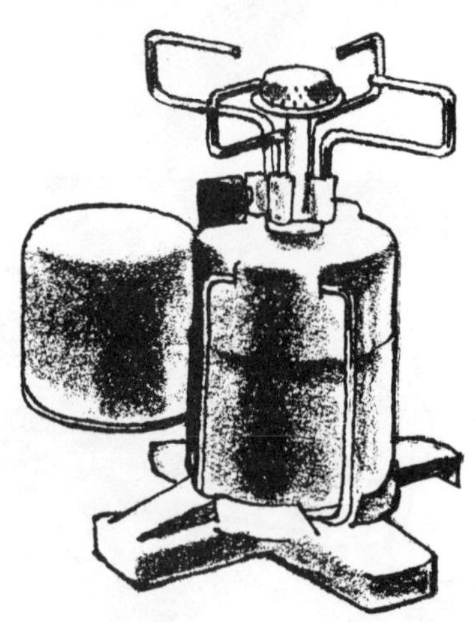

Butane Bleuet stove

White gas Optimus backpacking stove

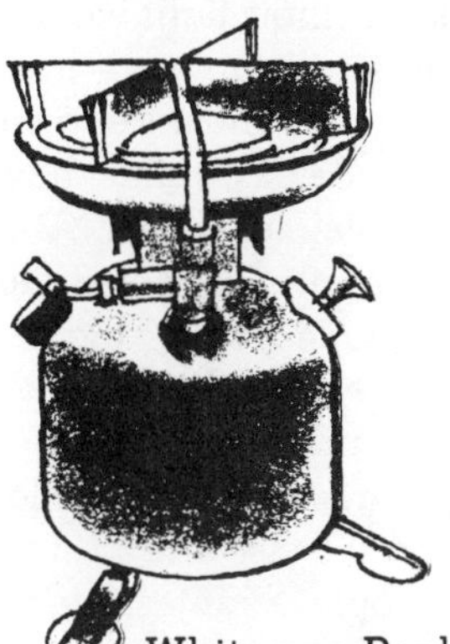

White gas Peak I backpacking stove

Two-piece MSR white gas stove

campers prefer carrying an eyedropper to avoid spilling fuel while pouring small amounts on the burner for preheating, and some prefer carrying a small vial of alcohol because it is less volatile than white gas.

Kerosene is the least popular fuel in America because it creates an unpleasant odor, is susceptible to smoking and staining pots and pans, and is the least volatile of the fuels, meaning it is the most difficult to coax into flame. However, it is worth noting that kerosene comes closer than any other stove fuel to being universal. It is available virtually everywhere in the world, whereas white gas is essentially a North American fuel and propane and butane cartridges are almost equally difficult to find. But kerosene is available on every continent of the world.

Alcohol is almost equally unpopular because of its low heat.

Heat: If you are going to camp in extreme cold—high elevations or stretching your camping season into the winter months—a tent heater will make your mornings and evenings more pleasant, although they add considerably to your weight and bulk and are not suitable unless you camp near your vehicle or haul everything into camp on pack animals.

Most heaters burn some form of gasoline, mostly white gas, which means they must always have ventilation to aviod the danger of suffocation. They are of more use in the heavier, nonporous tents than lightweight models that "breathe" well. Their usefulness in the latter models is limited because heat, like moisture, dissipates rapidly.

Another useful source of heat is hand warmers, which are only slightly larger than cigarette packs and provide heat for around an hour without flame.

Many winter fishermen and fall hunters consider them an essential part of their equipment because one in each pocket keeps hands from getting cold and stiff.

Generally speaking, however, if you have the proper gloves or mittens, you won't need them with you. But they are a good addition to camping equipment to warm cold or wet hands.

CAMPER'S SECRET

Plastic film cans make good, lightweight containers for salt, pepper, herbs, and spices. While traveling, keep the original snap-on lids with labels on the cans. Most camping supply stores now stock shaker lids that will fit over the cans.

CAMPSTOVE SAFETY

- Never fill a stove inside the tent or near an open flame.
- Keep fuel cans away from all sources of heat and in the shade.
- Never fill a stove inside the tent or near an open flame.
- Avoid cooking or heating with the stove valve wide open. It overheats the stove.
- Don't fill a stove while it is still hot.
- Don't carry fuel in plastic containers that can be punctured.
- Memorize the stove manufacturer's instructions and carry them with you.
- If you smell fuel—gasoline or natural gas—before lighting the stove don't light it until you have found the source.

UTENSILS

Since more and more camping areas—particularly national parks—prohibit campfires, it is best to assume you will have to carry your own stove. If you do camp in a place that has a good supply of firewood and you can build a campfire, you will probably still want to use your campstove so that your pots and pans stay clean and use the campfire for hot dogs, marshmallows, and evening cheer.

Nesting pots made of aluminum have long been standard camping utensils because of their light weight. However, recently campers—and even backpackers, who must fret over every additional ounce of weight—have been won over to stainless steel. Aluminum dents more easily, is more difficult to clean than stainless steel, and has a tendency to "shed" its black residue on your hands and towels. Although this microscopic residue is harmless and seldom alters the tase of food, it still bothers some people. Stainless steel is a bit heavier, true, but it is more durable and easier to clean and since it is always shiny, it looks cleaner.

In selecting your cooking gear and organizing menus, simplicity is the key word. Discipline yourself to use as few pots and pans as possible so that packing, unpacking, and camp tending will be less complicated. Some campers have learned to prepare meals for the whole party using only two modest-sized pots and a one-burner stove.

One essential that never should be overlooked is at least one high-quality cooler, preferably with a hard shell so it will take the normal wear and tear of camping. It is one of the most versatile of

Old-fashioned camping cook gear was heavy, often made of cast iron, but today's equipment, shown below, weighs only ounces and nests together to save space.

utensils—some are strong enough to use as a seat—and can double as storage on the way home.

Grills are not an essential if you plan to camp in organized areas with stone or concrete fireplaces at each campsite, but you should carry one along, if for no other reason than you will have a backup grill in case the one in the campground is broken, too large, or not even there.

They come in a variety of shapes and sizes. Often you can find one at a second-hand store that came out of a range oven, or you can use one from your barbecue set at home.

Outdoor equipment stores have smaller ones designed for campers that weigh only ounces and have their own carrying case to keep the soot away from other equipment.

The best are those with a fine mesh so you can cook hamburgers and hot dogs on them without losing the meat in the fire.

Griddles are a blessing for pancakes at breakfast and other uses and, like grills, don't take up much space. Some newer models are made of lightweight materials so you won't feel you're carrying a cast-iron stove top with you.

Ovens are available that fit above a burner of your campstove, and most will fold flat when not in use. Most models have a built-in thermometer.

Dutch ovens have long been a favorite for campers, who swear that they are not only versatile and easy to use, but also make food taste better. They are simple to use, and after you've practically buried one under coals for the specified time, the cooked meal that comes from it appears as if by magic.

The true Dutch ovens are sometimes difficult to find, and some outdoor supply stores don't seem to know the difference between them and ordinary kettles. A real Dutch oven is made of cast iron (some new ones are of aluminum, but purists stick with the heavier cast iron). They have three short legs about two inches long (so they can sit slightly above the fire) and a lid that is indented or flanged so hot coals can be placed on it. The lid has a handle, sometimes notched at the exact center of balance, and can be used for frying bacon or sausage and eggs.

New Dutch ovens must be "seasoned" before use. This means you should boil grease, lard, or suet in them, spreading it all over the inside, until it starts smoking. Then remove from the heat (your range oven at home is fine) and wipe it thoroughly—but leave a coating of grease. Never use strong detergent or scouring pads to clean it; you'll have to season it again. After each cleaning, lightly coat it with grease again to keep water from reaching the metal and creating rust.

Dutch ovens can be used for broiling, roasting, stewing, deep-fat frying, and just plain frying on the lid.

Like the crock pots used for home cooking, Dutch ovens are usually used for things like pot roasts and stews, and nearly every river guide in America has a Dutch oven stew he or she has named in honor of a river. One memorable dish, called the Salmon River Stew, has both beef and sausage patties mixed among the vegetables.

For these dishes, the oven and lid are preheated; then grease or suet is dropped in and heated until it begins to smoke and all parts of the oven are swabbed with grease. Then the roast or stew meat is seared on all sides. Add a cup of hot water slowly, salt the meat, and add vegetables.

Some cooks prefer cooking the meat

Dutch ovens are made of cast iron (some aluminum models are available) and have three legs and a recessed lid that holds glowing coals or can be cleaned and used as a skillet.

first, then adding the vegetables about an hour before the meal. Most, however, put the whole dish into the pot and forget about it while pursuing other activities.

There are two ways to heat the oven: over an open fire with coals added to the lid, or buried.

If you want to bury a Dutch oven, select a site where there is absolutely no danger of fire spreading or damaging the soil. A sand or gravel bar is best. Scoop out a wide hole, approximately a foot deep, and build a fire in it. When the wood turns into a bed of hot coals, put the oven into the hole, cover the lid with coals, and cover the whole thing with dirt, sand, or gravel. If using sand and/or gravel, cover it with a mound to preserve the heat. Better put some kind of "fence" around it, too, so people won't walk on it.

Buried ovens will cook the meal in four to eight hours, depending on the recipe and the intensity of the fire. After you've completed your day's activities, you can dig out the oven and speed the cooking along with a fresh fire.

Cooking above an open fire can be speeded along by keeping the fire stoked up. For this kind of cooking, use about the same amount of time you would if cooking on your range at home.

A Dutch oven filled with pot roast or stew is surrounded by coals, then buried for a day-long cook.

Remember, though, that Dutch ovens are heavy and, if you plan to camp out of sight of your vehicle, you'll probably curse the person who suggested it be brought along.

A twelve- to fourteen-inch Dutch oven will feed up to twenty persons if crammed full of food. A ten-inch size will feed up to ten persons, although campers seem to have heartier appetites than other people, even though some days

CAMPER'S SECRET

If you're thinking of trying a new dish on a camping trip, don't take chances: prepare it before the trip, preferably using the same kind of stove.

they may have done nothing more vigorous than washing the breakfast dishes.

In the recipe section of this chapter, the following dishes are specially designed for cooking in a Dutch oven: Beans, Pot Roast, Corned Beef and Cabbage, and New England Clam Chowder.

Pots and pans: Since most family cooks are rather particular about their cooking utensils, it is smart to invest in a set of nesting cookware and to keep camping equipment separated from the household equipment. Nesting cook sets come in a variety of styles, but most have at least three pots—ranging in size from three quarts to one quart or less. Other larger models are called nesting kettles and have capacities up to six quarts.

Nesting billies differ from other popular sets in that they are not designed to fit on a particular stove, as many cooking sets are. If you plan to car camp only, nesting billies are sufficient.

A typical nesting cook set designed to fit over a particular backpacking stove will include a base and a windscreen on which the three or four nesting pots fit. The whole thing can be placed inside the largest pot, stove included, making a small, compact package.

If you will be cooking over a campfire, almost any old pot and pan will suffice, provided weight is no consideration. You can shop around charity stores for this equipment or use your own second-generation kitchen ware. Usually a maximum of three cooking pots will suffice, and most experienced campers get along very well on two. Use the largest to keep water boiling and the other two for meals. In case of a larger group, all three may be needed. Your menu will dictate this.

After pots and pans you will need the same equipment you use at home—stirring spoon, spatula, good can opener, a sturdy pot gripper or pad, a portable grate, and a sharp knife.

In addition, each member of the party should have a plate, cup, knife, fork, and spoon. Plastic plates are fine, although some prefer using pie plates so they can also be used to cook a cake or cobbler. Cups may be plastic and metal, but some campers prefer to bring along their favorite mug on the grounds that drinks taste "funny" from plastic or metal. A set of tableware that clips together is a wise investment for each member of the party, and the truly fussy might want to label their own set to keep others from using it.

DOING THE DISHES

We have lived with automatic dishwashers so long that they are taken for granted by most Americans, and the thought of actually washing, rinsing, and drying dishes makes many beginning campers—especially children—consider fasting on their camping trip. But doing the dishes is a fact of life in campgrounds and one of those chores that is worse to ponder than to perform.

First, each member of the party should be responsible for his or her own plate, cup or glass, and eating utensils.

CAMPER'S SECRET

Charcoal is always welcome for camp cooking, and you can carry in rinsed-out milk cartons, then burn the cartons to help start the charcoal.

That way, since most camping meals don't require more than one or two pots, the person assigned to the dishwashing duty won't have that many things to wash after each meal.

If you are carrying only one large pot (which is unlikely with the nesting sets being so lightweight and practical), boil water in it, add detergent, scrub each item, and place it on a clean flat surface. When everything is washed, bring another pot of water to boil, then rinse each item thoroughly.

One of the best ways to let things dry is to buy or make a wide-mesh bag for each person that will hold a plate, cup or drinking glass, and all eating utensils. Once washed, equipment can be placed in the mesh bag and hung from a nearby tree or bush to air dry. You can carry a dish towel, which can double as a hot pad, but it isn't really necessary if you can set the utensils out to dry.

To do the washing, you can take along a small plastic bottle filled with dishwashing soap if you want, but those scrubbing pads preloaded with detergent are usually sufficient. Use one for the really dirty jobs, such as scrubbing soot off the bottom and sides of pots placed over the fire, and the other for the insides of pots, pans, and eating utensils. One each of these pads is usually sufficient for a weekend trip (double that for a week's trip).

Since liquid detergent is usually concentrated, a small plastic bottle of it doesn't take much room and comes in handy for soaking pots when the cook chars the stew pot. Buy detergent that is not poisonous; most outdoor stores have biodegradable detergent in small plastic containers or in scrub pads.

Sand from the beach or river bank is a good scouring agent, too, especially when mixed with a bit of detergent. Do not scrub and rinse directly into the stream or lake, however. Although the residue may be biodegradable, you and your neighbors won't like the idea of sharing dishwashing soap, uneaten food, and dirty water. If you aren't at a campground with a wash area and running water, move well away from your camp and your neighbors' campsites into the woods to dump the dishwater. Obviously it should not be dumped on a trail or anywhere someone might step on it.

Most scrapings from pots and pans will burn in the campfire, but be certain it does burn. Don't clean the breakfast egg residue into the fire, then pour water on the fire, leaving behind a mess of blackened wood decorated with fried eggs.

If you camp in an area that does not permit open campfires, carry a sturdy, waterproof sack—preferably nylon—or one lined with a plastic garbage bag and pack out your food garbage with you to the nearest garbage can. Most camping areas that do not permit campfires are

CAMPER'S SECRET

Use sand to scour skillets and pots, or bring pads preloaded with detergents. Dishwater should be dumped well away from sources of water.

heavily used, and it doesn't take long for food scraps dumped in the woods to become garbage. It also attracts wildlife and encourages animals to depend on mankind for food scraps.

CAMPER'S SECRET

Freeze as much of your food as possible before the trip. This will both keep the food longer and help cut down on the amount of ice you will need on the trip

MENU PLANNING

Most campers prefer a hearty breakfast a no-bother lunch, and a hearty dinner. This not only gives the cook more time away from the stove or fire; it also helps guarantee that the cook will have lots of help during mealtime; catch the others when they're hungry and they'll be more willing to help instead of wandering off.

In theory, the stove or fire should not be necessary for lunch, although someone will almost invariably want hot tea, coffee, or chocolate. Some campers have the will power to fill a thermos with hot drinks and consume it only at mealtime, but most—particularly children—do not.

CAMPER'S SECRET

Since most cereal boxes aren't sturdy enough for camping trips, try using well-rinsed half-gallon cardboard milk cartons. These cartons also make excellent water containers, and when you're through with them, they burn easily.

Variety of dishes is important, and you can plan so that each day's menu is different. Eggs with sausage, bacon, or bacon bits can constitute breakfast on alternate days, as can omelets. Individual packets of hot breakfast foods, such as oatmeal, can be prepared along with toast made over the fire or stove.

Lunches should always be of the serve-yourself variety: canned meats, paté and "hand foods" such as gorp, high-energy tropical chocolate and fruit bars, pilot bread or crackers, peanut butter and jelly.

Dinner can be the main event: roasts, stroganoff, chili, chipped beef, spaghetti, and so forth. Vegatables—freeze-dried, dehydrated, or canned—should be part of dinner. Desserts can range from cakes baked in camp to pudding, fudge, or even a novelty item such as freeze-dried ice cream. There's no need to limit yourself to boring dishes while camping, as the recipes later in this chapter will attest.

The U.S. Department of Agriculture's guide to daily menu planning can be easily adapted to camping menus. This guide is divided into four basic food groups:

CAMPER'S SECRET

Make your ice chest more efficient during the heat of the day by wrapping it in a blanket, a sleeping bag, or some other insulating material.

Meat: Two or more servings, for energy and growth. This includes red meats, poultry, fish, eggs, cheese, and legumes.

Dairy foods: Two servings for adults, three to four for children and teens, for growth and body maintenance. This group includes milk, butter, cheese, and other such products.

Vegetables and fruits: One dark green or yellow vegetable and one citrus fruit or tomato.

Breads and cereals: Four or more servings, for roughage. Includes breads, breakfast cereals, macaroni, or noodles and rice.

SAMPLE THREE-DAY NUTRITION PLANNER

Breakfast	First Day	Second Day	Third Day
Protein food			
Cereal			
Fruit or juice			
Beverage			

Lunch	First Day	Second Day	Third Day
Main dish			
Vegetable or fruit			
Bread			
Dessert			
Beverage			

Dinner	First Day	Second Day	Third Day
Main dish			
Vegetable			
Salad			
Bread			
Dessert			
Beverage			

THREE-DAY MENU

BREAKFAST	LUNCH	DINNER
scrambled eggs hash browns hotcakes hot drinks citrus drink	cup of soup pilot bread or crackers powdered milk drink oatmeal cookies	beef stroganoff rice mixed vegetables brownies milk mix hot drinks

BREAKFAST	LUNCH	DINNER
granola cinnamon toast hot drinks citrus drink	cheese spread (in tube) pilot bread peanut butter milk drink coffee or tea	macaroni, cheese, and ham bits instant potatoes vegetables milk mix hot drinks

BREAKFAST	LUNCH	DINNER
pancakes with corn or berries boiled eggs toast and jelly citrus drink	canned meat bread or crackers fig bars fruit drink coffee or tea	vegetable and beef stew vegetable applesauce peach cobbler milk mix hot drinks

CAMPER'S SECRET

Here's a general "pinch guide" in case you forget your measuring cups:

2 pinches = 1/8 teaspoon
3 pinches = 1/3 teaspoon
4 pinches = 1 teaspoon
1 fistful = 1/4 cup

Better test these at home in case your fingers are smaller or larger than average. Also test them with your camping spoons and a drinking cup.

CAMPING RECIPES

Obviously your camping menu is limited only by individual tastes and your imagination. Some families pay little attention to menus, other than as a source of nutrition and body fuel, and stock up on the simplest items to prepare, treating eating as a necessity that must be tolerated, like sleep. But most campers like a bit of contemporary civilization combined with their outings, something like the British during the empire days when they dressed for dinner even if they were on the muddy banks of the Congo River.

Following is a group of recipes gleaned from a variety of sources to show you things that are possible with only a modest amount of advance preparation and a dependable cooler. These recipes range from the ordinary to the gourmet.

HOT CHOCOLATE MIX

1 lb. instant cocoa mix
6-8 oz. dry cream substitute
8 qts. (or 9 cups) dry milk
1/3-1/2 cup powdered sugar

Mix thoroughly. This fits into a 3-pound coffee can with a plastic lid. To prepare, use 1/3 cup mix to one cup hot water.

HUNGARIAN MEATBALL SOUP

Serves 4

1 lb. lean ground beef (uncooked)
1 cup bread crumbs
1 egg, beaten
salt and freshly ground pepper to taste
1 onion, sliced thin
2 tbsp. butter
2 tsp. paprika
2 cups diced peeled potatoes
garnish: sour cream or plain yogurt and chopped parsley

Combine the beef, bread crumbs, egg, salt, and pepper. Shape into 1-inch balls. In a saucepan, sauté the onion in butter until soft, then add the paprika and 4 cups water. Bring to a rapid boil, lower heat, and add meatballs and potatoes. Cover and simmer 30 minutes. Ladle into soup bowls, top each with a spoonful of sour cream or yogurt, and sprinkle parsley on top.

NUTTY OATMEAL

Serves 2

1 cup oatmeal
1/4 tsp. salt
1/2 cup nonfat dry milk
1 oz. chopped dates
1 oz. chopped nuts (unsalted)
1 oz. brown sugar

Combine oatmeal, salt, and dry milk, and boil in 2 1/2 cups of water. Remove from heat. Stir in dates, nuts, and sugar. Return to low heat and simmer about 3 minutes, stirring occasionally.

NEW ENGLAND CLAM CHOWDER

Serves 6

1 qt. clams with liquor
1/4 lb. salt pork, well rinsed
and diced or 2 slices bacon, diced
1 onion, chopped
3 potatoes, peeled and cut into
1/2-inch cubes
3 tbsp. butter
1 pint half-and-half or 1 cup each milk and heavy cream
salt and white pepper to taste

Combine clams and their liquor and 1 quart water. Bring just to a boil in a large saucepan. Drain clams and reserve liquid; chop clams coarsely and set aside. Fry salt pork or bacon just to release fat; add onion and cook until it is transparent. Add the reserved liquid and potatoes and cook covered until potatoes are tender, about 20 minutes. Stir in butter, half-and-half or milk-cream mixture, reserved clams, salt, and pepper. Heat through; do not boil. Serve with pilot crackers.

POT ROAST

While cooking breakfast, braise a roast on all sides, then place it in the Dutch oven with half a cup of warm water, carrots, potatoes, onions, garlic, cloves, salt, and pepper—and whatever else you want. But don't overfill; leave a space between the lid and contents.

When the campfire has died down to a nice bed of coals, dig the coals aside and put the oven in the middle, cover the lid with coals, and cover the whole thing with fresh dirt. Dinner will be served at dusk.

CORNED BEEF AND CABBAGE

Serves 4-6

1 4-lb. corned beef brisket, soaked in cold water to cover for 2 hours to remove excess brine, if necessary. (Can be done before leaving home.)
1 bay leaf
6 peppercorns
1 onion, stuck with 2 cloves
1 carrot, sliced
1 celery rib, sliced
2 sprigs parsley
1 cup apple cider
6 carrots, halved crosswise
6 new potatoes
1 small head cabbage, cut into 4–6 wedges
accompaniments: Dijon-style mustard, horseradish, gherkin pickles

Combine all ingredients, except cabbage, with water to cover and cook in Dutch oven for 8 to 10 hours. Remove corned beef and vegetables to a platter and keep warm. Add cabbage to pot, cover, and cook on high heat for 20 minutes. Place cabbage on platter with meat and vegetables. Serve with accompaniments.

BEANS

Serves 2

1 lb. navy or pea beans
2 chunks of bacon (salt pork)
1 tsp. salt
1/2 cup molasses
pepper to taste

Soak beans overnight, then bring to a boil and cook until the skins burst when spooned or blown. Pour off liquid and save. Drop the salt pork in, add salt, molasses, and pepper to bean water, then pour it over the beans and add the other piece of salt pork on top. Use the hole-in-the-ground Dutch oven method of cooking (see page 126). Cook 4 - 8 hours.

SWISS FONDUE

Serves 4–6

1 clove garlic
2 cups dry white wine
1 1/2 lbs. imported Swiss or Gruyere cheese, grated or cut into small cubes
2 tsp. cornstarch mixed with
1/4 cup Kirsch
pinch of freshly ground pepper
pinch of freshly grated nutmeg
French bread, cut into 1-inch cubes
accompaniments: green salad or fresh fruit for dessert; a dry white wine, such as Riesling

Rub the inside of a cooking pot with the garlic clove, then discard garlic. Heat the wine in the pot placed over fire and add cheese, stirring constantly until smooth. When bubbles begin to appear, add the cornstarch-and-kirsch mixture. Season with pepper and nutmeg. Fondue should be kept slightly bubbling while cooking. Spear bread cubes with a fondue fork, roll in fondue, and eat. Serve with a green salad or fresh fruit for dessert and white wine.

DONUTS

Open a can of ready-to-bake biscuits and cut center holes in each. Fry in shortening until brown, and fry the "holes" too. Drain on paper towels or napkins, then shake in a sack of powdered sugar, white or brown.

CAMPERS' FUDGE

Serves 4

2 cups chocolate chips
3/4 cup sweetened condensed milk
dash of salt
1 tsp. vanilla
1 cup chopped nuts

Melt chocolate over hot water; stir in remaining ingredients. Pour into buttered pie pan, cool, and cut into squares.

CAMPER'S SECRET

Fill quart or half-gallon cartons with water, staple them closed, and freeze them before the trip. They will stay frozen longer than ice cubes and take up less room. The cartons can be used for pitchers later.

SOME-MORES

1 box graham crackers
1 bag marshmallows
chocolate bars

This is a favorite for scouting groups and everyone else who camps. Place a chocolate bar on top of two separate graham crackers. Toast marshmallows until soft, then squash them between the chocolate-lined crackers. You'll undoubtedly want some more.

PEPPER CHEESE

8 oz. cream cheese or pot cheese
2 oz. dry white wine
1/2 pint heavy cream, whipped
1 clove garlic, crushed
1 tsp. finely minced fresh fennel or chervil or summer savory
4 tbsp. peppercorns, crushed

Mix everything except peppercorns in mixing bowl, using fork or electric beater. Chill for several hours, then flatten mixture. Spread the crushed peppercorn on a flat surface and roll cheese in it, pressing down so that the pieces of pepper are pushed into the cheese. Roll the cheese around so that a lot of pepper is coating it. Store in plastic container in cooler until needed.

TOMATO BARBECUE SAUCE

About 2 cups

1 medium onion, chopped fine
1 tbsp. salad oil
1 large clove garlic, minced or pressed
1/2 tsp. dry mustard
1/2 tsp. each salt and chili powder
2 tbsp. brown sugar
3 tbsp. cider vinegar
Worcestershire sauce to taste
3/4 cup each catsup and dry red wine

Cook onion in oil in a 1 1/2- to 2-quart saucepan until soft but not browned; stir in garlic, salt, chili powder, dry mustard, brown sugar, vinegar, Worcestershire sauce, catsup, and wine. Stir until sugar dissolves and mixture begins to boil; simmer for 3-5 minutes, then remove from heat.

Cool before using as marinade for uncooked beef, pork, or poultry. Or brush, warm, over hamburgers and hot dogs.

HOMEMADE GRANOLA

Serves 4

3 tsp. (or more) cooking oil
1/4 cup water
3 cups rolled oats
1/2 cup brown sugar
(or use 1/2 cup honey and
leave out water)
1/4 tsp. salt
1/2 cup wheat germ
3 tsp. soy flour

Combine wet ingredients, then dry, and combine mixtures. Roast in shallow baking pan at 210° F. for 25 - 30 minutes, or at 350° for 15 - 20 minutes. After roasting, you may add dried fruit.

For crunchier granola, use only 1 cup rolled oats and add 1 cup unsweetened coconut and 1 cut sesame seeds.

PRECOOKED GROUND BEEF

Many experienced car campers have learned that one of the best shortcuts for a two- or three-day camping trip is to precook two or three pounds of hamburger meat a few days before the trip, then freeze it until the camping trip.

Prepare it the same way you would for chili or shepherd's pie or for a pizza. Fry it in a big pan, adding a bit of water, and cook until totally brown. Drain the water and grease, and place in a plastic container and freeze.

It will keep in the camping cooler for several days, or as long as you keep ice in the cooler, and greatly simplifies your cooking. It can be used in a variety of recipes, including Pizza Buns (next recipe) and Chili Bean Soup (page 137).

PIZZA BUNS

Serves 4

4 tsp. butter
4 English muffins, split
1 medium onion, chopped fine
1 15-oz. can tomato sauce with bits
4-oz. can mushroom stems and pieces, drained
1 tsp. Italian herb seasoning
1/2 tsp. garlic salt
1 lb. precooked ground beef (see preceding recipe)
4 oz. Jack cheese, shredded

Melt 2 tsp. butter in a 10-inch fry pan, and toast muffin halves. Keep warm. Melt remaining 2 tsp. butter in pan and sauté onion. Then stir in tomato sauce, mushrooms, seasonings, and ground beef. Cover and cook, stirring occasionally, until heated. Spoon over muffin halves, then sprinkle with cheese.

COLESLAW

Serves 6-8

3 cups shredded cabbage (1/2 head)
1 cup shredded carrot
1/2 small red onion, peeled,
thin-sliced, and separated into rings
1/2 cup sour cream
1 tbsp. sugar
2 tbsp. tarragon vinegar
1/2 tsp. salt

Combine cabbage, carrot, and onion with sour cream, sugar, vinegar, and salt in large bowl. Toss the slaw to coat thoroughly, and chill until eaten.

BATTER FOR HOT BUTTERED RUM

1 lb. butter
1 lb. brown sugar
1 qt. vanilla ice cream

Soften butter and ice cream and mix thoroughly. Makes 4 pounds of batter that can be stored frozen and used as needed.

Servings: To 2 tablespoons (or more) of batter, add rum to taste and hot water. Children love the batter (without the rum, of course).

CHILI BEAN SOUP

Serves 4

3 tsp. butter
onions to taste
30-oz. can chili beans
28-oz. can tomatoes
1 lb. precooked ground beef
(see page 136)
2 tsp. dehydrated sweet pepper flakes
1 tsp. garlic salt
2 - 3 tsp. chili powder

Fry onion in butter. Add rest of ingredients, cover, and simmer for 15 minutes.

CORNED BEEF AND NOODLES

Serves 5-6

2 12-oz. cans corned beef
12 oz. egg noodles
3/4 cup catsup
2 tsp. Worcestershire sauce
dash of Tabasco sauce
2 tsp. vinegar
1 cup water
1 tbsp. minced onion (or instant onion)

Boil egg noodles according to package directions. Combine sauce items separately and simmer about 5 minutes. Then break up corned beef and add to sauce, and simmer about 15 minutes. Drain noodles and mix with sauce and beef.

BEEF FONDUE

Serves 4–6

1 1/2 - 2 lbs. boneless top round
unseasoned powdered meat tenderizer
2 cups salad oil
salt and pepper
Spicy Garlic Mayonnaise
(see next recipe)

Trim fat from meat and cut into small cubes. Sprinkle with meat tenderizer and let stand for about 30 minutes, following tenderizer directions. Heat oil in pot to about 360° F. If you're using a campstove, lower burner and keep oil hot; if you're carrying a canned heat burner, use that. Cook each piece of meat to taste; 20 - 30 seconds will brown the outside, leaving the meat rare inside. Transfer meat to serving platter, sprinkle with salt and pepper, and dip into Spicy Garlic Mayonnaise for serving.

SPICY GARLIC MAYONNAISE

About 1 1/2 cups

1 egg
2 tsp. Dijon-style mustard
1 tsp. paprika
1 clove garlic, minced or pressed
1/2 tsp. salt
2 tbsp. white wine vinegar
1/4 cup olive oil
3/4 cup salad oil
1 tbsp. each chili sauce and drained capers
1 tsp. Worcestershire sauce
2 tbsp. snipped fresh chives or thin-sliced green onions
2 tbsp. chopped sour pickle

Before leaving home, combine in blender egg, mustard, paprika, garlic, salt, vinegar, and olive oil. Cover and blend at low speed, then immediately pour in salad oil in steady stream, and whirl until thick and smooth. Add remaining ingredients. Cover and keep chilled, then store in cooler for camping trip.

TOMATO RAREBIT

Serves 4

1 can condensed tomato soup
1/3 can milk
1/2 lb. (approx.) American cheese (sliced or diced)
1 egg

Heat tomato soup, milk, and cheese until cheese is melted. Beat egg, stir slowly into the cooking mixture, and cook for one minute. Serve over toast or crackers.

MARINADES

A variety of marinades can be made at home or from mixes bought in grocery stores, and experimenting at the backyard barbecue will tell you which is the most popular with your family. Marinating is one of the best ways to make inexpensive meat palatable.

One of the simplest marinades is made of two parts cooking oil and one part vinegar. Other recipes with more flavor to them follow:

MUSTARD AND HERB MARINADE

Less than 1 cup

1/3 cup salad oil
1/2 cup dry white wine
1 tbsp. each red wine vinegar and lemon juice
1 large clove garlic, minced or pressed
1 1/2 tbsp. Dijon-style mustard
1/4 tsp. each salt and sugar
1/8 tsp. each thyme, oregano, summer savory, and tarragon
dash of white pepper to taste

Combine all ingredients in blender until smooth. Use for lamb or chicken.

TERIYAKI MARINADE

About 3/4 cup

1/2 cup soy sauce
3 tbsp. sugar
2 tsp. grated fresh ginger or 1/2 tbsp. ground ginger
1 clove garlic, minced or pressed
2 tbsp. dry sherry

Mix or shake ingredients well. Use for beefsteaks or chicken.

SHISH KEBAB

Serves 6

1 lb. stew meat
1 16-oz. can pineapple chunks
1 16-oz. can potatoes
1 onion
1 16-oz. can cherry tomatoes
1 green pepper

Cut meat into small chunks and place in mixture of two parts cooking oil and one part vinegar. Let it marinate at least 24 hours, either before leaving home or in a plastic jar with a screw lid, left in the cooler at camp until needed.

Place meat on skewer, green stick, or spit, alternating with pineapple and vegetables. Cooking time is approximately 15 minutes over normal fire or coals. You can brush kebabs with barbecue sauce if desired.

SNACK ITEMS

Gorp: Acronym for Good Old Raisins and Peanuts. Also added are items such as M&M candies because they keep longer. Can be purchased as a mix or made at home, substituting other nuts, your choice of raisins, and other items.

Fruit: Usually freeze-dried or dehydrated for ease of handling and long life.

Trail cookies: These are high in nutritional value and proportionately expensive. Oatmeal cookies with fruits added are a good homemade substitute.

Party mix: This can either be purchased ready made or made at home. Combination of pretzels, Cheerios, nuts, and Wheat or Rice Chex, baked in the oven, and packaged in individual portions or in a large container.

Pilot bread: Large round crackers that don't crumble as easily as saltines.

EXTRAS

Bacon bar: A compressed bar of precooked bacon, excellent for adding to scrambled eggs or munching for snacks. Has long life.

Meat bars: Made of a variety of meats in same way as bacon bars, and won't spoil easily.

Pemmican: This old Indian recipe of meats and berries has been updated into individually wrapped bars that can be eaten like candy, or in cans and plastic containers to be used as a spread.

WILD FOODS

Every region of the world has its own wild edibles there for the taking—salad greens, nuts, fruits, roots, mushrooms, berries. Since the species vary so much from region to region, we won't attempt a complete listing here, but accurate and detailed guides to these wild foods are available in most outdoor equipment stores and regular bookstores. Look in the regional section of bookstores.

One word of warning: be absolutely certain you know your mushrooms before supplementing your food supply with those found in the forest. A few species are poisonous and some are deadly. Be

CAMPER'S SECRET

You can prepackage your camp syrup by combining one-half cup each of white and brown sugar with a dash of cinnamon. In camp, add one-third cup of water and simmer until dissolved.

certain you can identify them beyond a doubt.

Here are a few of the more common plants found in most parts of North America that can liven up your camping meals:

Cattail: One of the most abundant of the aquatic plants, it is also one of the most popular for wild-food gatherers. Just below the leaves is the tender shoot that tastes something like a cucumber and may be eaten raw or cooked. The root is also quite tasty and can be cooked like a potato, fried, baked, boiled, or mashed.

Chickweed: This bane of lawn owners can be partially controlled by eating it. Its leaves and tender stems can be used as a salad or a cooked green.

Dandelion: Another lawn pest, this one has been popular not only for eating but as an ingredient for wine. The leaves may be eaten as a salad, but the younger leaves should be picked to avoid the bitter taste that comes from older leaves. The roots can be chopped and roasted until hard and brittle, then used as a coffee substitute.

Dock: This common weed is known also as sheep sorrel, Indian tobacco, and curly dock. The leaves can be cut and boiled for about five minutes and taste a bit like asparagus.

Lamb's quarters: Usually found in drier climates, this plant is best as a cooked green. Some people call it pigweed, but it should not be confused with the true pigweed, which doesn't taste good at all.

Stinging nettle: Few things are more uncomfortable than a patch of your skin that has brushed a stinging nettle. In spite of this, by carefully pick-

CAMPER'S SECRET

Jar-lid rings can be used to hold the shape of poached eggs. Place the rings in two inches of water in the Dutch oven, then break eggs into them.

Dandelion

ing some of the leaves (preferably with gloves) and boiling the leaves a minute or two to remove the stinging oils, you can sample one of the tastiest of the wild plants. Boiled, it tastes much like spinach.

Bear grass: Usually found in higher elevations, even alpine, bear grass does grow in lower elevations and is easily identified by its tall spike of white flowers. The root is excellent eating and is pulled from the ground and peeled. It can be eaten raw or added to a stew like any other vegetable.

Arrowhead: Sometimes called a duck potato or tule potato or by its original Indian name of Wapato, this aquatic plant has underwater tubers that are excellent eating. It tastes something like potatoes and was used as a medium of exchange by many Indians.

CAMPER'S SECRET

Before the trip, fill plastic jugs with fruit drink or lemonade and freeze them. On the trip, these will help keep everything else in the cooler cold and will give you cool, refreshing drinks for two or three days.

CHAPTER 9

CAMP ACTIVITIES

Camping should be an active form of recreation, not just changing the places you sit on a weekend or vacation. Part of your preparation for the trip should include planning activities at the campsite that you normally wouldn't do at home. Camping is hardly the place to catch up on homework or correspondence, but it is a perfect opportunity to learn more about the outdoors.

The relatively simple act of going somewhere, pitching a tent, cooking meals, sleeping, breaking camp, and returning home is not sufficient cause for many families to consider camping as a goal in itself. However, there are many activities people engage in that could be enhanced by camping. These include hunting, fishing, bird watching, rockhounding, bike touring, canoeing and kayaking, cross-country skiing, geological field trips, photography trips, horseback treks, and summer festivals.

In each of these activities, and many others, basic camping skills and good lightweight equipment are essential to getting the most out of the trips. The more expert you and your family become at camping, the more adjunct activities you will find.

An example is the variation on the license-plate game children love to play while traveling. Instead of tallying up the number of licenses from other states, take along a guidebook that tells you how to identify species of plants, trees, flowers, and birds of the region you're camping in. Keeping a tally of species is a good way for the whole family to learn more about nature. One of the best

series of books on identification, with regional divisions, is published by the Audubon Society.

If you live in the city, the combination of too much light at night and air pollution shuts off your view of stars. But out in the woods they shine vividly from the deep sky overhead. Most bookstores and some outdoor supply stores sell star identification books or charts that can keep the children busy most nights attempting to identify the important stars and constellations. A side benefit of this is that they also learn the basics of celestial navigation.

Another practical activity is learning to use a compass and topographical maps. The basics of navigation can be learned by any child old enough to read and do simple mathematics. Learning to read the colored topographical maps is intriguing to children, for it appeals to their innate imagination. Often it is easier for them to mentally transfer those squiggly lines on maps to the surrounding terrain than it is for adults. As with learning to use the compass and the basics of celestial navigation, knowing how to read the topographical maps may one day be a life saving skill.

Navigation is fun and lends itself well to campground games. The child who thinks math is the pits will often be the one who pores over the maps and compass by the hour. A topographical map is as real as a photograph, and a compass is in some ways the ultimate toy. The stars, too, are real and fun to discover. Once a child learns to identify the major constellations, it helps make those nights a little less overwhelming.

Obviously not all campground recreation must be in the educational field; schoolchildren will quickly rebel if camping begins to sound suspiciously like a school field trip.

So don't hesitate to bring along cards, popular games, small chess or backgammon boards (some are tiny and magnetized), or whatever the family likes to play at home. These help ease long afternoons when the rain is falling or evenings around the campfire or in the light of a lantern.

CAMPER'S SECRET

When you purchase maps for camping trips, whether in the wilderness or along major highways at established campgrounds, consider buying two of each map: one for use on the trip and another for posting on a wall at home with the route marked in colored ink. As you return to the same area and visit different places, use another color ink. Such maps are nice mementos of the trips, and they also help teach map reading to children.

TOPOGRAPHICAL MAPS

The accompanying illustration shows the basic components of these maps that are the standard for outdoorsmen all over North America. The only way to become accustomed to them is to use them in the outdoors. Nothing replaces practical applications, and while you can pore over them for hours at home to select potential campgrounds or hiking trails, still you must consult them in the actual terrain to learn how to read them at a glance.

These "topo" maps are color-coded. Features such as buildings, roads, railroads, mines, windmills, churches, and schools are all printed in black. Water features are in blue. Vegetation is in green, and all the elevation indicators such as contour lines, altitude markings, and benchmarks are in brown.

Nearly all maps have detailed instructions printed on them along with a key to the symbols.

The standard map for outdoors enthusiasts is on the scale of 1:24,000, which means one unit on the map (this may be a kilometer or a mile or whatever has been chosen and noted) equals 24,000 of the same units in actuality. For example, a mountain that is shown on the map no larger than a pencil eraser will actually be 24,000 times as large.

Each topo map also shows the declination correction factor, which is vital in using your compass. This will be explained in the next section.

Topo maps are available at nearly every major outdoor supply store, at map stores, and by mail order from the U.S. Geological Survey. (See page 213 for addresses.) If you know the specific map or maps you need by their quadrangle names, fine. But it is best to write and ask for an index map, which shows the whole United States, then breaks it

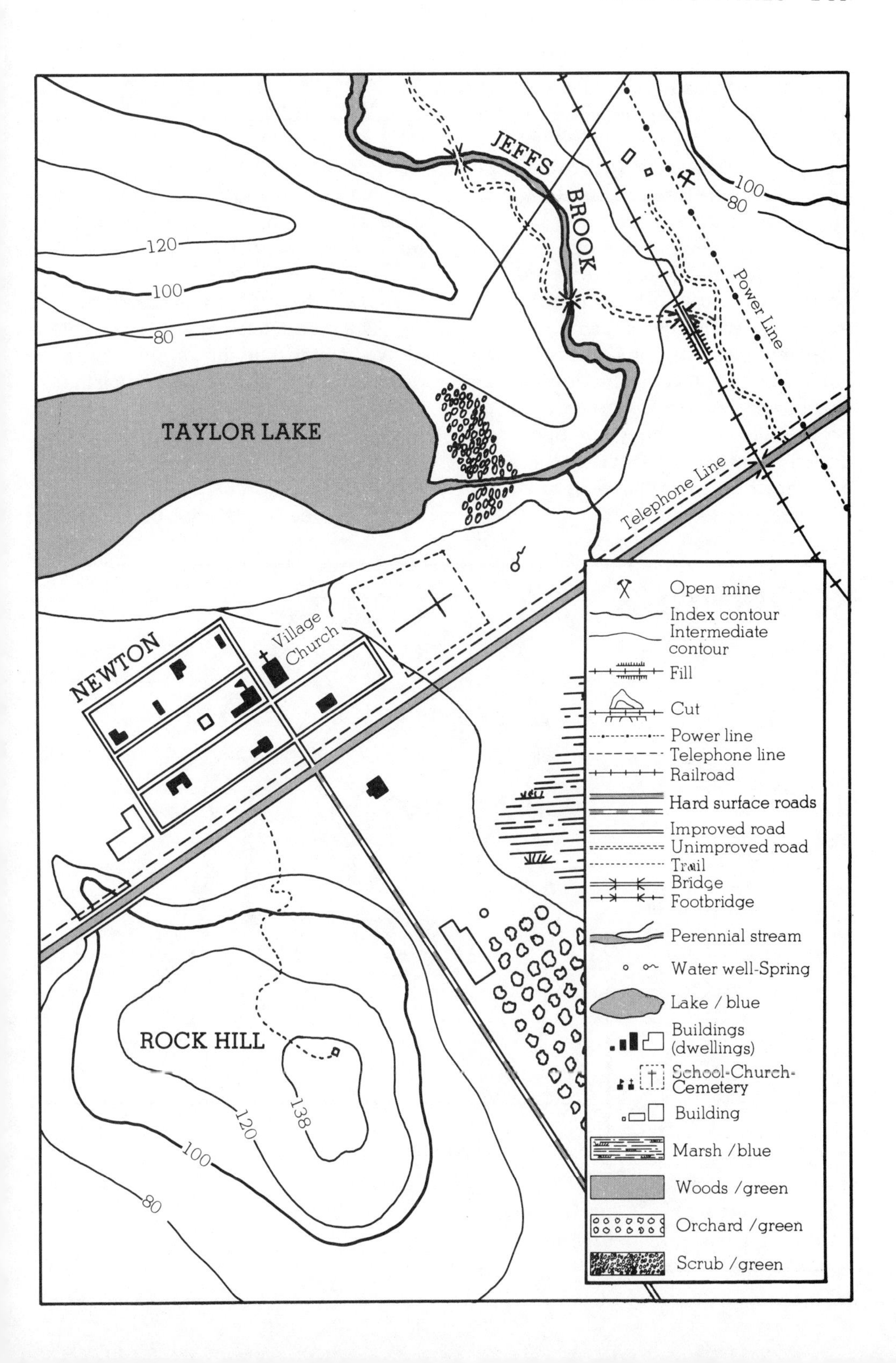
JEFFS BROOK
Power Line
100
80
120
100
80
TAYLOR LAKE
Telephone Line
NEWTON
Village Church
ROCK HILL
138
120
100
80
Open mine
Index contour
Intermediate contour
Fill
Cut
Power line
Telephone line
Railroad
Hard surface roads
Improved road
Unimproved road
Trail
Bridge
Footbridge
Perennial stream
Water well-Spring
Lake / blue
Buildings (dwellings)
School-Church-Cemetery
Building
Marsh / blue
Woods / green
Orchard / green
Scrub / green

down into increasingly small portions, or quadrangles, so you can order exactly what you need.

Almost identical maps are available for all of Canada through the Geological Survey of Canada.

It is essential to keep topo maps up to date. National Forest roads are continually being built, obliterated, or extended. Buildings shown on maps burn or vanish. New ones are built. New landmarks, such as microwave towers and power transmission lines, are built. So it is best to order your topo maps directly from the USGS if possible, since outdoor stores might have an older edition that hasn't sold out.

THE COMPASS

The compass is a needle magnetized on one end and balanced on a pin so it can swing freely. The magnetized end will always point toward the magnetic north pole unless it is deflected by local ore deposits (a rare occurrence) or metallic objects you are carrying.

The magnetized needle is mounted above a dial that shows all the major directions and often is marked off in the 360 degrees of a circle as well. Covering most compasses is a plastic plate that can be turned, and it is printed with an arrow marked north.

A compass does not point toward

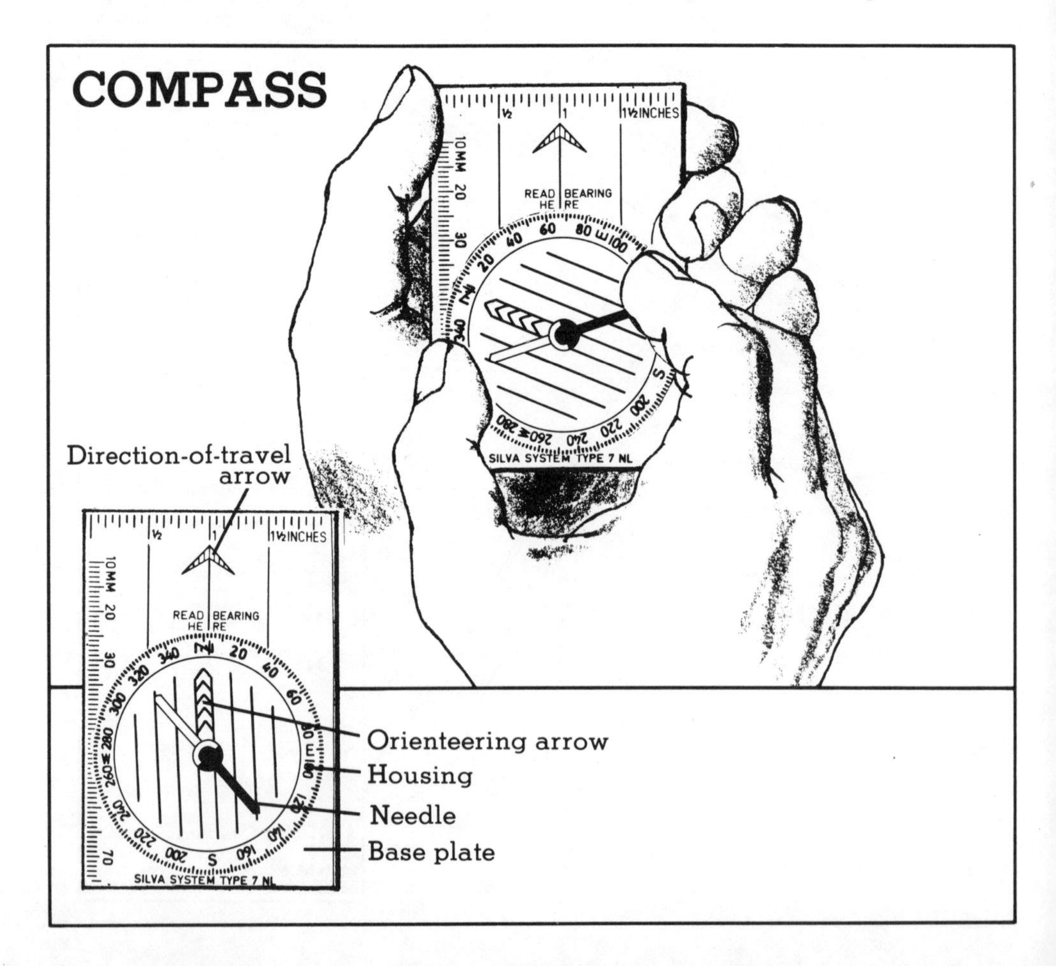

the North Pole. It points toward the magnetic pole, which is approximately 1,000 miles from the North Pole in northern Canada. This magnetic field tends to wander slightly, but not enough to make a lot of difference in your direction finding. At the most, it wanders only a degree.

All maps are based on the North and South poles. They are the only constants from which cartographers can operate with total accuracy. This means that your compass and the maps you carry do not match: whereas the longitude lines on the map point north, your compass will point a few degrees in another direction, toward the magnetic pole.

The angle of difference between the North Pole and the magnetic north pole on your campass is called declination. This varies in different parts of North America, and all U.S. and Canadian Geological Survey maps have the angle of declination printed on them.

This is where the plastic covers of the best compasses are needed. When you plan to use your compass on a trip, you must first consult the map of the area you will be traveling and find the angle or degrees of declination on it. Then you hold the compass with the arrow pointing to north on the dial, count the degrees of declination in the direction away from true north, and turn the upper dial to that point. Thus, the printed arrow will always point to the North Pole while the magnetized needle points to magnetic north.

This is the basis of all navigation by map and compass. The rest is applied mathematics and is relatively simple, provided you learn navigation in easy stages, starting in your own backyard or a local park. If you think you may need to use a combination of map and compass on a trip, it is vital that you practice with them before the trip, and that practice should be thorough and often enough so that you are as comfortable with the compass as you are with street maps.

In most family camping situations, you will not need the compass because you will probably want to stay near the major trails and campgrounds. But these family camping trips are excellent opportunities to teach everyone in the family the basics of wilderness navigation.

As your family becomes more adept at using the compass and maps, you may want to invest in more complex compasses. There are several different kinds available, each with its own uses. Of the five basic types, the **fixed-dial** compass has already been described. Then there's the **lensatic** compass equipped with a dial that pivots beneath the magnetized needle and has an azimuth scale that also is adjustable.

The **cruiser,** designed for forestry engineers and timber cruisers, has a dial with the degrees printed counterclockwise. It is of little use for recreationists.

The **sighting** compass is a sophisticated and highly accurate hand-held compass with a sighting lens in the case that magnifies the dial to within one-half-degree readings. They are useful for canoeists who want to maintain a totally accurate course and for other similar uses.

The **orienteering** compass is a sophisticated design that can be laid on maps for plotting routes. It has a transparent plastic plate for a base that is marked on one side with the map scale in millimeters; another side has the scale in inches. The orienteering compass is a good second choice for the family who wants to combine U.S. Geological Survey map reading and compass use in the same learning process and should

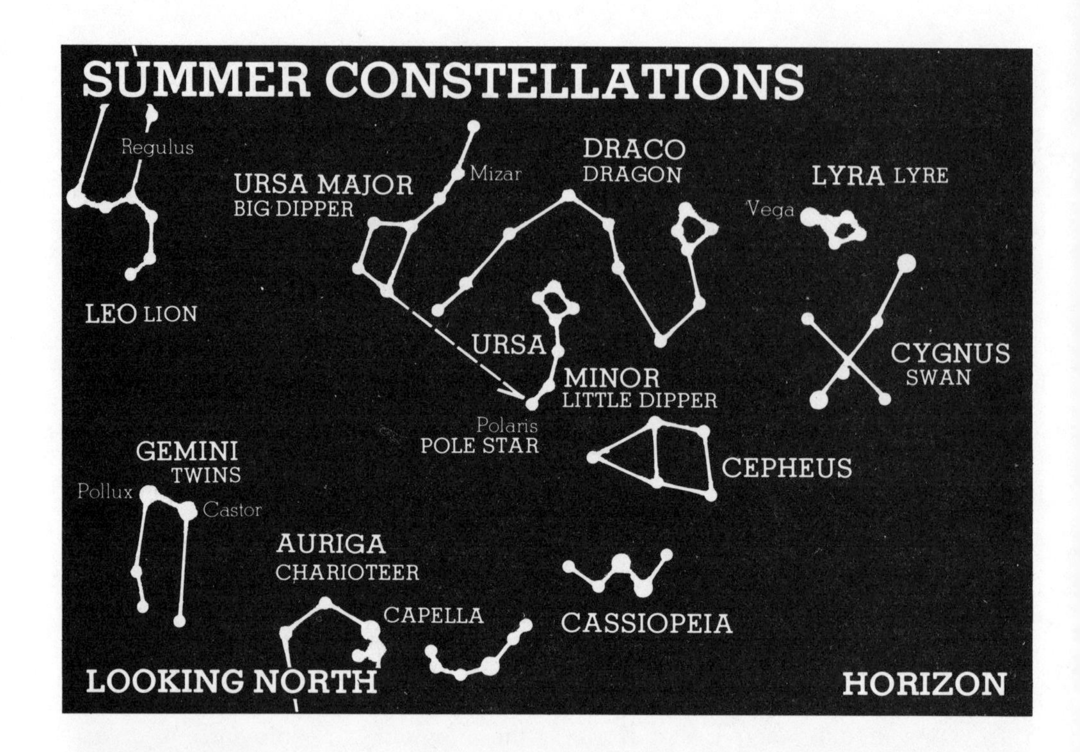
SUMMER CONSTELLATIONS
Regulus
URSA MAJOR
BIG DIPPER
Mizar
DRACO
DRAGON
LYRA LYRE
Vega
LEO LION
URSA
MINOR
LITTLE DIPPER
CYGNUS
SWAN
Polaris
POLE STAR
GEMINI
TWINS
Pollux
Castor
CEPHEUS
AURIGA
CHARIOTEER
CAPELLA
CASSIOPEIA
LOOKING NORTH
HORIZON

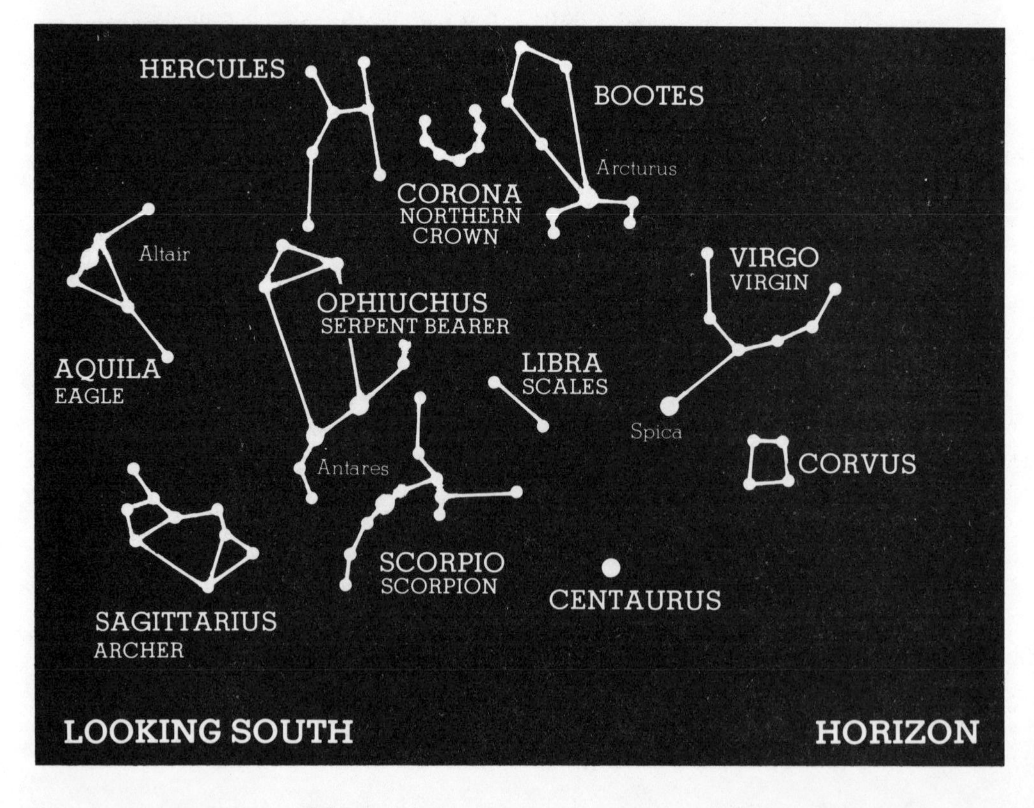
HERCULES
BOOTES
Arcturus
CORONA
NORTHERN
CROWN
Altair
VIRGO
VIRGIN
OPHIUCHUS
SERPENT BEARER
AQUILA
EAGLE
LIBRA
SCALES
Spica
Antares
CORVUS
SCORPIO
SCORPION
CENTAURUS
SAGITTARIUS
ARCHER
LOOKING SOUTH
HORIZON

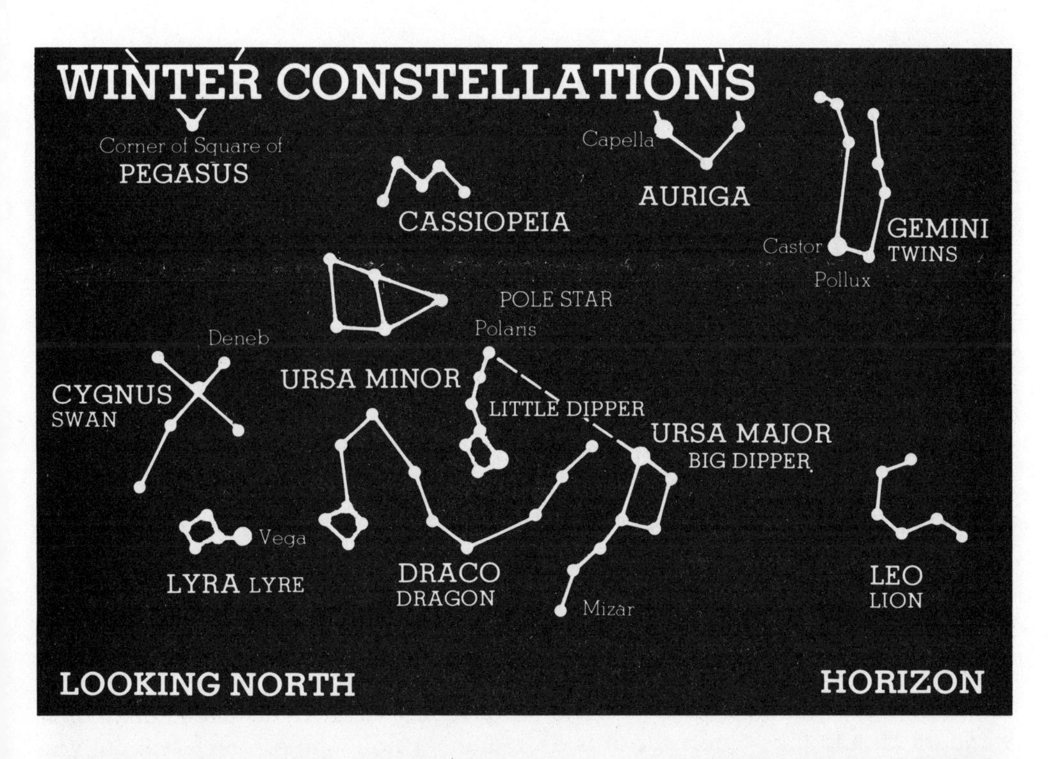
WINTER CONSTELLATIONS
Corner of Square of
PEGASUS
CASSIOPEIA
Capella
AURIGA
GEMINI
TWINS
Castor
Pollux
POLE STAR
Polaris
Deneb
CYGNUS
SWAN
URSA MINOR
LITTLE DIPPER
URSA MAJOR
BIG DIPPER
Vega
LYRA LYRE
DRACO
DRAGON
Mizar
LEO
LION
LOOKING NORTH
HORIZON

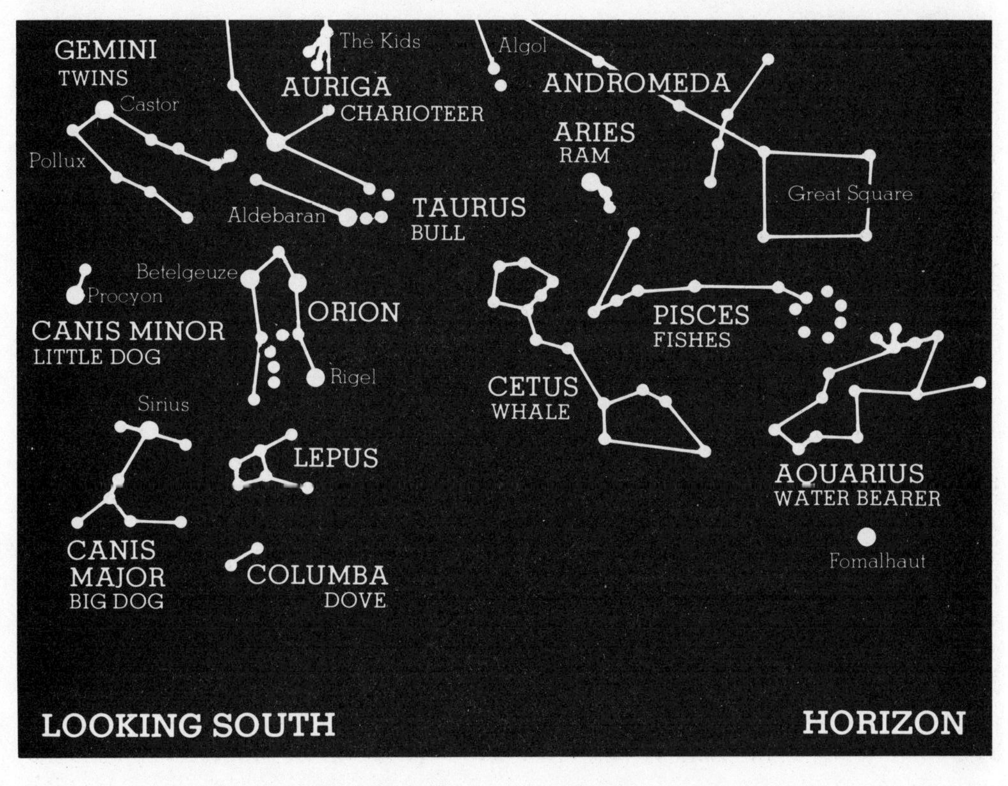
GEMINI
TWINS
Castor
Pollux
The Kids
AURIGA
CHARIOTEER
Algol
ANDROMEDA
ARIES
RAM
Aldebaran
TAURUS
BULL
Great Square
Betelgeuze
Procyon
CANIS MINOR
LITTLE DOG
ORION
Rigel
PISCES
FISHES
CETUS
WHALE
Sirius
LEPUS
AQUARIUS
WATER BEARER
CANIS
MAJOR
BIG DOG
COLUMBA
DOVE
Fomalhaut
LOOKING SOUTH
HORIZON

be considered as the second step up in compass purchases.

ORIENTEERING

This sport has become one of the most popular forms of recreation in many parts of North America, and especially in Europe. Essentially, it involves someone laying out a course with compass and topographical map for others to follow. The game is one of precision because you must follow the compass and map exactly, and it is one of speed because the first participant to complete the course wins.

This can be adapted for a campground game with checkpoints near enough to the camp so that nobody wanders off into the wilderness. It is a technical version of a scavenger hunt and one of the best ways to become intimately familiar with maps and compass.

A modest version of orienteering can be developed for the whole family as a learning tool. Simply lay out a course along the road to camp and have someone in the group read off the landmarks, follow the compass readings, and alert the driver which landmarks to watch out for next.

Another good way to become well acquainted with both map and compass is to find a camping spot in an open valley surrounded by peaks or hills with some other landmarks on the topo map. One member of the group can go out alone and establish checkpoints along a route, such as hiding a coin (or any recognizable item) at the base of a tree. Obviously the course must be simple for the beginners, and it should specify how many feet or kilometers each leg of the course is. This also teaches campers to estimate how many paces they must take for specified distances, such as one hundred yards or one hundred kilometers.

Another exercise can be conducted in wide-open country the first time, then in timbered and rugged terrain after a bit of practice. This is how to find a spot, such as a campsite, by making progressively smaller boxes on an imaginary map.

Assume you know roughly which direction you've walked from camp, but you know you can't reach it by walking a straight line. So you estimate how far you've walked. Then, from your present position, begin walking at a right angle from the direction you think camp lies. Walk at least half that distance at the right angle, then make a right-angled turn and walk that distance again, then another right angle, which will put you in a straight line from where you started.

Now, walk toward the starting point, but stop three-quarters of the way there (if you've walked one mile, stop at three-fourths) and take your right-angle turn inward again. If you don't see the camp on this leg, make another right-angle turn three-fourths of a mile along, then repeat it until you've made another three-fourths box.

Now cut your distance of the next leg another fraction.

Be sure that while performing this exercise you follow the compass route as closely as possible and allow for detours around rock outcroppings and other obstacles to your direct route.

Unless you are totally lost, or you didn't bother to mentally note landmarks around your campsite, you should pass it before the series of boxes you're walking come to the center. If you don't find your campsite in this fashion, begin another series of progressively smaller boxes farther along the backward route.

Obviously, an experienced outdoors person will mentally establish a series of landmarks along the route away from camp. However, hikers sometimes get caught in a dense fog that hides everything within a few feet.

It is virtually impossible for anyone to walk in ever-smaller circles, but with a compass to establish a route and reasonably accurate estimates of distances walked, this will bring you back to camp.

WATCH COMPASS

A good trick for everyone in the family to learn (although your chances of having to use it should be slight; always carry a compass) is how to find true north by using your pocket or wrist watch.

First, if you are on daylight saving time, move the watch back an hour to sun time.

Then get a piece of straight stick—a wooden matchstick will do fine. Place the watch on a level surface, such as a stump of rock, with the sun hitting it. Hold the stick so that it is in the center of the watch, and turn the watch until the shadow of the stick is along the hour hand.

If you are performing this exercise between 6:00 A.M. and 6:00 P.M., **south** not north will be directly between the hour hand and 12:00 on the watch in the angle formed by the two hands.

If you're doing this between 6:00 P.M. and 6:00 A.M., **north** will be between the hour hand and 12:00.

The formula to remember is **N** for North and Night; **S** for South and Sun.

This is accurate to within eight degrees, assuming your watch is set properly.

Another method of finding directions with a watch is to put a straight stick in the ground and mark the end of its shadow at 10 A.M., then again at 2 P.M. (You can use other times, but each must be the same number of hours away from noon.) Measure the distance between the two points, divide by half, and that point will be true north.

CAMPER'S SECRET

Here's how to measure the distance between you and that last bolt of lightning. When the lightning strikes, count off the number of seconds between the strike and the arrival of the thunder, and figure four and a half seconds per mile. If you don't have a watch, count slowly, "one thousand, two thousand," etc.

NATURE SCAVENGER HUNT

A scavenger hunt is always popular with young children and can be used as a method of getting them acquainted with a camping area. Adults should go out first and scout the area not only for items for the children to find, but also for potential dangers. After scouting the treasures—draw up a list that looks something like this:

a pine cone off the ground
a bird's feather
a wild flower
a dead leaf
a living leaf
one piece of litter
one piece of moss off a tree
one piece of moss off the ground

Make sure first, though, that no poison plants such as ivy or oak are in the area and do not let children stray beyond the campground perimeters. If there are several children in the group, divide them into teams.

The winner gets a prize, an extra dessert treat or whatever you choose.

KNOTS

It isn't of great importance in the field to know the language of knots, but when you look through a selection of knots, you will find them divided into knots, bends, and hitches.

A knot is used to tie a bundle of something and it makes a loop or a noose or knot in the rope.

A bend ties two ropes together.
A hitch ties a rope to something—a tree, a ring, an oar, or whatever.

Square knot: The most common, after the slipknot you use to tie your shoes, is the square knot. It won't slip if tied correctly, and it can be untied with considerable ease.

Bowline: This is used for strong loops that won't jam or slip. To tie it, form the loop, then run the free end through, around the standing part, and back through the loop.

Sheet bend: This is used to join two ropes of different sizes. The free ends must be on the same side of the knot or it will slip. This knot will jam on heavy loads, however, and the Carrick bend should be used for those.

Carrick bend: Virtually jamproof, this can be used to join two ropes of different sizes when a strong pull, as with a winch, is needed. To tie, make the loop with one rope, then interlace the other rope as shown. Since there will be considerable slippage until the knot is tight, allow plenty of length on each end.

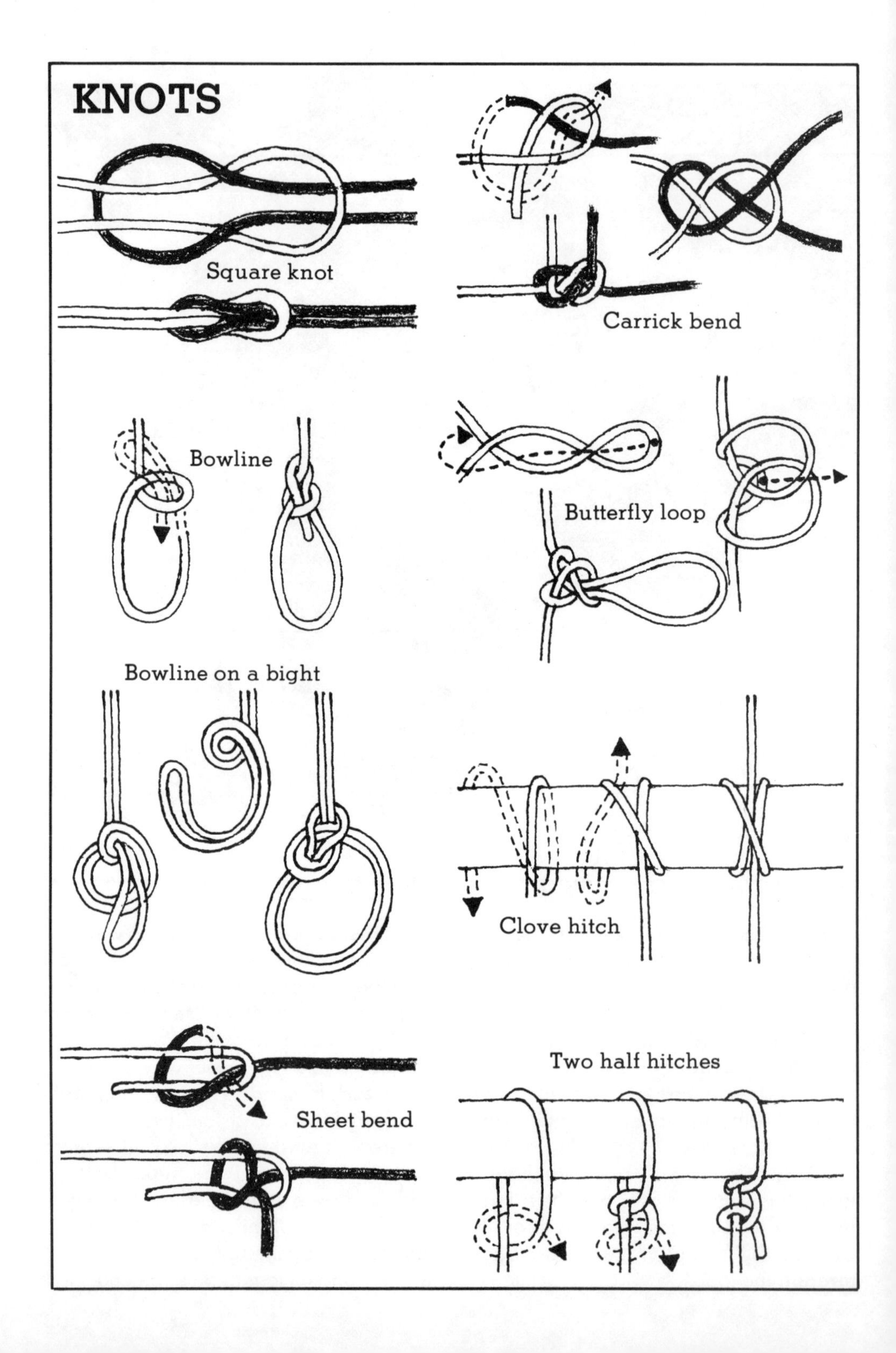
KNOTS
Square knot
Carrick bend
Bowline
Butterfly loop
Bowline on a bight
Clove hitch
Two half hitches
Sheet bend

Butterfly noose: This is a loop in the middle of a rope and will not jam. To tie it, twist a loop, then fold up the lower part of the loop and push it through the center opening as shown.

Clove hitch: This is a quick and secure method of attaching a line to a tree or post. It will, however, slip unless pressure is kept on the end. To avoid this, take a half hitch with the loose end around the taut part of the rope.

Half hitch: Half hitches and double half hitches are popular among sailors and cowboys for making quick, temporary knots.

SURVIVAL SKILLS

A number of wilderness survival skills can be taught as a normal part of the camping experience, and while some of this lore is absorbed almost by a process of osmosis, there are many things that can be a part of the camping trip.

Some old legends never die, such as the alleged ability to find north by seeing which side of the trees have moss growing on them. This may be true in some areas of North America, but not all. Some shaded forests have almost equal amounts of moss on all sides. In general, the south side of hills and mountains have less plant growth than the shadier and damper north side. In some areas of extreme glaciation during the ice ages, the hills have been scoured in a general north-south direction.

Animals are no more eager to waste energy than people are, so well-worn game trails usually follow the path of least resistance.

When you are traveling any distance in the outdoors, teach your children—and learn this yourself—to stop often and look behind you so you will recognize the route on your return. Things always look different when approached from the opposite direction. It is especially important to stop and study the lay of the land when you come to a fork in the trail. Memorize the intersection by looking backward at it.

Some outdoors enthusiasts insist they have body chemistry that makes them know instinctively where north is. This may be true in some extremely rare cases, but it is more likely that they have traveled in the outdoors so long that they subconsciously note wilderness signs and know their direction. Without a compass, nearly everyone traveling on strange terrain will walk in a wide circle.

If you are camping where wildlife is abundant, or if you're camping in the off season when snow is on the ground, a book on animal tracks is a good investment. You can also identify animals by their droppings. Soon you will learn to tell approximately how long ago the animal was there by a variety of signs, such as water seeping into the tracks, or crushed grass still returning to its original position.

CAMPER'S SECRET

Know what kind of wood you can use before leaving the trailhead on your trip. More and more federal lands are prohibiting campfires, and restrictions are being imposed to preserve what little dead wood is left so it can rot back into the soil to help new trees grow. Never cut a live tree or trim off the branches.

ANIMAL
TRACKS
Deer
Raccoon
Opossum
Porcupine
Cottontail rabbit
Skunk
Fox
Rat
Mouse
Deer
Raccoon
Porcupine
Opossum
Cottontail
rabbit
Skunk
Fox
Rat
Mouse

USING KNIVES, AXES, AND SAWS

Most campfire injuries are related to fires and tools such as knives and axes, and safety while using tools must be emphasized at every opportunity. Learn how to fold a knife so that you don't catch your finger. Learn how to use one so that there is never a danger of its slipping and slicing either you or someone near you. Always whittle away from your body. Do not let children play with bowie knives by throwing them against stumps or trees. Not only is it rough on equipment, it is also dangerous and can damage living trees. Emphasize always that knives and axes are tools, not toys.

The use of an ax should be approached with the same respect an animal tamer uses when entering a cage full of tigers. Because of the force required to make an ax do its job, its potential for serious injury is greater than that of almost any other campfire tool.

The illustrations on pages 161-164

THINK SAFE

Probably more camping accidents occur with knives and axes than any other tools, so it is extremely important to set strict rules for the use of each. If your companions—young or old—are not experienced with knives and axes, establish rules early in the outing. Wood-chopping areas should be set aside some distance from the campsite, and you can borrow an idea from the Boy Scouts which involves putting a "fence" around the wood-chopping area with a rope. Children should use knives and axes only with supervision.

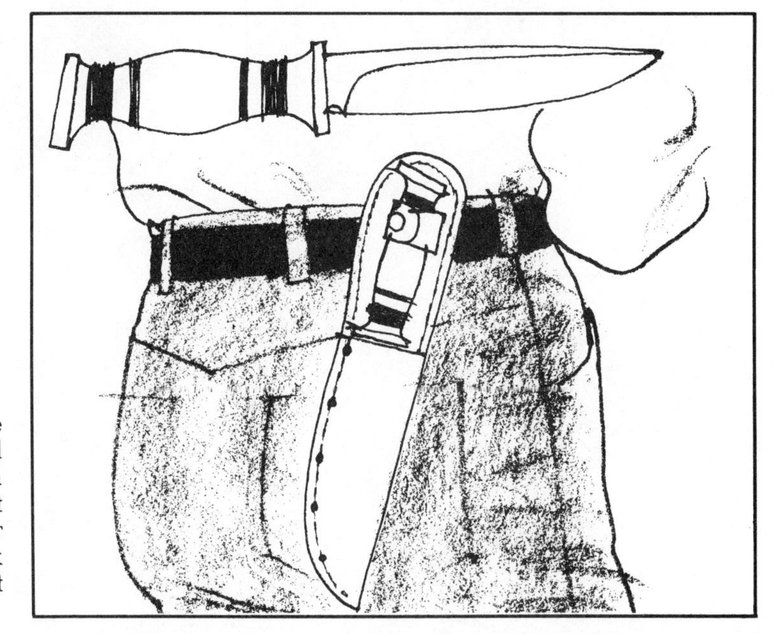

A sheath knife should be placed in your belt in such a way that it won't interfere with your movement or stick you when you sit down.

Camping knives range from two-blade models to the multiblade-and-tool Swiss army knives.

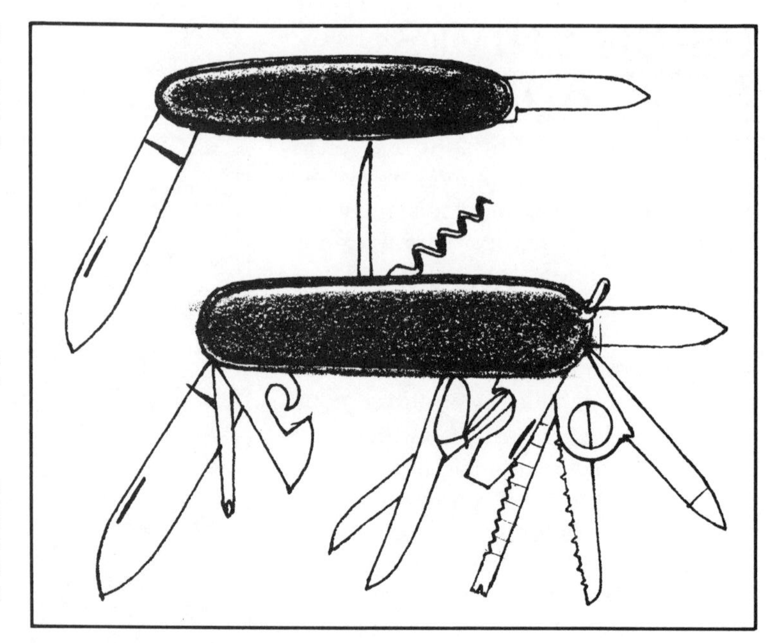

Always whittle away from your hands and body.

Carry a whetstone to keep knives and axes sharp, and keep your fingers out of the way when sharpening.

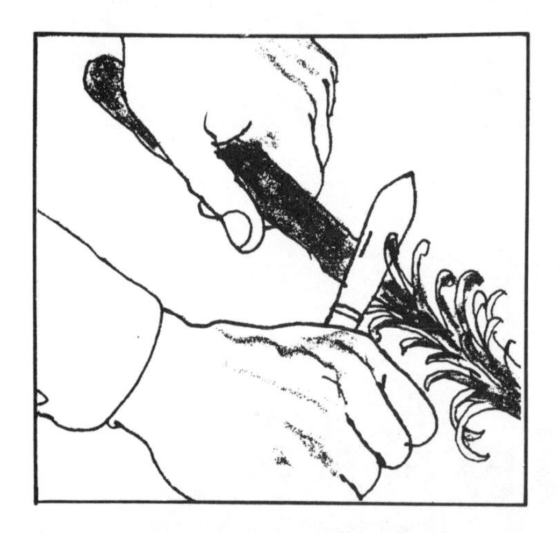

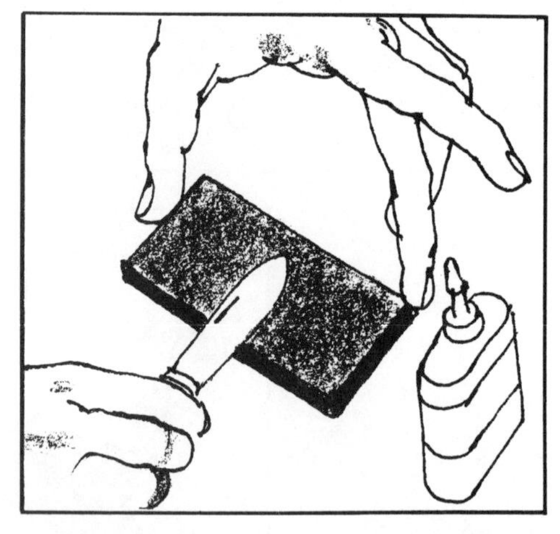

CAMPING ETIQUETTE

The more children know about outdoor lore, the more they will enjoy camping. It is also the perfect place to teach them the basics of outdoor good manners and environmental responsibility. After you have camped at several different sites and the children have seen the evidence of thoughtless campers who came before you—litter, fire-blackened boulders, trees scarred by carved initials or ax marks—they will take more interest in keeping campsites cleaner.

show examples of how to use an ax properly in most camping situations. In all cases, always carry an ax or hatchet in a sturdy sheath.

In many campgrounds you won't really need an ax or hatchet, and a folding saw will be sufficient for all your uses, plus a sharp knife for making shavings, and tinder for starting campfires.

A variety of folding saws are on the market, all considerably lighter than the smallest ax or hatchet. One of the most popular is the Sven saw, which folds into a compact tube that can be tucked away in a pack or in the truck of your car without danger of snagging something.

Other saws are available, including the kind found in many survival kits, which is simply a thin wirelike piece of steel with teeth and with a ring on each end for you to hold.

Most environmental protection

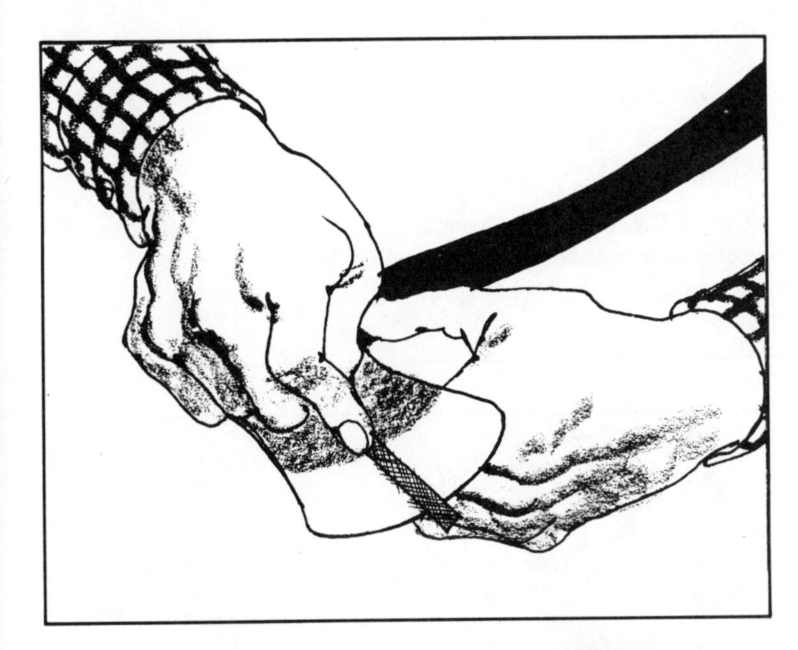

A mill file and double-grit whetstone are needed to sharpen axes and hatchets, the file to knock off the burrs and the heavier whetstone to dress and sharpen the hard steel.

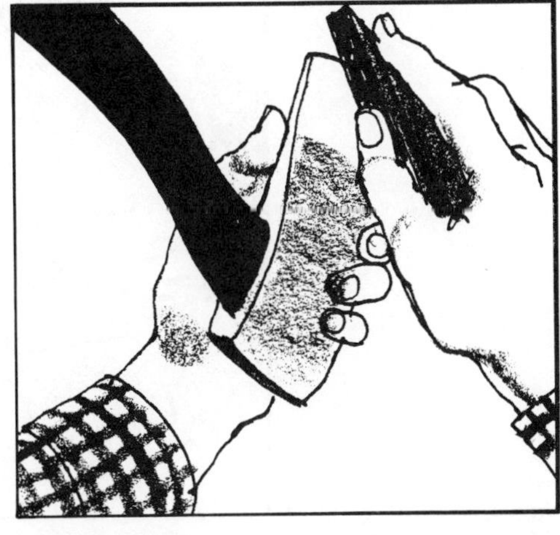

MAKING A HARDWOOD WHISTLE

Many of us remember that great sense of accomplishment when we made our first successful whistle from a hardwood branch. Where tree growth is abundant, you can cut a branch and, usually after a few false attempts, make a one-toned flute.

The branch section should be free of knots or other imperfections. Cut it into about a six-inch length, then gently tap the bark all around until it loosens from the hardwood beneath.

Then cut a notch in the middle, as shown in the illustration, and a notch for the mouth piece. You may not get it right the first time, but after a series of minor adjustments with a sharp knife, you will note that the closer you get to the right cuts, the closer the sound of your blowing through it comes to a whistle.

You can wedge the head of an ax or hatchet between two pairs of stakes to anchor it, placing a piece of firewood between the stakes for more security.

When cutting a stick, always cut at an angle to reduce the distance it will fly. Make the first cut deep enough to hold the wood with the ax; then hold the stick with one hand and swing it downward on a chop block or log to complete the cut. This gives you much more control over the flying wood.

Use the same technique for splitting. Chop into the end of the stick, leave the ax in, and bring it down on the chop block.

organizations strongly discourage the use of an ax or hatchet by campers. They argue that, as with owning a firearm, you feel compelled to use it whether it is needed or not. This is particularly true with children.

On the other hand, if you check into a Forest Service campground and find the wood supply consists of pieces of wood as thick as a runner's thigh, you will wish you had brought along an ax.

WILDCRAFTING

A pleasant way to combine camping with a chance for school-age children to earn extra income is to camp in areas known for having marketable plants and parasitic growths. In those parts of the country where these things grow in profusion, it is possible for a family not only to recover the costs of camping trips, but also to turn a tidy profit. Some students have paid most of their way through college by "brush-picking" or other forms of wildcrafting.

Here are some examples of forest products for which there is usually a market:

For use in floral arrangements, wreaths, etc:

- evergreen huckleberry
- salal
- sword fern
- scotch broom
- false boxwood
- dwarf Oregon grape
- galax
- leucothoë
- mountain laurel
- mistletoe
- holly
- evergreen boughs

Used by pharmaceutical manufacturers for medicines:

- virginia snakeroot
- cascara bark
- quinine conk

Collecting seed cones for timber companies is another source, but quite specialized.

If you plan to undertake this activity, check first with the local Forest Service office or other governmental agency for information on which forest products have a market value, how to go about getting permission to harvest them, and where to sell them. Some timber companies assign portions of their land to individuals who earn a comfortable income from this line of work.

Most of this activity is in the Pacific Northwest, from northern California to upper British Columbia, and in the Appalachian Mountains.

CAMPER'S SECRET

One chore most children enjoy performing on long trips is making a chart to show which oil companies will honor your credit cards. The names of many oil companies change from state to state, and the various brands under the same parent company are shown on the reverse side of your credit cards. Have children list your credit cards across the top of a sheet of paper, then the different brand names by state beneath. It gives them something useful to do and greatly simplifies your search for a service station.

CHAPTER 10

FIRST AID AND SAFETY

Cuts, scratches, minor burns, insect bites and stings, and other minor injuries that are common at home and in the backyard are to be expected sometimes while camping, too. Be pleasantly surprised if you don't have to open the first-aid kit on a family outing. No matter how careful you are, someone will pick up a hot dish, step on a sliver, get scratched on a sharp rock, or cut with a knife or on the sharp edge of a grill.

Most camping injuries are in this minor category, and you should be prepared for them with a well-stocked first-aid kit. Be sure to take an inventory of the kit's contents before each outing in case someone raided it for a cut at home.

Although serious injuries are unlikely to occur, it is wise to know in advance how to deal with them. Many families, as a matter of course, take the Red Cross first-aid classes offered in most community centers or at work. The Red Cross has made inroads into the public's general medical ignorance with these courses, which are very reasonable in cost and taught by experienced personnel. It will give you a feeling of security both at home and in the outdoors, and this alone makes the time invested well worth it.

Treatment of injuries continues to change and improve. Some of the old ways have been proven to be almost as damaging as the injury itself. For example, we used to be taught to treat frostbite by rubbing it with snow. Now we know that not only should the frostbitten area not be rubbed, it should be warmed instead of chilled. There are many examples of this in first-aid history, so you should be certain your home first-aid manual is up to date.

A BASIC FIRST-AID KIT

No two first-aid kits are alike because each group of campers has its own particular needs, such as prescribed drugs and personal preferences. However, here is a basic list of first-aid equipment that should always be with you when away from home, either on an extended trip by car or public carrier or while camping.

Medicine

Prescription drugs with dosage clearly labeled on the container. Each drug in a separate container. Wrap dosage instructions in clear plastic tape to prevent smearing or erasing.

aspirin—at least 12 tablets, more if you expect to need them

salt tablets—24, to prevent heat cramps

antacid—at least 6 tablets, in case the cook blunders

antihistamine—at least 6 tablets, for insect bites, stings, colds, etc.

> **CAMPER'S SECRET**
>
> When you need a cold compress to reduce swelling, don't forget to look in your ice chest. It sounds insultingly simple, but when you're in pain and the others are worried, it's easy to forget the obvious.

Bandages

bandage strips—a dozen one-inch strips for minor lacerations

butterfly bandages—6 or more, for closing lacerations

Carlisle battle dressing—one or two, for large wounds

moleskin—a 6-inch-square sheet or more, for blisters

sterile gauze pads—for large wounds

tape—2-inch roll, for holding bandages on or for sprains

triangle bandage—for supporting an injured arm or covering a large dressing

elastic bandage—one 3-inch size; know how to use it properly

Tools

needle, to remove splinters, etc.

razor (disposable is fine)—for shaving hairy areas before taping

safety pins—3 or 4 for mending clothing, holding bandages, etc.

oral fever thermometer

Liquids

tincture of benzoin—1-oz. bottle, to help hold tape to skin

antibacterial soap—1-oz. bottle, for mild antiseptic cleansing of cuts

Suggested Additions

coins for telephone calls
small first-aid book
thread for clothing repairs
sunscreen to prevent sunburn
anesthetic lotion for use if sunburn prevention fails
Caladryl lip balm

You should also carry a snake-bite kit when traveling in areas where poisonous snakes are common—but you must know how to use the kits properly and be prepared to do so. Most kits are equipped with a small knife for lacerating the bitten area and a small suction

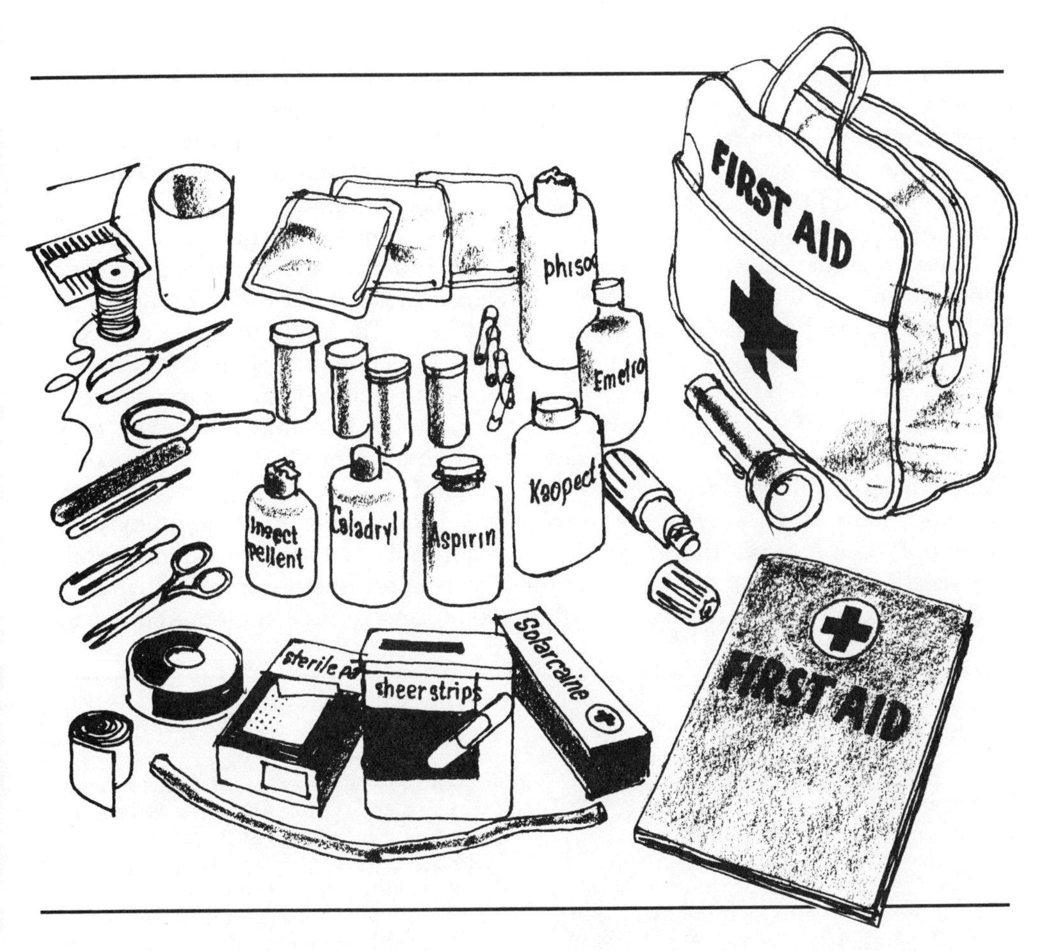

device for removing as much of the venom as possible. Some also have antivenin to counteract the poison. Check with your family physician for the latest information on snake-bite treatment. Very few people are bitten by snakes each year, and deaths are very uncommon, except from bites of the coral snakes in the southeastern United States. Obviously, as with all camping accidents, taking preventive measures is the wisest course of action; everyone should be trained to watch carefully for snakes.

Burns: For minor burns, immerse the burned area in cold water to relieve the pain, then carefully wash the area with a liquid soap using sterile cotton soaked in water that has been boiled and cooled. Cover the burn with a layer of sterile dressing, preferably coated with petroleum jelly so it won't stick to the burned skin. Then cover this with a larger, thicker bandage that applies only a moderate amount of pressure to the injured area. Leave for approximately a week without removing to avoid the possibility of infection.

In cases of minor first-degree burns, bandaging isn't necessary except to prevent further injury to the area. Burn ointments, similar to that recommended for sunburn, can be applied to superficial, cooking-type burns.

If the burn covers a large area, however, or destroys the skin, take the victim to a doctor or hospital to avoid further damage. Keep all clothing, jewelry, and other foreign matter from touching the skin and keep the victim as immobile as possible so additional damage to the tissue won't occur.

Heat exhaustion and heat stroke: The best treatment obviously is to take preventive measures. An adequate intake of salt and fluids will prevent most from occurring.

Heat exhaustion is caused by prolonged physical activity in a hot climate and is caused by the blood vessels in the skin becoming so dilated that they rob the supply to the brain and other vital organs. The result is similar to fainting and usually isn't serious. The victim will feel faint and often has a rapid heart beat. Sometimes the victim will also feel nauseated and have a headache. Rest, along with an intake of salt and fluids, is the best treatment.

Heat stroke (or sunstroke) is in many ways the opposite of hypothermia and can be fatal if not treated quickly and properly. The body must be cooled as rapidly as possible, either by immersion in tepid—not cold—water, application of wet towels or blankets over the entire body or, best of all, with towels soaked in alcohol.

The injury may come quickly. The victim will be normal one minute and in a very short time become confused, irrational, and uncoordinated. At this stage,

WATER SAFETY

It is always surprising to find how many people never learn to swim or bother to learn or have their children learn the rudiments of water safety. Several thousand people drown each year, and most of those are in still water rather than swift rivers or far out to sea. Most occur within sight of the beach, and most could be prevented if more people were at ease around the water.

Water safety should be an important part of your outdoor experience, not only the ability to swim but also the ability to stay afloat in water for long periods.

One way to do this is testing yourself at the local swimming pool or lake. Try the lifesaving techniques, such as using your clothing for life preservers. Most cloth is airtight when it is wet, and pants or long slacks make good life preservers. Tie the ends of the legs together and, while treading water, swing the pants over your head with the opening down. This will at least partially fill the legs with air, and you can inflate them the rest of the way by blowing into them. By keeping the opening below the surface, you'll be able to float for several minutes, or longer if the material is closely woven.

No matter how well you swim, you should never be out in a boat without a life vest on, and children should grow up taking these vests for granted.

Your local Red Cross chapter will offer lifesaving classes that include the accepted methods of artificial respiration for drowning victims. The odds are against your ever having to use this knowledge, but the peace of mind you'll have from the knowledge is well worth the small amount of time and the modest expense involved.

the victim will usually have a body temperature of 105 degrees Fahrenheit or higher and no sweating at all.

The treatment must begin immediately. Sunstroke is one of the most dangerous accidents because death or brain damage can occur if the body's temperature isn't brought down to at least 102 degrees rapidly. The victim will not recover for some time, and the temperature may fluctuate up and down for a few days afterward.

OUTDOOR PESTS

Pesky critters that buzz through the air or crawl about are a natural part of outdoor experiences, from backyard picnics to wilderness treks. Perhaps as much as fear of discomfort, the dread of these pests makes many novices nervous about camping. Some people have an almost irrational fear of harmless pests, and there apparently is little that can be done to remove the fear, other than wearing it away by finding places to camp with a minimum of pests, and spending as little time as possible talking about it. Like fear of the dark, talking about it can only make it worse; treating it like the normal thing it is will gradually erode fear into an easy acceptance.

These outdoor pests come in almost as many varieties as there are campers. Rather than go through the whole encyclopedia of things that bite, sting, or stink, let's consider the most common pests and how to coexist with them.

Ticks live mainly in hardwood forests and open range where they have cattle available. They are seldom in the high country above timberline, and they do not like cold, damp climates. If you are camping in an area where ticks may be found, be sure to check your body at least once a day, especially around your boot tops, belt line, armpits, crotch, beneath folds in your skin, etc.

If you wear high-top boots, tuck your pants legs in them if possible or into your socks, then spray your pants legs and socks with an insect repellent. This may not keep all of them off, but it will help.

When you find a tick on your body, remove it immediately and carefully. If it has only its head buried in your skin, gently ease it out with your fingers or tweezers. If it doesn't dislodge easily, apply a lighted match or cigarette to its body. Then kill it.

If the tick has burrowed into your skin deeply, apply heat to it. Sometimes a bit of alcohol or stove fuel (white gas or kerosene) will remove it.

Should you accidentally leave the head burrowed in your skin, remove it

CAMPER'S SECRET

Some outdoor pests are more of an irritant than a major problem, such as the slugs and big snails common in some areas of the country. They are disgusting to touch, leave a trail of slime behind them as they inch along and, if they are accidentally stepped on, leave a mess on your shoes and the ground. They can be killed by sprinkling salt on them, but you'd have to carry bags of salt to build a salt fence around the campsite. No final solution is known, although some people put out saucers of beer to attract them to their death by drowning.

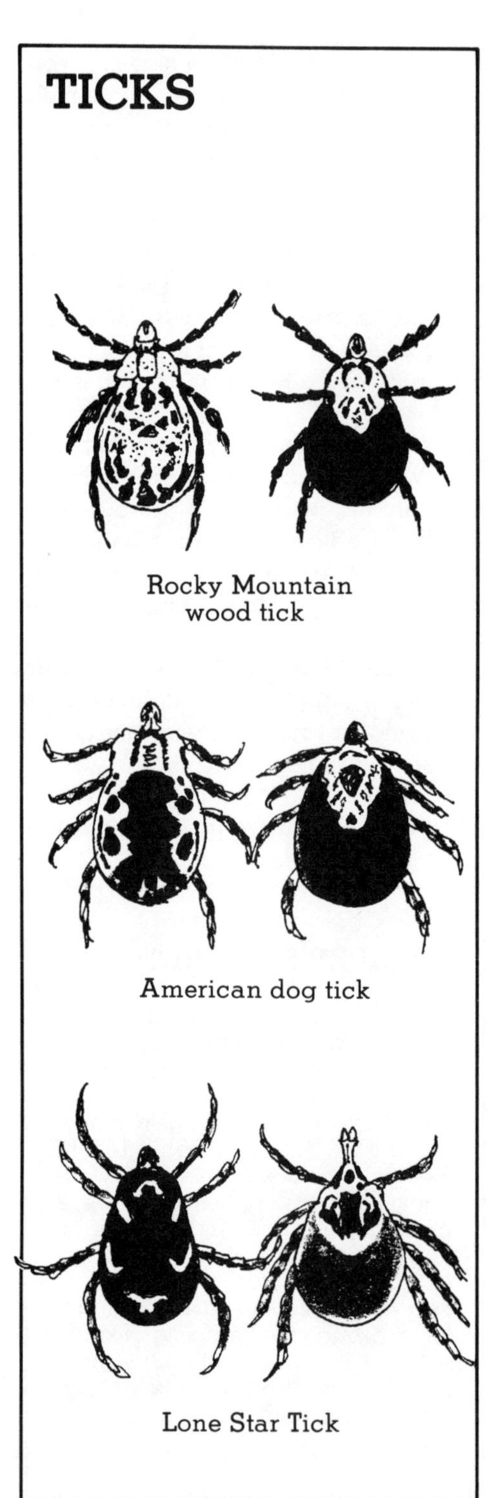

Rocky Mountain wood tick

American dog tick

Lone Star Tick

immediately with a bit of minor surgery. Sterilize a sharp knife and make a small incision, then treat the area with an antiseptic. In all cases, an antiseptic should be used.

A few diseases are occasionally transmitted by ticks, even though the incidence of Rocky Mountain tick fever and other afflictions is on the decrease. So take no chances.

CAMPER'S SECRET

One method of keeping flies and stinging insects such as hornets and bees away from the camping area is to pour a small amount of syrup or honey in a dish and place it well away from the campsite. It won't attract all of them, but it will help.

Spiders: Thanks to a host of horror movies, spiders elicit a stronger fear among us than they actually rate. Although all spiders are venomous, only a tiny fraction of the species have teeth strong or long enough to penetrate human skin. And of these, only two are likely to be a problem for humans. Fortunately, few live beyond urban areas because their food supply is the flies, gnats, and roaches that also depend on man's garbage for food.

The black widow is the most dangerous, although almost every fatality from one's bite is among very small children or the elderly; individuals in good health almost always recover. Black widow spiders are about one-half inch long with an almost spherical body. The underside has a red-orange hourglass marking. After a person is bitten by a black widow, severe pain usually starts within an hour in the area of the bite, then spreads to the rest of the body. The victim is usually quite ill with abdominal pain and spasms, and often shock.

Antivenin is available but should be administered by a physician. Most treatment, such as hot baths, is directed at relieving the muscle spasms. But the victim should be taken to a doctor or hospital as soon as possible. The pain and prostration caused by the bite usually disappear in two to four days, but the victim will usually be weak for a few weeks afterward.

The other dangerous spider is the brown spider, called the brown recluse or violin spider. It isn't often fatal, except to very small children. The brown recluse is slightly smaller than the black widow and has an oval body with a dark violin-shaped spot on the front portion of its body. Its bite causes a severe inflammation in a small area around the bite. Over the next two weeks, the resulting blister usually ruptures and the skin turns dark and falls away, leaving a sore that usually heals with some scarring.

The most feared spider, because of its ghastly appearance, is the large tarantula, native to the southwest desert and parts of the South and Midwest. Its bite is no more damaging than a bee sting; it just looks awful.

When walking through underbrush, where spiders build their webs between branches of trees or bushes, carry a walking staff or simply a stick to beat the webs away before you get to them.

When picking up a stick or rock, first roll it over with your toe or a stick.

Carry insect spray designed to kill spiders and use it on areas with lots of dead wood or rocks near your campsite.

Always check your sleeping bags and clothing for spiders or insects before using them.

Scorpions can inflict a bad sting. Since they are primarily nocturnal, especially in desert areas, you probably won't see them unless they find their way into your clothing or personal effects. So always shake out clothing, bedding, and towels before use in areas where scorpions are common. Be careful when lifting rocks or old boards, where they like to hide.

Bears: Although they hardly qualify as pests, owing to their size, many campers fear bears without knowing the difference between species. Except for Montana and Wyoming, particularly in Glacier National Park and Yosemite National Park, the grizzly bear is not common in the contiguous United States. They are common, however, throughout western Canada and Alaska and should be treated with great respect. The major safety precaution with grizzlies is guarding your food supply. Park rangers have been trying for decades to educate campers not to leave food and garbage around so they won't attract bears to their camping area. If you are camping in an area known to have grizzlies, keep your food suspended at least ten feet above the ground from a rope hanging from a tree branch. Do not keep it in your tent. Do not sleep near the campground garbage dump.

For the rest of North America, the smaller black bears are even more frightened of us than we are of them. Their major contribution to camping problems is that they, too, are given to raiding the food larder. Thus, the necessity of carrying a sturdy container, preferably plastic, in which all your food will fit and a rope to sling over a high tree branch.

The prime rule of safety with bears is to be certain they know you are in the area. Many hikers carry small bells on their packs that aren't loud enough to irritate other hikers, but let the bear know they're around. Some people rattle a small stone in a tin can as an alert for bears that might be feeding in a nearby thicket. Others say that a thin willow whip will create a noise that frightens bears away.

Whatever you use, it is important that bears not be surprised by your presence, and particularly sows with cubs cavorting around them. Sow bears will take drastic measures to protect their cubs, and if you find yourself near a cub, depart immediately. Cubs are as much fun to watch as a pup or kitten, but their mothers can become quite cranky if they fear their offspring are in danger.

Mosquitoes are the most common of all pests and are found almost everywhere you are likely to camp. Insect repellent, netting on your tent, and long pants and long-sleeved shirts are the best insurance against bites. Mosquito coils, rings of a very slow-smoldering and evil-smelling chemical, will clear them out of cabins and keep them out as long as the coil burns.

Chiggers and noseeums are among the most irritating of all pests because they are so difficult to see, and their bites cause itching. As with ticks and mosquitoes, the best approach is prevention: apply insect repellent to your boots and pants legs. Obviously, camping in shorts gives them a bigger banquet table.

Ants seem to be almost everywhere, especially where food is kept. Other than the irritating prospect of finding them in your cup or in your bedding, they're of little consequence. You can usually keep them away by hanging food from tree branches. Check the campsite before settling down for a long stay in case there is a hill of ants nearby. If so, let them have the area.

Bees and hornets: Virtually everyone has been stung sometime by a bee or hornet. Although painful, the stings can be treated easily with a paste of baking soda and water. Some people who become allergic to their stings should be treated with antihistamines carried for those purposes.

A word about insect repellents: Some people swear by massive doses of various vitamins as the best insect repellent; others say eating garlic by the pound will keep them off your body— and your companions at their distance as well. There apparently is no single insect repellent that works best for everyone, and you'll have to experiment with two or three of the major brands before you find which is best for you. They come in spray cans, squeeze bottles, lotions, and sticks, and have brand names such as Off, Cutters, Jungle Juice, and Mirehol.

Warning: Do not let insect repellent spray hit your tent wall or any other synthetic material. Some varieties will weaken or destroy these fabrics. Read the manufacturers' warnings for each tent, and ask outdoor-equipment stores about other items, such as rain gear.

Skunks: When you come across a skunk in camp or on the trail, let it have the right of way. It won't spray you with its pungent and clinging urine unless it is frightened or cornered. With justification, skunks are not very frightened of people, but they don't lurk around camp very long if they know people are wandering around as well. If one does spray you, about the only thing you can do is bury your clothing and take a series of baths to remove the stubborn odor. Vinegar is one agent that will eventually remove the odor; tomato juice is another.

Porcupine: These slow-witted camp robbers cause no trouble if you leave them alone. They usually come into camp in search of salt, and that salt may be on your clothes or backpacking equipment. They can be frightened away and are harmless unless you make the mistake of touching them. Dogs are the biggest problem with porcupines because they never know enough to leave them alone, and a dog with a nose full of porcupine quills is in agony until they are removed. Usually they have to be removed by a veterinarian.

CAMPER'S SECRET

It is hardly a secret that you should always carry some emergency provisions and tools in your car. One of the most important is a shovel, either a short-handled and sharp-pointed one or a folding shovel so popular in the military services. This can be used around the campground for digging latrines, for fire safety, and for similar uses. You should also carry emergency flares that are easy to ignite, and a block of wood or something similar to block your wheels in case you have to change a tire or do some other automotive surgery. Another convenient tool is a small winch that can be used to extract your car when it is stuck in the sand or mud. A portable winch with strong cable or cord plus a shovel will get you out of most situations provided you have something strong, like a tree or another vehicle, to winch toward, and also provided you don't dig the car in too deeply before giving up.

TREATING PROBLEMS

Following are methods of treating some of the most serious wounds.

Stretchers can be made of a variety of things at hand in case a member of your group must be carried for some distance. One of the most simple consists of two poles, made from saplings cut and with the branches trimmed off, and two or three jackets or parkas.

The parkas should be zipped up or snapped, however they are closed, and the poles inserted through the arms and the bottom of the parka. Two of these will usually suffice for an average-size adult, but more can be added if necessary.

If you are carrying the victim over rough terrain, you may need to lash the victim in, particularly if he or she is unconscious or very weak.

Make the victim as comfortable as possible by using other articles of clothing or sleeping bags as padding.

Shock is usually caused by a sudden reduction in the volume of the victim's blood, either as a result of severe bleeding or when the blood serum is rushed by the body's healing mechanism to a severely burned area. It can also result from dehydration when the body's fluids are lost through vomiting or diarrhea.

When one of these causes occurs, the arteries constrict to divert the blood supply to the vital organs, and the heart begins pumping faster to circulate the blood that remains or to help in the healing process.

A victim of shock is pale, with skin turning cool, first in the extremities and later the trunk. As shock becomes more serious, the patient will sweat and complain of being cold and frequently thirsty. The pulse is often rapid and breathing fast and shallow as the blood pressure lowers.

It is best to treat potential causes of shock quickly to prevent shock from occurring, or at least to minimize it. Bleeding should be controlled immediately, because it is the loss of blood, not the initial injury, that causes shock. In the case of burns, fluids should be given to minimize the loss of other fluids to the injured area.

The victim should be made to lie down and keep both feet elevated about twelve inches above the level of the head. Hot water bottles, heated stones, or body-to-body contact beneath plenty of cover should be administered to help raise the body temperature. Keep the victim prone until treatment has been completed and he or she appears to be in stable condition.

In all cases—even when someone has suffered only mild shock—he or she should be taken to a doctor as soon as possible after being stabilized.

Fractures are usually quite easy to diagnose, although they sometimes can be confused with severe sprains in joints. In either case, the victim should be immobilized and taken to a doctor for treatment.

> **CAMPER'S SECRET**
>
> A small vial of lighter fluid can be carried and used to remove adhesive tape without irritating the skin.

If the fracture isn't compound (i.e., if the bone isn't protruding through the skin), the area of the break will be painful to the touch; swelling and discoloration will set in soon, and a lump or crook will show in the bone.

Even if these symptoms are not evident—though they usually are—the victim should lie down and you should make a splint to cover the injured area so more damage won't occur. Do not make an attempt to set the broken bone; that is the job of a doctor. But do make every attempt to prevent more damage through the construction of a splint.

Splints can be made from a variety of materials at hand, from small saplings cut for the purpose to pieces of boards or tent poles. For them to work properly in immobilizing the fractured area, splints must anchor the limb both above and below the fracture.

The splint should be padded between the rigid surface and the places it touches the body. Wrap it firmly at once, though not so tight that it interferes with blood circulation.

If the fracture is below the knee, the victim can move on the uninjured leg with assistance, but this should not be permitted unless necessary. Fractures of the knee and above require that the victim be carried on a stretcher, as do fractures of the hip and pelvis or any spinal injury.

Arm, wrist, and shoulder fractures and dislocations should be immobilized, also. Fractures of the arm should be splinted and the arm immobilized against the body trunk with a sling.

Severe cuts: It has been only a short time since doctors have recommended that we forget about learning the so-called pressure points throughout the body where you can apply pressure to stem the blood flow through arteries to injured areas. Now even tourniquets have fallen into disfavor because, like pressure points, they can lead to more damage if not used properly.

Pressure directly to the wound is the only safe and effective means of stemming blood loss. This collapses the severed blood vessels so that clotting can occur.

Arterial bleeding, identified by bleeding in spurts as opposed to steady bleeding from veins, is the most serious kind and can also be controlled by direct pressure, although it sometimes takes longer. If the bleeding doesn't stop after several minutes of holding gauze bandages on the wound, wrap the injured spot tightly, but not so tightly that the blood supply is shut off to the rest of the limb. If the skin beyond the bandage turns dark, or if the patient complains of a tingling or numbing sensation, the bandage is too tight.

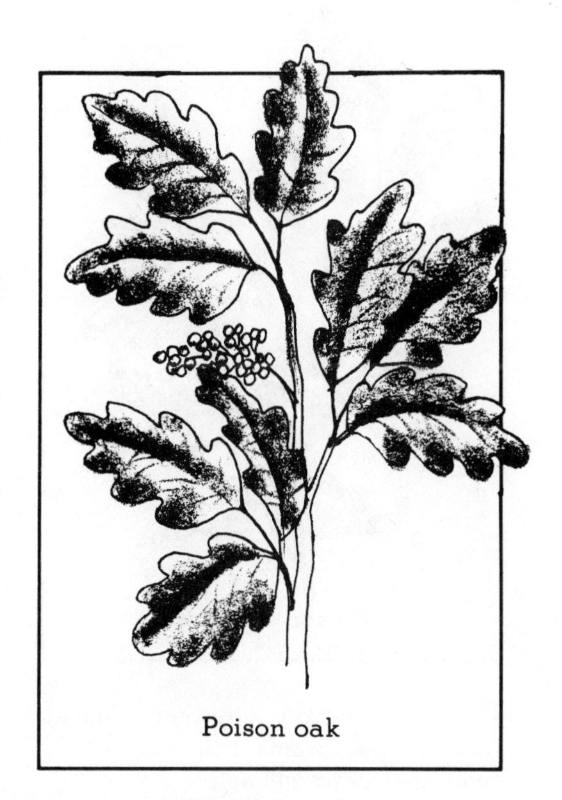

Poison oak

POISONOUS PLANTS

Before taking your first camping trip, be certain everyone in your group can identify the major poisonous plants in the camping area.

Poison oak and poison ivy are the most common. They are almost identical, except that ivy is a vine and the oak variety is a bush. Each has a cluster of three leaves and clusters of white berries.

Poison sumac closely resembles the more common staghorn sumac except that its leaves have smooth instead of jagged edges and its berries are white or gray instead of red.

Nettles, which can be boiled and eaten, are another irritating plant that can cause a rash, especially if you scratch the area they have touched.

Scratching should be avoided because it can cause infections or scarring. Calamine lotion helps relieve the itching, and a prescription drug, Syn-

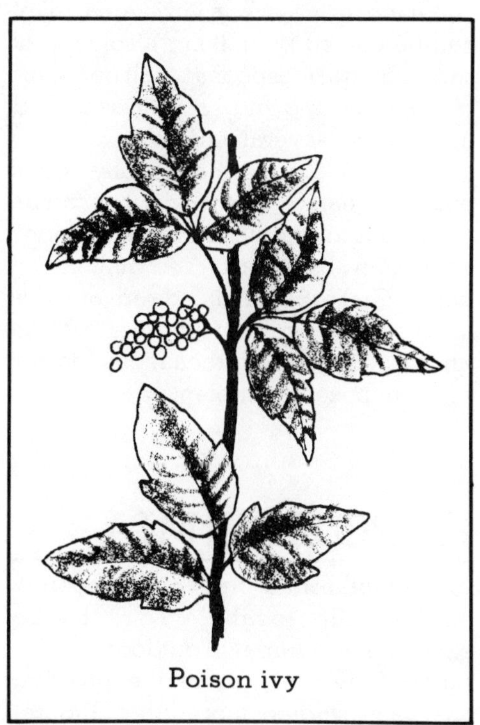

Poison ivy

Poison sumac

alar, or its weaker cortisone relatives such as the creams and lotions sold over-the-counter will help.

If the rash is over a large area, relief can be gained by making a solution of saltwater (two teaspoons of salt per quart of water) and applyng a compress soaked in this water several times a day.

No area of America is totally free of outdoor creatures and plants that can be an irritant to people, and this includes your own backyard. Yet millions of campers take these minor disadvantages for granted and, by recognizing them and taking normal precautions, do not let them become problems.

HYPOTHERMIA

Hypothermia has only recently entered our vocabulary, replacing exposure as a threat to outdoors enthusiasts. Fortunately, since the problem has been studied thoroughly and received much publicity, fewer and fewer deaths have occurred from this major problem.

Essentially, hypothermia is the loss of body heat. When body temperature drops, more than three degrees below the usual 98.6 degrees Fahrenheit, certain changes occur, and the lower the body temperature drops, the worse these changes become. Unless reversed, hypothermia can lead to death.

However, as stated above, most outdoors people have been indoctrinated against this, and in most cases it is no more complicated than telling your children to wear warm coats when they go outside in the winter. The problems come when people go on hikes or other outings on a warm, sunny day, and then a cold rainstorm comes and they are not prepared for the cold and wetness.

Thus, the importance of having everyone in a camping group always carry rain gear, a warm sweater or jacket, and the other Ten Essentials (see CHAPTER 3) when leaving the campsite. If you are camping in higher elevations, the weather can turn cold faster than at or near sea level. Many, many campers have never given a thought to hypothermia because they are so well equipped. To repeat, always carrying the proper clothing is the best prevention.

While prevention is relatively simple, except in the case of campers who might be stubborn to the point of stupidity (and you should find someone else to camp with in this case), a great deal of research is still being conducted to test various ways of speeding the recovery of hypothermia victims.

Basically, hypothermia is broken down into several stages. The first stage, which usually sends people in search of a warmer coat or wool socks, is shivering.

The next stage is violent, uncontrol-

Hypothermia victims go through four stages of the condition unless treated promptly and properly.
Shivering
Uncontrolled, violent shivering and difficulty with speech
Erratic movements, stiff muscles, and inability to think clearly
Irrational behavior, unconsciousness, and death

It is essential to treat the hypothermia victim quickly.
Get the victim warm and into dry clothing
Stoke up a big fire.
Walk the victim around to force his or her body to rebuild its own heat sources.
Feed the victim hot drinks.
Once the danger is past, keep the victim warm and comfortable.

Wind Chill Chart

COOLING POWER OF WIND EXPRESSED AS "EQUIVALENT CHILL TEMPERATURE"

WIND SPEED MPH	TEMPERATURE (°F)																				
CALM	40	35	30	25	20	15	10	5	0	-5	-10	-15	-20	-25	-30	-35	-40	-45	-50	-55	-60
	EQUIVALENT CHILL TEMPERATURE																				
5	35	30	25	20	15	10	5	0	-5	-10	-15	-20	-25	-30	-35	-40	-45	-50	-55	-65	-70
10	30	20	15	10	5	0	-10	-15	-20	-25	-35	-40	-45	-50	-60	-65	-70	-75	-80	-90	-95
15	25	15	10	0	-5	-10	-20	-25	-30	-40	-45	-50	-60	-65	-70	-80	-85	-90	-100	-105	-110
20	20	10	5	0	-10	-15	-25	-30	-35	-45	-50	-60	-65	-75	-80	-85	-95	-100	-110	-115	-120
25	15	10	0	-5	-15	-20	-30	-35	-45	-50	-60	-65	-75	-80	-90	-95	-105	-110	-120	-125	-135
30	10	5	0	-10	-20	-25	-30	-40	-50	-55	-65	-70	-80	-85	-95	-100	-110	-115	-125	-130	-140
35	10	5	-5	-10	-20	-30	-35	-40	-50	-60	-65	-75	-80	-90	-100	-105	-115	-120	-130	-135	-145
40	10	0	-5	-15	-20	-30	-35	-45	-55	-60	-70	-75	-85	-95	-100	-110	-115	-125	-130	-140	-150

WINDS ABOVE 40 HAVE LITTLE ADDITIONAL EFFECT.

LITTLE DANGER

INCREASING DANGER (Flesh may freeze within 1 min.)

GREAT DANGER (Flesh may freeze within 30 seconds)

DANGER OF FREEZING EXPOSED FLESH FOR PROPERLY CLOTHED PERSONS

Source: National Weather Service, U.S. Dept. of Commerce

Summer* Weather Chart

Regions	Normal Monthly Precip. (in.)	Normal Daily Max. Temp.	Normal Daily Min. Temp.	Normal Daily Mean Temp.	Average % of Poss. Sunshine
Pacific N.W.					
Seattle	1.08	73.8	53.7	63.8	62
Spokane	.58	81.9	54.0	68.0	77
California					
Bay Area	.03	71.6	54.3	63.0	65
Los Angeles	.02	75.8	63.2	69.5	83
Interior	.05	91.3	56.9	74.1	96
Colorado					
Denver	1.29	85.8	57.4	71.6	72
MN/WI					
Minneapolis/ St. Paul	3.05	80.8	59.6	70.2	67
IL/MI					
Chicago	2.73	82.3	59.9	71.1	68
Detroit	3.04	81.6	62.1	71.9	65
N. Atlantic States					
Boston	3.46	79.3	63.3	71.3	67
Mid-Atlantic States					
Wash. D.C.	4.67	86.6	67.6	77.1	63

*Figures are for August, normally the warmest, sunniest month. Source: U.S. National Oceanic and Atmospheric Administration, as reported in "Statistical Abstract of the United States," U.S. Dept. of Commerce, Bureau of the Census. Temperatures recorded at airports, in Fahrenheit degrees.

lable shivering and difficulty in speech.

Then, unless checked, the shivering decreases and muscles become stiff. The victim makes erratic movements and can't think clearly.

If the condition is allowed to continue, the victim becomes irrational and loses contact with reality. Unconsciousness is the next step. Death is the last.

As with any response to one's environment, hypothermia affects each individual differently. One camper may be shivering miserably while the other, dressed in identical clothing, will be standing around with coat unbuttoned, complaining that he or she is too warm. It isn't a sign of weakness to get cold easily, it is simply due to the physiological differences among us.

It is essential to act quickly. Watch other members of your group for the first signs of shivering and emothional withdrawal. Get them into dry, warm clothing, feed them hot drinks and high-energy foods, such as chocolate energy bars, and make them walk. Stoke up the fire. And remember that this is no time to tease or ridicule the sufferer.

Some researchers on the subject say it is best to keep the chilled person moving about, building up body heat with warm drinks (nonalcoholic) and energy-producing foods, and making the body restore its own internal heat. Most researchers say this is the most import part of the recovery process: to help the body restore its own heat, by exercises, warm liquids, and high-energy foods.

Barring dunkings in cold water, few campers will suffer from the more advanced stages of hypothermia if they follow the basic rules of dressing properly for the weather conditions.

CHAPTER 11

STRETCHING THE SEASON

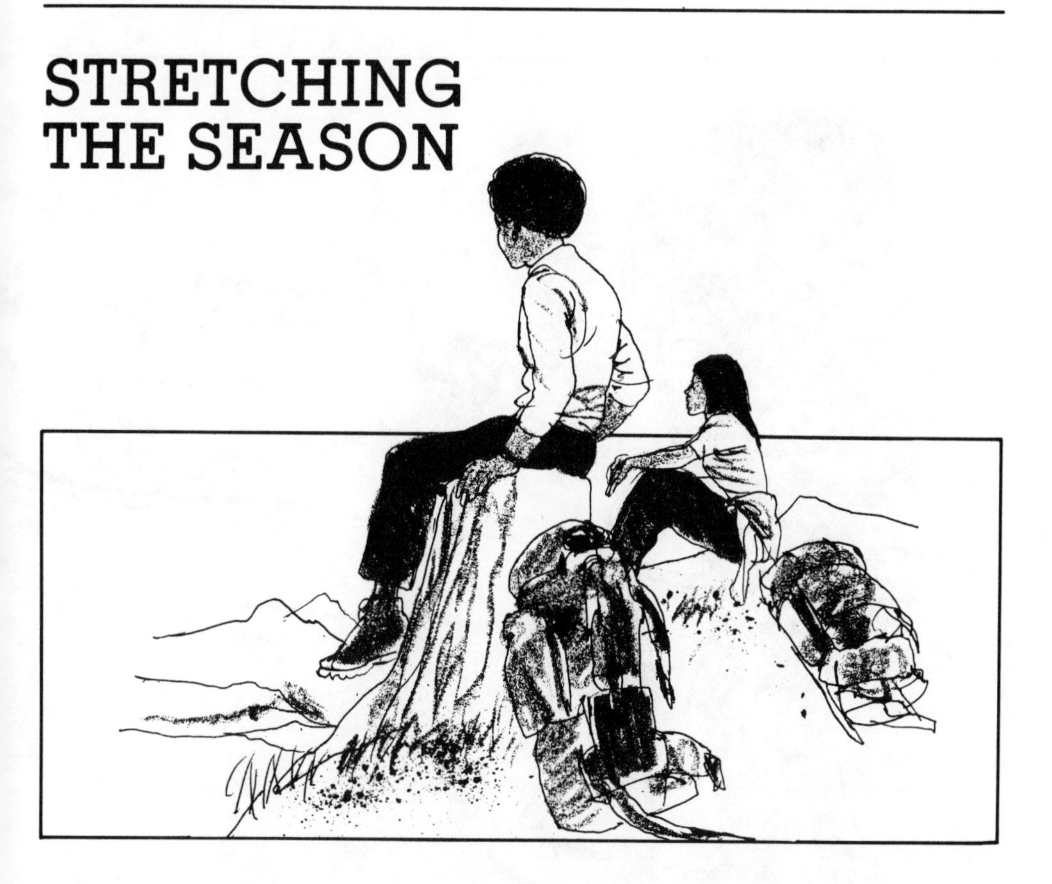

After you have been camping a few times during the summer months and find that it isn't nearly as primitive or as uncomfortable as you feared, you will also discover that with only a modest addition to your equipment list, and a few precautions, you can extend your camping season by several months.

One definite bonus is that before June 15 and usually after September 15 of each year, the camping sites that were crowded during the summer are virtually vacant. School activities keep most families home on weekends nine months of the year, and the majority of campers have put away their equipment until the summer months return.

Winter camping will be discussed later, but for spring and fall camping, with little prospect of snow, your summer equipment of tents and sleeping bags will be sufficient. Following are suggestions on how to stretch the season for your basic equipment.

Sleeping bags: Most are rated to about freezing temperature, or 30 degrees. You can add another ten to twenty degrees by wearing dry woolen clothing to bed, by adding a wool blanket over the bag, or by wearing waffle-weave underwear at night. Some bags can be extended even more by inserting a lightweight goose-down liner or one made of flannel. When you buy your sleeping bag, ask about these options so you won't have to buy an additional bag for colder camping.

Clothing: For spring and fall camping, an additional layer is usually sufficient. The days are usually warm, but the

parka, will not have the insulation in the hood. Thus, a cap that covers your ears should be worn. Some people do not like hoods unless they are absolutely necessary, so a wool watch cap or balaclava is a necessity.

Gloves or mittens: You might need a pair of each, with liners. Many ski gloves come with removable liners that reflect heat. Mittens can be leather with wool liners or down-insulated with lightweight liners.

Rain protection: Rain is usually more common during the spring and fall than midsummer. Although rain gear should always be taken along on camping trips, it is especially essential when stretching the season. In addition to a raincoat or poncho, rain pants should also be worn. True, no totally satisfactory

nights have to be taken into consideration. Sometimes this is only a matter of wearing a heavier wool shirt and pants or a goose-down jacket or parka.

Footwear: Your shoes or boots should be leather instead of tennis or running shoes, and you should be able to wear two pairs of socks in them. Down booties are excellent for sleeping and tent wear.

Head wear: A cap or insulated hood should be worn. Sometimes your lightweight parka, essentially a wind

rain gear has yet been developed. It is all bulky, tends to overheat when you exert yourself, and makes a lot of noise when walking. But the alternatives of cold and wet discomfort make it essential. A dining fly is also excellent protection against the sun and rain.

Since you must expect rain (and you can be pleasantly surprised if it doesn't come), the best campsites to choose are those above the level of nearby streams, with a slight slope to them so you can avoid seeing your campsite turn into a mudhole.

Food: You will be burning, and needing, more body fuel to keep warm, so your diet should be heavier in carbohydrates and hot drinks. Additional meats such as jerky should be added to your larder, and additional packets of hot chocolate are useful for both heat and energy.

Without getting too technical or scientific, this increased need for fats and carbohydrates means you should stock up on items that many reducing diets would ban: hash-browns, pancakes and syrup, cookies, bacon bars, jerky, and so forth. Gorp with an ample supply of chocolate candy and peanuts is another good source of carbohydrates (the

CAMPER'S SECRET

Spare mitten liners can be made of old wool socks with thumb slots sewn in.

chocolate and raisins) and fats (the peanuts).

Tents: Since few tents are designed to hold heat inside, you can't depend on them for comfort—only for shelter and privacy. You'll have to depend on your clothing and sleeping bag for extra warmth. You can improve your tent so that it isn't quite so crowded by carrying along an eight-by-ten-foot, or slighly smaller, waterproof tarp. You can construct an awning over the tent entrance, or a simple A-frame extension of the tent, and store your gear in the shelter to expand the roominess of your tent.

The same tarp can be used as a cover for the kitchen area by stretching it overhead as an awning or as a wind break.

It is essential that the tent be kept clean, especially if it is cold and raining, and this A-frame tarp over the entrance will allow you to remove your wet rain gear and muddy boots before entering the tent.

Each of the major basic tent designs discussed in CHAPTER 4 have modifications for use in cold-weather activities such as cross-country skiing and mountaineering. Some cold-weather variations available include the following:

Vestibule: This is an extension over the entrance to give extra protection from the wind, rain, and snow. Some models offer it simply as an extension of the rain fly, others as a part of the entire tent. You can store extra gear in the vestibule, or use it for the camp kitchen since the danger of damage from fire or spilled food is less.

Tunnel entrance: Because all zippers are susceptible to freezing, many cold-weather tents have a tunnel entrance with a drawstring at the end. Some have a drawstring at the outer end and a zipper over the inside entrance. Both designs offer a safe entrance in bad weather.

Frost liner: This is a detachable inner wall, often made of cotton, that collects the body moisture that would otherwise condense and form ice on the tent wall and ceiling. The liner can be detached and the ice shaken off at regular intervals. For ease of removal, most are held in place by ties rather than straps or snaps.

Cook hole: Some tents designed for hard use have a zippered hole in the floor that can be removed so that cooking inside presents less of a danger. Spilled fuel or boiling food goes onto the ground or snow beneath instead of the tent floor. As with other modifications, this appears in tents designed for hard use and is not a standard feature of all tents.

WINTER CAMPING

As you become more experienced in the woods, and have stretched the camping season closer and closer to both ends of winter, you probably will become intrigued with the prospect of camping in the dead of winter. The rules of keeping infants happy apply to winter camping—make sure you're dry, warm, and well fed. All gear additions refer back to these three basics.

Generally speaking, most of your summer camping equipment can be adapted, or added to, for winter camping. However, you will want to invest in special winter tents, heavier sleeping bags with a lower comfort range, and special clothing if you plan to make it a regular sport.

Another new factor to contend with

is transportation. You won't want to "post-hole" your way through deep snow back into the forest, so you will most likely take up winter camping in connection with active winter sports such as cross-country skiing or snowshoeing.

The site you select for your camp should be on a flat clearing with protection from the wind. A good source of water should be nearby. Melting snow for camping water supplies is a tedious process and involves the use of more stove fuel than is necessary.

Avoid camping on hillsides, or at the foot of a bank or hill, because of the danger of avalanches or small snowslides.

If possible, avoid camping beneath evergreen trees because they often dump big loads of snow that can demolish your campsite and collapse your tent. If you do camp beneath a tree, shake off the snow or use a long pole to beat it off before setting up camp.

It is best to clear the snow from your tent site several feet beyond the tent and the cooking area. Tents and kitchens set up on bare ground do not tilt wildly with the wind. Depressions beneath the tent can be stamped level with snow. If you sleep on a closed-cell foam pad, you won't be bothered by melting that's taking place beneath you.

Before leaving on the trip, check the waterproof bottom of your tent to be certain it doesn't have pinprick holes through which snow melt will invariably seep. Sealing compounds and patches are available at nearly all outdoor equipment stores. Check with the manufacturer's specifications for the best sealer. You can inspect the tent floor by holding it up to a bright light

Special equipment includes snow scoops to clear the campsite and tent stakes specially designed for anchoring lines in snow. Anodized-aluminum poles are almost essential for your tent so the snow and ice won't stick to them

For information on sleeping bags, see CHAPTER 5. You will need to extend the comfort range downward by either buying a heavier bag or adapting your summer-weight bag for winter use. This can be done by wearing heavy-duty underwear, such as high-quality wool or goose down. Your bag should be equipped with an insulated hood. You can also sleep with a wool cap or, in extreme temperatures, a down face mask that covers your entire head except for eye holes and slits for your nose and mouth.

Don't sleep in the buff. The nylon of the bag will be incredibly cold against your unprotected skin.

Zipping two bags together in cold weather is also an invitation to cold sleeping. It is almost impossible to keep cold air from seeping in around the shoulders.

The layering system of dressing is extremely important in winter camping, and careful shopping for clothing can give you a wide selection of warm clothing that individually isn't very heavy. A good camping wardrobe would include:

- net or wool underwear
- wool shirt
- wool sweater
- wool pants
- jacket or parka
- Dacron, nylon, or silk undersocks
- heavy wool outer socks
- waterproof boots with felt liners
- gaiters
- lightweight, water-repellent wind parka
- wind pants to match
- mittens with wool liners
- leather gloves with liners
- wool cap, balaclava style
- sunglasses

Individually, each of these items offers some protection from the cold and wet, and when you have them all on (with mittens and liners instead of gloves), you can comfortably survive temperatures well below freezing.

Your food should lean heavily toward carbohydrates and fat, to help your body produce heat, and lots of liquids.

Winter camping presents special hazards not covered in the chapter on first aid and safety, including thin ice and wet feet.

Ice: Many people fall through thin ice every winter. Avoid ice whenever possible, and search for shallow parts of streams to make your crossing. Learn to "read" ice and where to expect thin ice. It is usually thickest where streams are broadest and the water slower. On lakes it is thicker along the shore than farther out. If it is blue in color, it is usually thin.

Wet feet: If your feet get soaked in cold weather, stop immediately to build a fire and change socks. Frostbite and freezing occur rapidly, and the faster you get your feet dry and warm again, the less the danger.

Writer and guide Andy Russell tells

CAMPER'S SECRET

Everyone likes a bargain, and free campgrounds are one of the best bargains of them all. One couple who travels several weeks each year in their motorhome found a clever way to camp free, and often were the only campers in the park. They selected small towns along rivers where fishing prospects were good, and then talked to the local police. Frequently the police were happy to have someone in the town's park all night to help prevent vandalism. The couple had a CB radio in their motorhome, which they could use to summon the police if necessary. In several summers of free camping, they never had to use it.

of falling through the ice into a stream in the middle of an Alberta winter. He built a fire in record time and stripped off his woolen clothing that was already freezing stiff. When his companion arrived, his first sight was Russell standing stark naked in a snowbank, flailing his woolen long johns against a tree. With some justification, Russell's companion thought for a moment he was traveling with a demented woodsman. But Russell, having beaten the ice out of his woolen clothing, put it back on, and warmed himself by the fire, suffered no ill effects.

Obviously you should never be more than shouting distance from camp without carrying the Ten Essentials (see CHAPTER 3) and some kind of emergency shelter with you. In some parts of the country, blizzards can blow up almost instantly, and you may have to stop and hole up wherever you are to sit it out. If you have shelter and warm clothing and can build a fire and brew yourself a warm drink, you can sit out a storm in comfort.

The primary rule for winter camping is always to be prepared for the worst. Learn to live with winter instead of trying to force your schedule or your energies against it. You'll lose more such battles than you'll win. If the weather is foul, stay in camp. If bad weather is predicted, stay home. The more you understand winter camping, and your own limitations as well as your equipment's, the more you will enjoy this form of camping.

CHAPTER 12

WHERE TO NEXT?

One pitfall of instructional books on the outdoors is that the reader will become so overburdened with information that the purpose of the book—enjoyment of the outdoors—will be forgotten amid the rules, suggestions, charts, choices, and author's preferences (or prejudices). As more than one bewildered novice camper has said, books tell you how to do everything except have fun; they get you all dressed up with no place to go.

Thus we have departed slightly from the standard how-to approach in this chapter to give you an idea of what awaits the properly outfitted and educated camper. While most of the information in this book has been written on the assumption that the reader knows no more about camping than the average American knows about tribal customs in the Borneo highlands, by now you should feel a bit more comfortable in the maze of knowledge and equipment that technology and regulations have forced upon us.

Most campers who go to established campgrounds and camp within sight of their car soon get the urge to move on into the wilderness; to buy a pack and a pair of hiking boots, stock up on freeze-dried food and strike out through the forest or across the desert. Or they want to tackle a wilderness river, with either a guide service or a group of experienced whitewater people. The innate curiosity that has sent man all over the world and space soon becomes an urge you will want to follow.

This is the primary reason that

throughout the book we have continually nudged you toward purchasing lightweight equipment so that you won't have to duplicate tents, stoves, or sleeping bags when you want to try a different kind of trip. Buying the best and lightest equipment gives you a freedom of movement and choice you don't have if you buy according to price or convenience alone.

The basic camping equipment is universal in that you can use the same equipment in all seasons, all terrains, and all outdoor sports. Simply take the basic equipment and add the seasonal or terrain requirements to it. Winter camping requires additional clothing and warmer sleeping bags, for example, but the layer system of dressing takes care of most of the clothing, and you can increase the comfort range of your sleeping bag by adding a liner or one of those half-bags called bivouac bags for your lower body and a goose down-filled parka for your upper body.

CAMPER'S SECRET

Some of the best places for day trips and one-night camping trips are federal and state game refuges—but not during hunting season, of course. These refuges are usually near a source of good water, and camping areas are not difficult to find. Check with your state department of game or federal Fish and Wildlife Service for maps, because these areas are frequently overlooked by other campers.

This same basic outfit is all you will need for long river or lake trips, with the addition of waterproof duffel bags and life jackets.

And it can be used with equal success traveling in recreational vehicles. You may want to add a collection of books, maps, and food items such as corn-on-the-cob and any other kind of food you prefer, but you'll be glad you own sleeping bags that compress well: you can simply fold them into the stuff bag every morning.

After a few outings with your new equipment and your new form of recreation, you will quickly become accustomed to the outdoors, and if you have children they will soon be insisting that you take them on some real adventures. They will want something to tell their friends when school starts, something beyond the car camping that "everyone" does. They will want to go backpacking, or on a two-week bicycle trip. They will want to run a whitewater river, or take a long canoe trip that may involve portaging from lake to lake. Once you have started this form of recreation, the natural progression is for each trip to be a bit more rugged—and interesting—than the last.

All of us have our dream trips, those

journeys we would love to take if we had the time, and as soon as possible and before the children leave home for good.

Here are some examples of trips taken by one family that can be duplicated or applied to other parts of North America. These were taken in the American West and northern Canada and the Yukon Territory where some of the most beautiful wilderness in the world can be found. And all of these trips were taken with the basic equipment found in this book.

HIKING

Most of the Olympic Coast of Washington is enclosed in the coastal strip of Olympic National Park and is not touched by highway, logging road, commercial development, or campground. It is wild, subject to torrential rainfall and overcast days as well as beautiful sunny weather that exhibits the offshore rocks, the rugged headlands, and many natural arches carved out of the shoreline by wind and sea. The series of headlands can be walked around only at low tide; otherwise hikers must scramble over them on steep trails that often require dangling ropes for safe climbing.

The best time to make the trip is during a series of extremely low tides so you will have an opportunity to walk around the tide pools to examine the starfish, sea anemones, sea urchins, hermit crabs, barnacles, the hundreds of varieties of seaweed, and the brilliant-colored algae that stain the rocks several shades of green.

The National Park Service has been insistent on keeping the coast wild and has not established campsites anywhere along the way. Obviously, the most popular campsites are near the small streams that trickle down from the Olympic Mountains. Much of the water is stained brown from the decaying wood and other plant life it flows across. It must be boiled for purity, then left to stand a few more minutes so the sediment will collect on the bottom of the pot or pan.

Hikes along the coast can last as long as you want, from a single day to a week or longer, and it is possible to walk up to fifty miles along the beach in one direction with only one brush with civilization along the route. The hike is a combination of long nights with the sound of the surf against the rocks and on the sand and gravel beaches, and occasional heavy surf when a storm blows in off the Pacific.

One point of interest is the former archaeological dig on the northern end of the hike at Cape Alava. Here the Makah Indians lived for centuries in a village protected from the storms by a long shelf of a beach and a small, tree-topped island just offshore. The Makahs were famous whalers, and the archaeologists found a treasure trove of artwork in the village. The village was subject to infrequent but disastrous landslides from the cliff above, and each slide covered several of the cedar houses, burying the houses and occupants and keeping the artifacts intact beneath the solid covering of clay and soil.

Many hikers along the beach trail do not wear standard hiking boots. Instead, they prefer rubber boots for wading through the surf when it is necessary, and walking shoes or running shoes on the sand and gravel.

All the basic camping equipment discussed earlier is suitable for the trip, especially the lightweight equipment plus a standard backpack.

Another hike, which some refer to

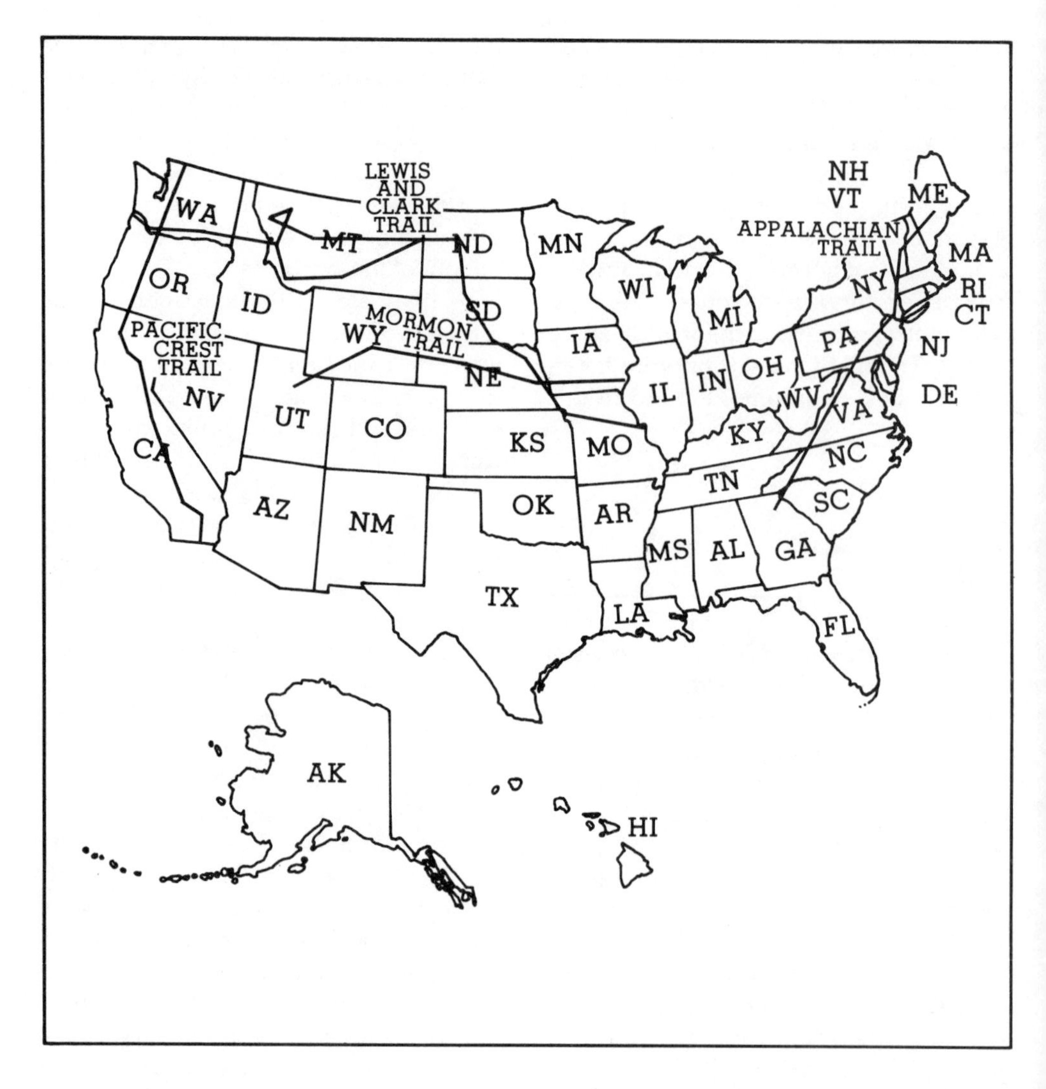

as a "classic" hike, is over the Chilkoot Trail from near Skagway, Alaska, to the headwaters of the Yukon River at Lake Bennett, British Columbia. The thirty-four-mile trail was the only overland portion of the route from the United States to the gold fields along the Klondike River near Dawson City, and during the gold rush of 1897 - 98, more than 20,000 persons walked the Chilkoot carrying their equipment from tidewater to the headwater lakes. The Northwest Mounted Police of Canada required each person to bring in a year's supply of food, clothing, and tools, and this translated to roughly a ton of gear per person.

The gold rush ended quickly, but during the two years that people—men, women, children, and assorted pets and work animals—traversed the trail, an amazing civilization rose and fell in that period. Six different towns sprang up almost overnight—Dyea at tidewater, Canyon City at the head of canoe navigation on the Tiaya River, Sheep Camp just below timberline, The Scales at the foot of the incredibily steep pass, Lindeman City at the lake of the same name, and Lake Bennett, where most of the Klondike stampeders spent the winter and spring building more than 7,000 boats and rafts while waiting for the ice to clear from the lakes and Yukon River so they could continue on north to the Klondike.

Today a hike over the trail is like walking through one of the world's longest museums. Bits and pieces of all the towns remain beside the trail, and one of the greatest of all engineering feats of the day is shown there, too. This is the aerial tramway that was built from Canyon City all the way over the summit to a small lake on the Canadian side, a total of ten miles. Some of the supports for the cable still stand, and sprockets, lengths of cable, and other parts lie beside the trail. At Canyon City are the remains of the steam power plant needed to operate the tramway.

Shoes, cookstoves, broken lanterns, collapsed cabins, lost or abandoned tools, a cache of canvas-and-wood collapsible canoes, harness, wagon parts and hundreds of other artifacts are still lying where they were dropped during the stampede.

The hike takes most people about five days to make, not because it can't be done in less time but because they want time to stop and explore the ghost towns and the other remnants of the gold rush. Excellent camping sites are found on both sides of the summit, and at the end of the trail on Lake Bennett, hikers can catch the White Pass & Yukon Route noon passenger trains that stop at Bennett Station to serve lunch to passengers and train crews. There they can continue northward to Whitehorse, the Yukon's capital, or go back over White Pass to Skagway, where they began.

OTHER HIKES

Two great Lower 48 trails: Each spring—sometimes more like late winter—several people carrying enormous loads in their packs strike out on an expedition that must be high on the dream list of every dedicated backpacker. These people are hiking one of the two great trail systems in America—the Appalachian and the Pacific Crest trails.

These are two of the premier trails in America, and fortunately they can be hiked in small, manageable portions instead of everyone having to take on the entire route.

The Appalachian Trail follows the crest of the Appalachian Mountains through 14 states. It begins to the south at Springer Mountain in northern Georgia and ends some 2,000 miles later at Mt. Katahdin in central Maine. It passes through Georgia, North Carolina, Tennessee, Virginia, West Virginia, Maryland, Pennsylvania, Vermont, New Hampshire, and Maine.

Since the Appalachians are not an unbroken range, the trail occasionally is forced to follow a highway short distances, which helps its hikers take breaks to repair shoes, purchase more food, and contemplate whether they want to continue or not.

Although it has been hiked in three months, perhaps less by some youthful charger, the average seems to be four months.

The Pacific Crest Trail is slightly longer, 2,400 miles, and traverses considerably more wilderness than the Appalachian. It goes through higher mountain ranges, across the Great American Desert, and hikes usually begin and end in winter.

The trail begins to the north in British Columbia's Manning Provincial Park, which adjoins North Cascades National Park across the international boundary in Washington. Glacier Peak Wilderness, skirts the edge of Mt. Rainier National Park, and crosses the Columbia River into Oregon at Bonneville Dam.

It continues along the ridges of the Cascades through Oregon into California, where the Cascades end and the Sierra Nevada Range begins. It swings southwest from the Sierra almost to Los Angeles, then swings east almost to the Joshua Tree National Monument, and stays in the desert until it reaches the Mexican border 30 miles east of San Diego.

Neither trail is easy to hike, and each has its own unique problems to conquer. Each has sections that are easy to hike; each has very tough stretches.

THE RIVERS

America has an abundance of great rivers for great float trips, ranging from the wilderness rivers of the West to the tamed giant rivers such as the Missouri, Mississippi, and Ohio and the smaller rivers with relatively short stretches of wilderness characteristics and great whitewater.

Some can be navigated by a novice in an inflatable raft or even a rented houseboat. But the most "interesting" rivers, meaning those with whitewater rapids and wilderness settings, should be attemped only by the very experienced or with a guide service that specializes in these rivers.

You can be as versatile on your river trips as you want. The large inflatable rafts that are so popular in many of the wilder rivers can carry enormous loads of people and gear, and you don't have to take freeze-dried food along unless you want to.

A memorable trip can be a gourmet float trip with the greatest concoctions you can dream up that are stored in the ice chest in one of the inflatables. You can carry tents large enough for standing, and you can take a folding table and folding lawn chairs if you want.

But in order to get the best "action" from the river it is best to take smaller inflatables that will crash through and over the rapids to give you a more thrilling ride. Some outfitters specialize in renting the small one- or two-man inflatable canoes with an experienced boatman as a leader to keep the incidence of dunkings to a minimum.

While traveling in these smaller craft, your lightweight camping gear will be especially appreciated because you can load all your belongings into a single waterproof duffel bag, and your

cameras, binoculars, and other similar items in one of the surplus military ammunition cans that are purchased by river outfitters by the hundreds.

Some of the best and longest rivers for whitewater trips are in the West, including Colorado, Wyoming, Montana, Utah, Idaho, Washington, Oregon, and California.

Smaller rivers with rapids are found in the South and Northeast, and a few are scattered through the Midwest. Some of the wildest, in terms of population centers, are in Alaska and western Canada.

One of the longest of the wilderness rivers is the Yukon, which begins in northern British Columbia on the Yukon border and winds its way north to Dawson city, Yukon, before turning west across Alaska, then southeast to its final run to the sea. The best portion for an interesting trip is from headwaters just south of Whitehorse, Yukon, to Dawson city, some 500 miles away. This is the last easily accessible place along the river for several more hundred miles. Rental canoes can be picked up at Dawson City, or you can have your own shipped back out more easily there than anywhere else along the river.

THE LAKE VOYAGES

Few manmade lakes are really suitable for wilderness-type trips because most are so thoroughly developed and heavily used. However, they are good places to become familiar with your equipment and for your family or group to become accustomed to water travel.

The best natural lakes—almost the only natural lakes in America—are in the northern states where the glaciers from the ice ages scoured out trenches that became lakes when the glaciers receded back into Canada and the Arctic.

Wisconsin, Minnesota, Michigan, Pennsylvania, New York, New Hampshire, Vermont, and Maine have the bulk of natural lakes that are not overcrowded with speedboats all through the summer. Each state has its own good selection of lakes for canoe trips. Occasionally you will find a group of lakes near enough together for you to portage from lake to lake, sometimes enabling you to make a complete circle in your trip.

These thousands of lakes continue on north into Canada, and the maps are dotted with them all the way from below the American border to the Arctic Ocean and Hudson's Bay.

One major canoe course is in the Voyagers National Park in northern Wisconsin along the Ontario, Canada, border. Strung all along that border are numerous lakes, many connected by shallow, slow streams that make it an ideal trip for canoeists.

RV TRIPS

Not all great trips require constant use of muscle power for transportation, and many families like to go on long trips in RVs when distance is the primary consideration.

One of the greatest trips for a family in an RV is to follow the Lewis and Clark Trail from St. Louis to the Pacific Ocean near Astoria, Oregon. The purists can point out that since Lewis began his trip in Pittsburgh, Pennsylvania, and met Clark along the Ohio River, that the trip really began on the East Coast. Thus, to follow the entire route, you would start at Pittsburgh, follow the Ohio River to the Mississippi, then go north along it to St. Louis (where historians generally say the expedition really began).

From St. Louis, the party followed the Missouri River into southwestern Montana, then across Idaho, down the Snake River between Washington and Oregon, and at last to the Pacific.

The Lewis and Clark Trail is perhaps the best route for seeing the northern portion of the Great Plains, the Rocky Mountains, and the Pacific Northwest. The Missouri River flows through Missouri, Kansas, Iowa, Nebraska, South Dakota, North Dakota, and Montana.

From there, the explorers crossed a low pass into central Idaho near Salmon, went on north again back into Montana, and crossed back into Idaho over Lolo Pass and followed the Clearwater River down the Snake River at Lewiston, Idaho. Then the route goes through the southeast corner of Washington, and between the two states along the Columbia River to the sea.

One of the bonuses of this route is that when you reach Montana, you aren't far from two of the most popular parks in America—Yellowstone to the south and Glacier to the north.

All along the route are other sites of historical and scenic interest, such as old army posts, trading posts, Indian reservations, and some of America's most beautiful scenery.

RVs are also an excellent way to travel through some of the more beautiful scenery that does not lend itself to camping. This is true of much of America's coastline which has ample commercial campgrounds, but few (and quite crowded) places suitable for pitching tents. Most of these can be reached only by hiking some distance, and you have to plan your trip quite carefully and take risks of finding available campsites. Also, you must take a chance and leave your car behind at the trailheads.

Thus, camping in these areas in RVs is much simpler and safer.

Again, all the basic equipment you have bought for the family camping trips is suitable for RVs, even those which have microwave ovens, televisions, food processors, and all the other modern conveniences. The children can get away from the adults (or vice versa) by pitching a tent just outside the RV door. The lightweight sleeping bags take up much less room than the heavier so-called slumber bags, and you'll invariably find good uses for the other equipment you bought for car camping.

In the following section are listings of all the major public agencies involved in camping. Also, you will find some basic information on different regions of America and other things of interest to campers.

Enjoy.

PART IV
RESOURCES

CHAPTER 13

DESTINATIONS

PRIVATE OR GOVERNMENT CAMPGROUNDS

Most families develop their own list of favorite camping sites over a period of time, some of which may be nothing more exotic than a level spot on a bank overlooking a stream. Yet, owing to the demand for such private sites, you will likely spend a portion of your time in either privately owned or government campgrounds.

Some privately owned campgrounds are part of nationwide chains, each site operated by a franchise owner. Most of these chains have well-defined guidelines for the franchise owner so that you know what to expect each time you pull into the office and register for a stay. These welcome everyone from car campers with their own tents to motorhome campers.

They will have a combination office, grocery store, and game room. They will also have shower rooms and a coin-operated laundry. Campsites for self-contained RVs will have an electrical outlet, a water faucet, and a sewage hook-up. You must carry your own electrical cord—and always carry a variety of plug adapters, because there is no standard plug; some require two-pronged plugs, others three-pronged ones, and some oversized plugs similar to those on an automatic dryer or range.

You must also carry your own section of water hose and sewage hose. If you don't want to be hooked to the sewage system, you can use the RV's holding tank until you prepare to leave the following morning. Nearly all privately owned campgrounds have a special place set aside for dumping sewage, usually near the entry area.

These campgrounds are usually near interstate and major highway interchanges, and an increasing number of car campers use them for their convenience. Renting a tent site is considerably cheaper than an RV site with the electrical, water, and sewage hookups, and it is an inexpensive way to travel across country, much cheaper than staying in motels and eating all meals in restaurants.

A few campgrounds have tents for rent, but these are very few in number, and you should take no chances. Always carry your own tent.

Government campgrounds are usually less expensive than privately owned ones, but they are also usually more primitive, less convenient to find, and farther away from population centers. The larger ones, however, sometimes have a small store and service station nearby.

These campsites range from small county parks to those developed by various state agencies, and federal government parks developed by the Bureau of Land Management, the Forest Service, and the National Park Service.

Canada has a wide selection of excellent parks in the national system, Parks Canada, and each province has its own selection of parks.

At the most primitive end of the scale, you will find a leveled area with firepits or concrete blocks with steel grates, a few garbage cans, and an outhouse. A water supply is usually nearby, either a stream or lake, or a well with a hand pump. Unlike the larger or better-developed parks, these frequently have no time limit on your stay and are often free.

Most states have a good selection of state parks with a resident ranger and well-developed campsites that include water and electrical hookups. These will also have recreation areas where you can play volleyball and badminton, pitch horseshoes, or enjoy other forms of entertainment. Nearly all will have picnic tables at each site. Usually you must make reservations well in advance for campsites, although a few continue to operate on a first-come, first-served basis. During the peak summer season, it is obviously best to arrive at these first-come, first-served campgrounds just before checkout time so you can claim a campsite, then walk back to the ranger's office to pay.

The same policy prevails at most Forest Service campgrounds, and many have a self-payment system.

The Forest Service has established a rating system on a 1-to-5 scale, with the most primitive campgrounds rated 1. This system is as follows.

1. Very primitive: These are primarily trail camps in wilderness and primitive areas, and usually are reached only on foot or horseback. Although some have stone or concrete firepits, usually they are little more than clearings on high ground. They will be shown on the maps you should purchase from the ranger station or district headquarters.

2. Rugged: Only a bit more developed than primitive sites, these can usually be reached by rough roads in vehicles that are not towing trailers. No established campsites are marked, and usually no outhouses or other such structures are present.

3. Rustic: These usually appear primitive but have a few amenities, such as outhouses. Campsites are usually laid out so that you'll have only two or three per acre, giving you plenty of room for privacy. Trails and roads into them are usually maintained well enough for towing a trailer to them and are either all-weather gravel or dirt.

4. Modern: Hard-surfaced roads lead into these, and you'll find good latrines and fire facilities. Usually these have three to five campsites per acre and often have some kind of natural screening between sites. These campgrounds are often near a recreational area, such as a stream or lake. Nature trails are frequently established nearby.

5. Ultramodern: These are the top of the Forest Service line, with laundry facilities, community showers and bathhouses, flush toilets, electrical and water hookups, landscaped grounds, and a ranger assigned to the campground.

Most Forest Service campgrounds are free, except the "ultramodern" ones, which charge a nominal fee.

You are advised to obtain information on campgrounds in all national forests you plan to visit, and do so well in advance, since many operate only on the reservation system. A list of the major national forests throughout the United States is in the next section.

Most major national parks have well-maintained and well-patrolled campgrounds, usually with a daily schedule of activities available to visitors. These range from organized nature hikes with a park naturalist to evening slide shows and lectures on the park. Usually the national park campgrounds are on a par with the best of the privately owned campgrounds. Privacy is often a low priority because the parks are so heavily used that the park planners must crowd as many campsites into a limited area as possible. In exchange for your fee, you receive some security, the use of the bathhouse and flush toilets, and the convenience of being near the attractions of the park itself. Nearly all of these operate on a reservation system, and your stay will be limited to a specific number of days.

ADDRESSES

After you have gone camping a few times and made some long trips to sample a variety of campgrounds, you will soon develop a sixth sense to detect areas where you will want to camp in the future, as well as places you will want to avoid. You will learn which public land agencies have the best facilities for your requirements, and which national forests have a reputation for taking good care of their campgrounds.

Soon you will find yourself going to the same general areas again and again, in part because you know what to expect from previous visits, and also because we are creatures of habit and like familiar surroundings.

Following is a list of addresses for information on national parks, national forests, and state camping areas in the United States. National park maps are free; national forest maps, which are excellent and beautiful enough to frame for the recreation room, cost $1 each. Most state agencies have free maps that range from full-color detailed maps to rough reading of main roads.

The best sources for information on these camping areas are three books published by Rand McNally & Co.:

Campground and Trailer Park Guide, revised annually.

National Park Guide, by Michael Frome, also revised annually.

National Forest Guide, by Len Hilts, revised as necessary.

Outdoor equipment stores and bookstores in each area will stock regionally published books, but the three above will serve you well.

Although there are several national campground organizations, one has outrun all the others in establishing a chain of franchised campgrounds with uniform standards throughout all of North America, from the Yukon to Mexico. This is Kampgrounds of America (KOA). Its directory of campgrounds is free at any KOA franchise, or available for $1 from KOA, Inc., Billings, MT 59224.

For information on other privately owned and operated campgrounds, contact the National Campground Owners Association, 804 D Street N.E., Washington, DC 20002.

One of the best sources of information on all campgrounds is the American Automobile Association; your membership in any state association entitles you to the detailed guides of any state or area of North America.

Still another good campground guide is that published annually by Woodall Publishing Co.

The best source for information on camping in Alaska, British Columbia, and the Yukon Territory is the annual **Milepost,** published annually by Alaska Northwest Publishing Co.

Many other campgrounds can be found by contacting state and federal fish, game, and land management agencies in the areas you plan to visit.

Several large timber companies also offer free camping and provide free maps showing locations. Some major utility companies with holdings along artificial lakes for hydroelectric projects also have free campgrounds.

EASTERN NATIONAL PARKS

North Atlantic Region
National Park Service
15 State St.
Boston, MA 02109

Mid-Atlantic Region
National Park Service
143 South Third St.
Philadelphia, PA 19106

National Capital Region
National Park Service
1100 Ohio Drive, S.W.
Washington, DC 20242

Southeast Region
National Park Service
1895 Phoenix Blvd.
Atlanta, GA 30349

MIDWESTERN PARKS

Midwest Region
National Park Service
1709 Jackson St.
Omaha, NE 68102

WESTERN PARKS

Western Region
National Park Service
450 Golden Gate Ave.
San Francisco, CA 94102

Southwest Region
National Park Service
Old Santa Fe Trail
P.O. Box 728
Santa Fe, NM 87501

Pacific Northwest Region
National Park Service
Westin Bldg.
Seattle, WA 98121

Rocky Mountain Region
National Park Service
655 Parfet St.
P.O. Box 25287
Lakewood, CO 80225

NATIONAL FORESTS

Eastern Region
National Forest Service
633 West Wisconsin Ave.
Milwaukee, WI 53203

Southern Region
National Forest Service
1720 Peachtree Rd., S.E.
Atlanta, GA 30309

Alaska Region
National Forest Service
RR 4, Box 1628
Juneau, AK 99801

California Region
National Forest Service
630 Sansome St.
San Francisco, CA 94111

Intermountain Region
National Forest Service
324 25th St.
Ogden, UT 84401

Northern Region
National Forest Service
Federal Bldg.
Missoula, MT 59801

Pacific Northwest Region
National Forest Service
319 S.W. Pine St.
P.O. Box 3623
Portland, OR 97208

Rocky Mountain Region
National Forest Service
11177 West 8th Ave.
Box 25127
Lakewood, CO 80225

Southwestern Region
National Forest Service
517 Gold Ave. S.W.
Albuquerque, NM 87102

EASTERN AND SOUTHEASTERN STATE CAMPING AREAS

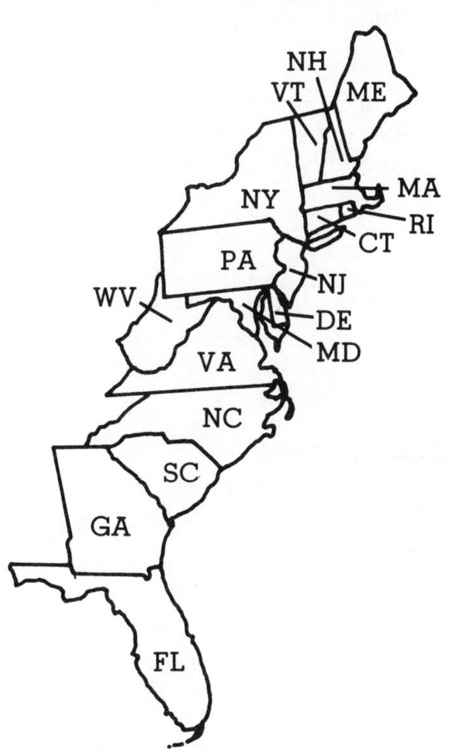

MAINE
Bureau of Parks & Recreation
State Office Bldg.
Augusta, ME 04333

NEW HAMPSHIRE
Office of Vacation Travel
P.O. Box 856
Concord, NH 03301

VERMONT
Dept. of Forests, Parks and Recreation
Montpelier, VT 05602

MASSACHUSETTS
Dept. of Environmental Management
Div. of Forests and Parks
100 Cambridge St.
Boston, MA 02202

CONNECTICUT
Dept. of Environmental Protection
Office of Parks and Recreation
165 Capitol Ave.
Hartford, CT 06115

NEW YORK
Bureau of Communications
N.Y. State Office
of Parks and Recreation
Empire State Plaza
Albany, NY 12238

NEW JERSEY
Dept. of Environmental Protection
Div. of Parks and Forestry
CN404
Trenton, NJ 08625

PENNSYLVANIA
Bureau of State Parks
Dept. of Environmental Resources
P.O. Box 1467
Harrisburg, PA 17120

RHODE ISLAND
Div. of Parks and Recreation
83 Park St.
Providence, RI 02903

MARYLAND
Maryland Park Service
Tawes State Office Bldg.
Annapolis, MD 21401

DELAWARE
Dept. of Natural Resources
and Environmental Control
Div. of Parks & Recreation
P.O. Box 1401
Dover, DE 19901

VIRGINIA
Div. of State Parks
1201 Washington Bldg.
Richmond, VA 23219

WEST VIRGINIA
Parks and Recreation Division
Dept. of Natural Resources
State Capitol Bldg. No. 3, Rm. 311
Charleston, WV 25305

NORTH CAROLINA
Div. of Parks and Recreation
Dept. Natural Resources and
Community Development
P.O. Box 27687
Raleigh, NC 27611

SOUTH CAROLINA
Div. of Parks and Recreation
and Tourism
Rm. 30, Box 71
Columbia, SC 29202

FLORIDA
Florida Dept. of Natural Resources
Bureau of Education Information
3900 Commonwealth Blvd.
Tallahassee, FL 23200

GEORGIA
Office of Information
Dept. of Natural Resources
270 Washington St., Rm. 817
Atlanta, GA 30334

SOUTHERN STATE CAMPING AREAS

ALABAMA
Alabama Dept. of Conservation
Div. of State Parks
Administrative Bldg.
Montgomery, AL 36130

ARKANSAS
Dept. of Parks and Tourism
One Capitol Mall
Little Rock, AR 72201

LOUISIANA
Louisiana Office of State Parks
P.O. Drawer 1111
Baton Rouge, LA 70821

MISSISSIPPI
Travel and Tourism Dept.
Mississippi Dept.
of Economic Development
P.O. Box 849
Jackson, MS 39205

TENNESSEE
Dept. of Conservation
Div. of State Parks
2611 West End Ave.
Nashville, TN 37203

KENTUCKY
Travel
Frankfort, KY 40601

MIDWESTERN STATE CAMPING AREAS

MISSOURI
Missouri Dept. of Natural Resources
P.O. Box 176
Jefferson City, MO 65102

OHIO
Publications Center
Ohio Dept. of Natural Resources
Fountain Square Bldg. B
Columbus, OH 43224

ILLINOIS
Illinois Adventure Center
160 North LaSalle St.
Chicago, IL 60601

INDIANA
Dept. of Natural Resources
Div. of State Parks
616 State Office Bldg.
Indianapolis, IN 46204

MICHIGAN
Parks Div.
Dept. of Natural Resources
P.O. Box 30028
Lansing, MI 48909

MINNESOTA
Div. of Parks and Recreation
Box 39, Centennial Bldg.
St. Paul, MN 55155

WISCONSIN
Dept. of Natural Resources
Bureau of Parks and Recreation
Box 7921
Madison, WI 53707

IOWA
State Conservation Commission
Henry A. Wallace Bldg.
Des Moines, IA 50319

GREAT PLAINS STATES CAMPING AREAS

ND
SD
NE
KS
OK

OKLAHOMA
Oklahoma Tourism and
Recreation Dept.
Div. of Marketing Services
500 Will Rogers Bldg.
Oklahoma City, OK 73105

KANSAS
Parks and Resources Authority
503 Kansas Ave.
P.O. Box 977
Topeka, KS 66601

NEBRASKA
Game and Parks Commission
22000 North 33rd St.
P.O. Box 30370
Lincoln, NE 68503

SOUTH DAKOTA
Game, Fish and Parks Dept.
Div. of Parks and Recreation
Anderson Bldg.
Pierre, SD 57501

NORTH DAKOTA
North Dakota Park Service
Pinehurst Office Bldg.
1424 W. Century Ave.
Bismarck, ND 58502

SOUTHWESTERN STATES CAMPING AREAS

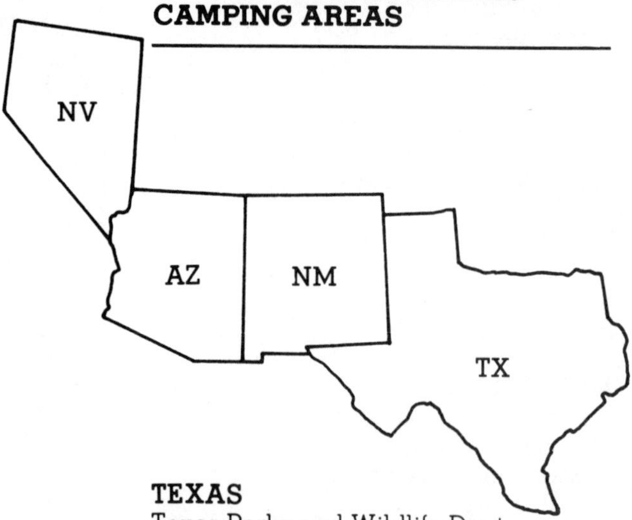

NV
AZ
NM
TX

TEXAS
Texas Parks and Wildlife Dept.
42000 Smith School Rd.
Austin, TX 78744

NEW MEXICO
State Park and Recreation Div.
Box 1147
Santa Fe, NM 87503

ARIZONA
Arizona Office of Tourism
3507 N. Central, Suite 506
Phoeniz, AZ 85004

NEVADA
Dept. of Conservation and
Natural Resources
Nevada Div. of State Parks
Capitol Complex
Carson City, NV 89710

ROCKY MOUNTAIN STATES CAMPING AREAS

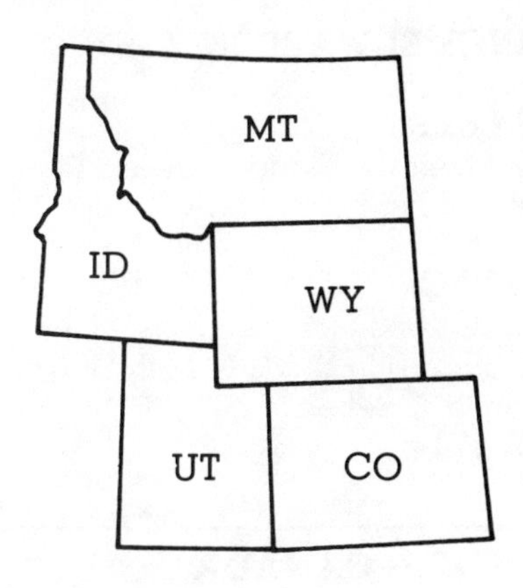

COLORADO
Div. of Parks and Recreation
618 Centennial Bldg.
1313 Sherman St.
Denver, CO 80203

WYOMING
Wyoming Recreation Commission
604 E. 25th St.
Cheyenne, WY 82002

MONTANA
Dept. of Fish, Wildlife and Parks
Parks Div.
1420 East Sixth Ave.
Helena, MT 56901

IDAHO
Parks and Recreation Dept.
Statehouse Mall
2177 Warm Springs
Boise, ID 83720

UTAH
Parks and Recreation
1596 West North Temple
Salt Lake City, UT 84116

PACIFIC COAST STATES CAMPING AREAS

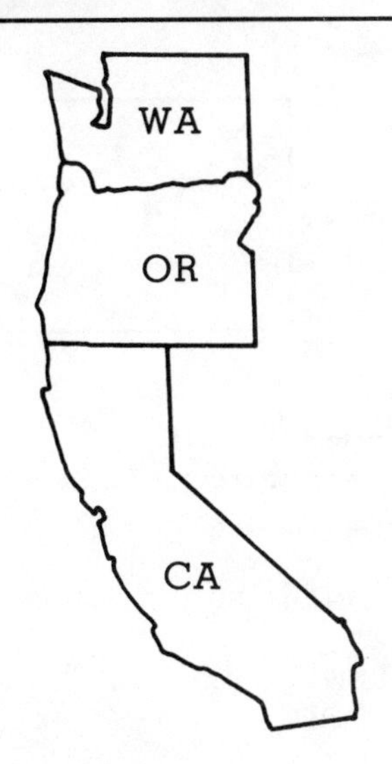

CALIFORNIA
Dept. of Parks and Recreation
P.O. Box 2390
Sacramento, CA 95811

OREGON
Travel Information
Oregon Dept. of Transportation
101 Transportation Bldg.
Salem, OR 97310

WASHINGTON
State Parks and Recreation Commission
7150 Clearwater Lane
Olympia, WA 98504

ALASKA
Div. of Parks
619 Warehouse, Suite 210
Anchorage, AK 99510

HAWAII
Dept. of Land and Natural Resources
State Parks Div.
1151 Punchbowl St., Rm. 310
Honolulu, HI 96813

CANADIAN NATIONAL PARKS

Parks Canada
Dept. of Indian Affairs
Centennial Tower
400 Laurier Ave. West
Ottawa, Ontario, Canada K1R 5C6

PROVINCIAL PARKS

EASTERN CANADA

Dept. of Tourism
Promotion Branch
P.O. Box 12345
Fredericton, New Brunswick
E3B 5C3

NEWFOUNDLAND
Dept. of Tourism
130 Water St.
St. John's, Newfoundland
A1C 1A8

ONTARIO
Ontario Travel
Queen's Park
Toronto, Ontario
M7A 2E5

PRINCE EDWARD ISLAND
Visitor Services Division
P.O. Box 940
Charlottetown, Prince Edward Is.
C1A 7M5

QUEBEC
Parks Quebec
150 St. Cyrille Blvd., E, 10th Floor
Quebec, Quebec
G1R 4Y1

WESTERN CANADA

ALBERTA
Travel Alberta
Alberta Tourixm and Small Business
Box 2500
Edmonton, Alberta
T5J 2Z4

BRITISH COLUMBIA
Ministry of Tourism
1117 Wharf Street
Victoria, British Columbia
V8W 2Z2

MANITOBA
Dept. of Economic
Development and Tourism
Travel Manitoba
Dept. 2044, Legislative Bldg.
Winnipeg, Manitoba
R3C OV8

NORTHWEST TERRITORIES
Travel Arctic
Yellowknife
Northwest Territories
X1A 2L9

SASKATCHEWAN
SaskTravel
Dept. of Tourism and
Renewable Resources
3211 Albert St.
Regina, Saskatchewan
S4S 5W6

YUKON
Parks and Historic Resources
Box 2703
Whitehorse, Yukon
Y1A 2C6

MAPS

For recreational maps of national parks and national forest service campgrounds, write or visit the offices nearest you, or write them at the addresses shown in the previous section.

For Geological Survey maps and the index, write any of the following offices, whichever is nearest you:

EAST OF THE MISSISSIPPI
U.S. Geological Survey
Map Distribution Center
1200 Eads St.
Arlington, VA 22202

WEST OF THE MISSISSIPPI
U.S. Geological Survey
Map Distribution Center
Federal Center Bldg.
Denver, CO 80225

CANADA
Canadian Map Office
Dept. of Energy, Mines and Resources
614 Booth St.
Ottawa, Ontario K1A 0E9

CHAPTER 14

FURTHER READING

Since the arrival of the first white men in North America, outdoor-adventure reading material has reflected Americans' great love of the outdoors; after all, we have more of it left intact than most other nations. Our national heritage is laden with great outdoor adventures that include journals of explorations toward the West, great sea voyages, reports by fur-trapping expeditions, first descents of rivers, and first ascents of mountains.

In addition to the great adventures, we have also had a number of great naturalists 'who described the world around us in terms we never tire of reading.

Since World War II Americans have been more and more interested in outdoor recreation, and a flood of how-to books, guidebooks, and other basic publications have come onto the market, many of which have remained in print for several generations of readers. All of this activity has made American campers among the most literate and well informed in the world.

Thus, the following bibliography is a mix-and-match of the outdoor adventure and the more straightforward instructional books.

Angier, Bradford. **Field Guide to Edible Wild Plants.** Stackpole, 1976. A heavily illustrated guide to wild plants, where to find them, and how to prepare them.

How to Stay Alive in the Woods. Collier, 1976. A basic book on survival that covers how to find food, shelter, and water and how to use both compass and celestial navigation.

Ferber, Peggy, ed. **Mountaineering: The Freedom of the Hills.** The Mountaineers, 1974. How-to for the advanced outdoorsman who will be doing some climbing as part of the outdoor experience.

Fletcher, Colin. **The New Complete Walker.** Alfred A. Knopf, 1974. An informative, often hilariously funny instructional book on backpacking.

From Katahdin to Springer Mountain. Rodale Press, 1977. A collection of stories of hikers along the Appalachian Trail, including grandmothers and young children.

Gibbon, Euell. **Stalking the Blue-Eyed Scallop.** David McKay, 1964.
Stalking the Wild Asparagus. David McKay, 1962.
Both books are standards by the American who single-handedly introduced wild foods to the nation.

Hillcourt, William. **The Official Boy Scout Handbook.** Boy Scouts of America, 1979. One of the best sources of outdoor how-to in existence.

Kjellstrom, Bjorn. **Be Expert with Map and Compass.** Charles Scribner's Sons, 1976. A basic guide to using compass and map.

Lathrop, Theodore G., M.D. **Hypothermia: Killer of the Unprepared.** The Mazamas, 1972. Basic information on hypothermia, how to avoid it, and how to treat it.

Leopold, Aldo. **A Sand Country Almanac.** Oxford University Press, 1979. One of the classics on enjoying the outdoors with an emphasis on environmental protection.

Manning, Harvey. **Backpacking: One Step at a Time.** Vintage Books, 1980. A guide to outfitting yourself for the outdoors with an emphasis on backpacking.

Olsen, Larry Dean. **Outdoor Survival Skills.** Brigham Young University Press, 1973. Food, shelter, water, and other information on surviving.

Rugge, John, and Davidson, James West. **The Complete Wilderness Paddler.** Alfred A. Knopf, 1977. A how-to book with excellent illustrations of how to outfit yourself for running whitewater rivers.

Rutstrum, Calvin. **The Wilderness Routefinder.** Collier Books, 1967. A book written with the wilderness traveler in mind, including celestial navigation and lots of personal observations.

Sutton, Ann and Myron. **The Pacific Crest Trail.** Lippincott, 1975. Not a guide to the trail (several of those exist), but a naturalist's description of the trail and the flora and fauna to be expected on the 2,400-mile trek.

Van Lear, Denise, ed. **The Best about Backpacking.** Sierra Club, 1974. Each of the chapters is by an expert on various subjects related to outdoor recreation.

Wilkerson, James A., M.D. **Medicine for Mountaineering.** The Mountaineers, 1979. Excellent information on all forms of outdoor survival and how to treat common ailments and injuries.

One of the greatest adventures in American history was the Lewis and Clark Expedition of 1804-6, mentioned earlier as an excellent car camping or RV route across the West. You can consult works on the adventure, beginning with the eight-volume set of the journals kept by Capt. Meriwether Lewis and Lt. William Clark. The set is available from Arno Press and from antiquarian book dealers. Two other works of value are:

Cutright, Paul Russell. **Lewis and Clark, Pioneering Naturalists.** University of Illinois Press, 1969. Easily the most readable account of the expedition, this is written without the usual political-economic overtones of most studies.

Satterfield, Archie. **The Lewis and Clark Trail.** Stackpole, 1978. A combination history-guide to the expedition and trail today.

In addition, see the books listed in the Acknowledgments for further reading suggestions on a variety of outdoor subjects.

INDEX

D

E

ECONOMIC DEVELOPMENT IN RETROSPECT:

The Italian Model and Its Significance for Regional Planning in Market-Oriented Economies

SCRIPTA SERIES IN GEOGRAPHY

Series Editors

Richard E. Lonsdale • Antony R. Orme • Theodore Shabad

Shabad and Mote · Gateway to Siberian Resources, 1977

Lonsdale and Seyler · Nonmetropolitan Industrialization, 1979

Dienes and Shabad · The Soviet Energy System, 1979

Pyle · Applied Medical Geography, 1979

Rodgers · Economic Development in Retrospect, 1979

ECONOMIC DEVELOPMENT IN RETROSPECT:
The Italian Model and Its Significance for Regional Planning in Market-Oriented Economies

Allan Rodgers
The Pennsylvania State University

1979
V. H. WINSTON & SONS
Washington, D.C.

A HALSTED PRESS BOOK

JOHN WILEY & SONS
New York Toronto London Sydney

V. H. Winston & Sons, a Division of Scripta Technica, Inc., Publishers
1511 K Street, N.W., Washington, D.C. 20005

Distributed solely by Halsted Press, a Division of John Wiley & Sons, Inc.

Library of Congress Cataloging in Publication Data:

Rodgers, Allan L.
Economic development in retrospect

(Scripta series in grography)
Includes index.
1. Italy, Southern–Economic conditions.
2. Economic development–Case studies.
3. Regional planning–Case studies.
I. Title. II. Series.
HC305.R553 338'.0945'7 78-20862
ISBN 0-470-26628-7

Composition by Marie A. Maddalena, Scripta Technica, Inc.

To Audrey

CONTENTS

ACKNOWLEDGMENTS

Space limitations do not permit listing all who provided assistance in my work on southern Italy. I literally owe thanks to dozens of government officials, planners, and university scholars who provided important insights into the southern development program and facilitated my understanding of that impoverished region. Over the years, they have granted me numerous interviews and supplied publications, unpublished data, and computer compilations which were fundamental for this book.

I must also acknowledge with gratitude the research grants received from the Guggenheim Foundation, the Italian Fulbright Commission, and the National Science Foundation, as well as from the Pennsylvania State University; these made it possible for me to live and travel in southern Italy for well over 3 years. My own institution, the Pennsylvania State University, also awarded me two sabbatical leaves and a number of grants for travel, research assistance, cartographic and photographic aid, computer time, etc.

Three organizations in Italy were most helpful and cooperative: the Cassa per il Mezzogiorno, the Association for the Development of Industry in the Mezzogiorno (SVIMEZ) and the Istituto Centrale di Statistica (ISTAT). A list of all in Italy who were of assistance over the years would be too lengthy, but those who were most helpful included Prof. G. Pescatore (former President of the Cassa) and his successor Av. A. Servidio, Dr. L. Pinto of ISTAT, and Prof. G. Tagliacarne, currently of SVIMEZ.

At the Cassa, I was assisted by Dr. C. Simoncelli and Prof. P. Vicinelli; at SVIMEZ by Drs. D. Cecchini, R. Cagliozzi, and S. Cafiero; at the Unione Italiana delle Camere di Commercio, by Dr. O. Cherubini; at IRI, by Prof. C.

Vanutelli and Dr. Del Canuto; at IASM, by Dr. C. Turco; at ANAS, by Dr. Ing. E. Scotto; at INAM, by Dr. F. Sartoretti; at INAIL, by Prof. M. Brancoli; and at ISTAT, by Drs. P. Quirino, E. De Angelis, and G. Marrocchi; Dr. Cipriana Scelba of the Italian Fulbright Commission was most helpful on many occasions.

My Italian colleagues, Profs. C. Muscarà of the University of Venice, A. Pecora of the University of Turin, C. Della Valle and O. Baldacci of the University of Rome, and D. Ruocco of the University of Naples, provided invaluable help. Last but not least in Italy, I wish to thank Dr. Segrè of the library of the Unione Italiana delle Camere di Commercio and the library staff of SVIMEZ.

In this country, I have made extensive use of the New York Public Library, Columbia University, and the Library of Congress. I greatly profited from participation in the Seminar on Modern Italy at Columbia University and the scholarly associations formed there. At Penn State, I owe very much to my colleague and friend Prof. Anthony Williams who came to my aid on innumerable occasions. I also owe a major debt of gratitude to a host of research assistants, most of all to Mr. James Meyer who prepared many maps for this book.

Finally, I cannot conclude this long list of acknowledgements without expressing overwhelming indebtedness to my wife Audrey. As editor, ever-ready listener and critical commentator, as one who shares my fascination for Italy, she, in a real sense, is a co-author of this book.

PREFACE

The process and character of industrial growth in less developed regions and the socioeconomic consequences of that experience are still imperfectly understood despite a rapidly burgeoning literature in the development field.[1] One such problem region is southern Italy, a classic region of sub-national underdevelopment. This study is primarily concerned with the evolving geographical pattern of industrialization,[2] under government planning and subsidization, in that region since 1951. The "South's" problems are replicated in many other segments of the developed world, including the United States. Here, I am deliberately excluding other problem areas often termed "depressed," i.e., those that formerly had a high level of economic development and are now classified as "unhealthy" because of changed economic circumstances.

The type of underdeveloped region which is my frame of reference is often populous and is commonly at an early stage of the demographic transition. It therefore has relatively high birth rates, low mortality levels, and a comparatively young population with a large dependent cohort. Yet such regions are typically further along in their demographic development than is common in the truly underdeveloped world. If these areas possess resources, they are either unused or underutilized. Their manufacturing activities are largely at pre-industrial levels, often dominated by artisan workshops, undercapitalized, and there is frequently a lack of local entrepreneurial initiative. Those branches that have developed are typically sectors which have been termed traditional or first stage industries. There is also a lack of modern technology, low literacy levels, and minimal worker skills (adaptable to factory-type industry). Such regions may also be disadvantaged because of their excessive dependence on unproductive agricultural systems and the inadequacy of their infrastructures. They are also often handicapped because of their distance and poor transport connectivity to major internal and international markets that possess sufficient demand for modern industrial products. Local demand is commonly far too limited to offset such locational disadvantages.

Examples of such underdeveloped regions within the developed world include Appalachia, the Maritime Provinces of Canada, Brittany and the Central Massif of France, the eastern Netherlands, south-central Spain, and northern Sweden. Southern Italy, which is also termed the "Mezzogiorno,"* unquestionably fits the characteristics noted above. This area includes the mainland south of Rome and the islands of Sicily and Sardinia. Its area encompasses nearly 43% of the nation's territory, and in 1951, at the start of the development program, it accounted for 39% of Italy's population (Fig. 1).

*The "Mezzogiorno" literally means "Land of the Midday Sun."

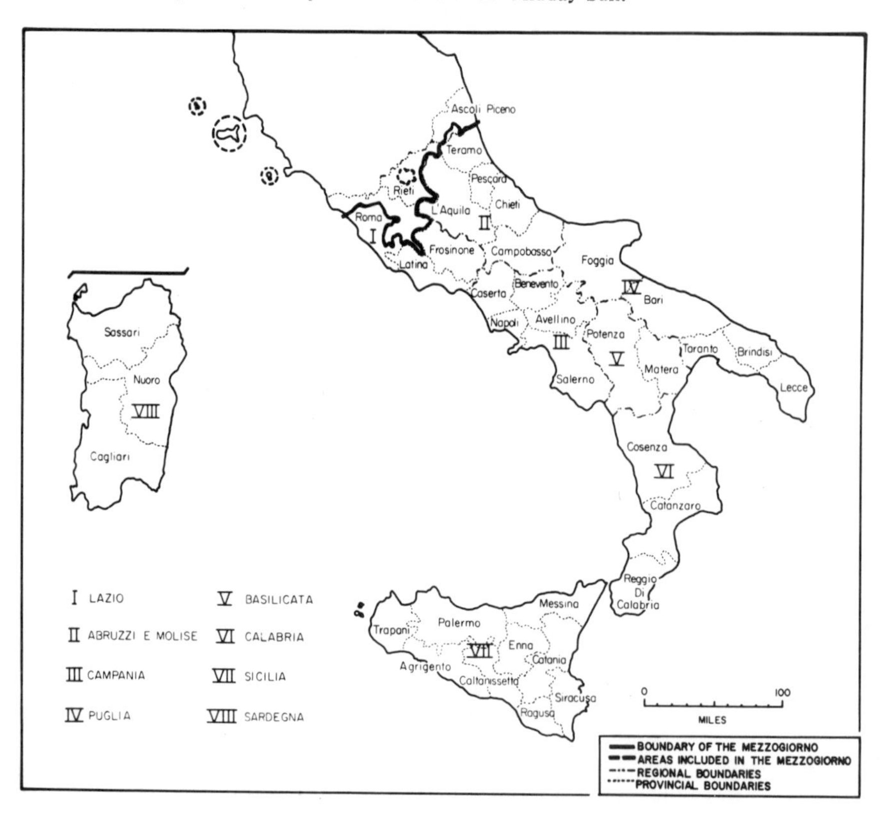

Fig. 1. Regional and provincial divisions in Southern Italy.

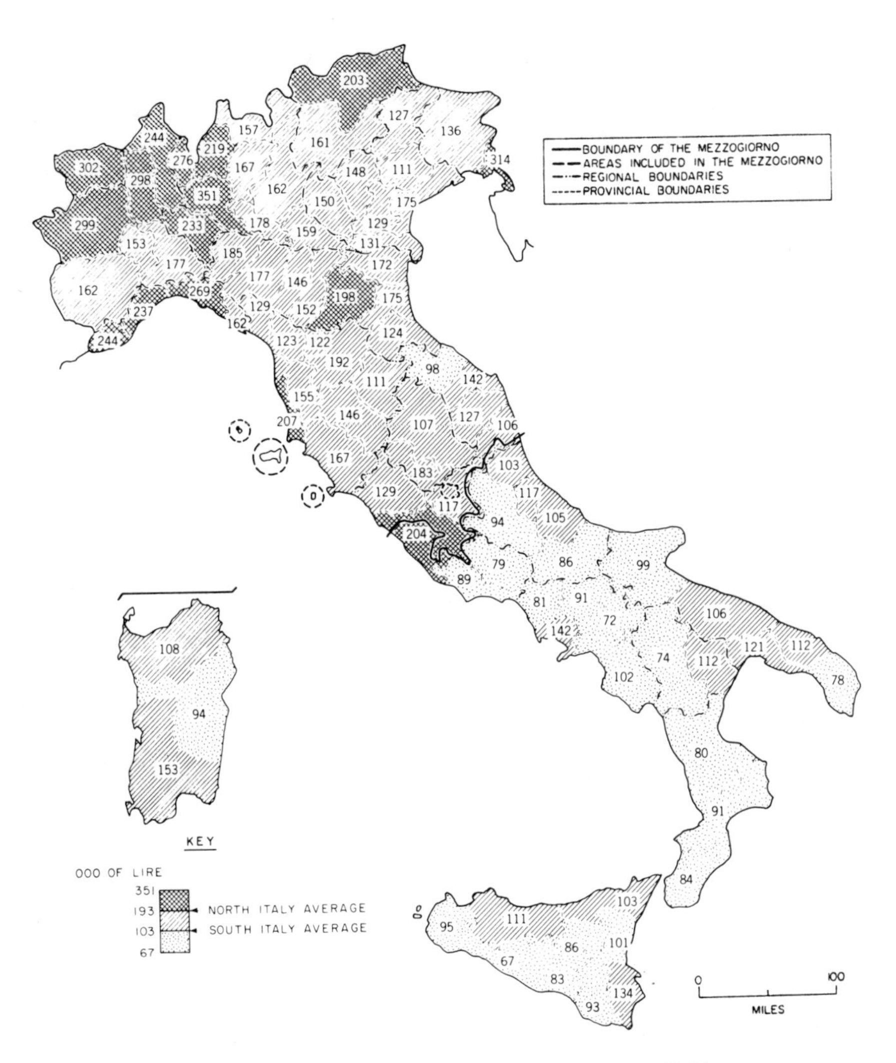

Fig. 2. Per capita income by province in 1951.

While Italy's per capita gross national product was less than one-sixth of that of the United States and roughly one-third of those of Sweden and the United Kingdom, these national values mask stark internal contrasts between the backward areas of the Mezzogiorno and northern Italy. Thus, per capita income levels in the thriving industrial centers of the Po Valley of the North were not strikingly below those of Western Europe; in contrast, per capita incomes in Southern Italy were half those of its northern counterpart. In the extreme case, as demonstrated on Figure 2, several southern provinces had values that were roughly one-fifth of that of Milan. Even the richest province in the Mezzogiorno—Naples—had a per capita income that was only about two-fifths of the average in the northern economic capital. But even these data understate inter-regional income differentials because of marked urban-rural contrasts within the southern provinces as well as those between the upper and lower classes of the cities themselves. The great majority existed at subsistence levels, some with incomes and living conditions resembling those commonly described in the truly underdeveloped world, while a small minority lived "apart" in relative luxury.

In addition to reviewing the origins of the "Questione Meridionale" or the problem of southern Italy, I am concerned with an appraisal of the impact of the development program for the Mezzogiorno on economic growth in that region and the ramifications of that performance on socioeconomic changes within the South. Although other variables such as population, agriculture, and transportation will be discussed, my principal focus will be on the industrial transformation of the region.

As an economic geographer, I am primarily interested in the locational or spatial facets of growth when viewed within the considerably broader framework of regional development theory. Our knowledge of that theory must be built on "in depth" probing and evaluation of representative regions throughout the world. Such probing and assessment is an essential task which serves to erect the foundation. I view the Italian experience as such a basic foundation because *the region constitutes a classic model for retrospective analysis of economic development* elsewhere. This model illuminates and enhances *our understanding of contemporary regional planning in advanced nations*, particularly in North America and Western Europe.

NOTES

[1] The first good review of the development literature treated from a spatial perspective was David Keeble, "Models of Economic Development," pp. 243–302, in Richard Chorley and Peter Haggett, *Socio-Economic Models in Geography*, (London: Methuen, 1968). The same volume contained F. Ian Hamilton's essay, "Models of Industrial Location," pp. 361–424. The latter treated geographical aspects of industrial growth. A more current review of the literature is found in B. S. Hoyle, *Spatial Aspects of Development*, (London: Wiley, 1974); a good review of the earlier growth pole literature was provided by D. F. Darwint, "Growth Poles and Growth Centers in Regional Planning," *Environment and Planning*, Vol. 1 (1969) pp. 5–32 and, more recently, by Antoni Kuklinski, *Growth Poles and Growth Centers in Regional Planning*, (UNRISD, Geneva and the Hague: 1972).

[2] In this book the term "industry" is used synonomously with the designation "manufacturing" and does not, unless indicated otherwise, include extractive industry, construction, electricity, gas and water utilities. The terms *South, Mezzogiorno*, and *Southern Italy* cover southern Lazio unless that area is specifically excluded by a footnote, or the "regioni" in a tabulation do not include either southern Lazio or the provinces of Latina and Frosinone. I have drawn freely in this study on my own publications, which include (Allan Rodgers): "Mezzogiorno: Regional Inequalities of

Industrial Growth within an Underdeveloped Region," in F. Ian Hamilton, *Industrial Change* (London: Longman, 1978), pp. 99-118; "Southern Italy: The Role of Governmental Locational Policy and Practice in the Development of Industry in an Underdeveloped Region," in Calogero Muscarà, ed., *Mezzogiorno e Mediterraneo* (Venice: Università di Venezia, 1976), Documento 4, pp. 47-81; "Migration and Industrial Development: The Southern Italian Experience," *Economic Geography*, Vol. 46, No. 2 (April 1970), pp. 111-36; "Industrial Location–Theory and Practice: An Italian Experiment," *Earth and Mineral Sciences*, Vol. 38, No. 7 (April 1969), pp. 58-60; "Naples: A Case Study of Government Subsidization of Industrial Development in an Underdeveloped Region," *Tijdschrift voor Economische en Sociale Geografie* (Netherlands Journal of Economic and Social Geography) (January-February 1966), pp. 20-32; and "Regional Industrial Development with Reference to Southern Italy," *Essays on Geography and Economic Development* (Chicago: University of Chicago, Research Paper in Geography No. 62, 1960), pp. 143-73.

There are many books, monographs, and articles on regional economic development in the Italian South. However, in my view, from the Geographer-Regional Economist-Regional Planner's context the best is Kevin Allen and M. C. Maclennon, *Regional Problems and Policies in Italy and France*, University of Glasgow, Social and Economic Studies, No. 19, (London: 1970). The Italian facet of this study was covered by Dr. Allen. He also co-authored with Andrew Stevenson a volume entitled: *An Introduction to the Italian Economy*, Glasgow Social and Economic Research Studies, No. 1, (London: Robertson, 1974). Also noteworthy and of great value for its perceptive critique of the development effort is G. Podbielski, *Twenty-Five Years of Special Action for the Development of Southern Italy*, SVIMEZ (Milan, Rome: 1978); note in particular its extensive bibliography covering the Italian literature. The most important Italian articles are reviewed in *Informazione SVIMEZ*, which is published on an irregular basis by the *Associazione per lo Sviluppo dell' Industria nel Mezzogiorno*. Monographs published under that sponsorship are by far the most useful of the voluminous literature on the Mezzogiorno.

Chapter 1

THE HISTORICAL CONTEXT

There has been considerable debate over the past decades as to the origins of the Southern Problem. The so-called "Meridionalists" of the South and their supporters, including the Swiss scholar Vöchting,[1] have argued that the North-South dichotomy was a product of "Unification." In contrast, another group of scholars, these mainly Northerners supported by several foreign economic historians, among them Clough[2] and Eckhaus,[3] has contended that this socioeconomic differentiation predated the creation of a unified state in Italy (1861). The weight of the existing evidence appears to support the latter school. What is more important, however, for our purposes, are the events following the unification of the nation, for there is no question that only limited industrial development had taken place in either area prior to the 1860s. Here the available statistics, which are admittedly limited, support the thesis that "modern," factory-type industry grew more rapidly in the North, particularly in the last decades of the 19th century. It is also clear that in contrast to northern Italy the industries of the Mezzogiorno prior to 1861 were mainly high-cost artisan establishments, which had only survived and prospered as a result of the protection of the tariff barriers of the Kingdom of the Two Sicilies. The advent of a period of free trade after Unification meant that these workshops could not compete with the flood of manufactured goods from abroad nor, under a unified economy, with goods from the North. In contrast, while neither area possessed a significant resource base, the North had a greater availability of social overhead capital and decided locational advantages such as market proximity and access to cheap water power. Northern industry was also more receptive to the diffusion of foreign technology, in part because of its location but also because of a generally higher level of entrepreneurial initiative. The comparatively earlier start of modern industry in the North was also reflected in the accretion of similar industries within the same provinces. These plants were able to take advantage of external economies and multiplier effects resulting from backward and forward industrial linkages. The textile and machinery sectors are often cited as examples of these stimuli in the industrial evolution of the Po Valley. In sum, the last two decades of the 19th century and the period before the first World War witnessed a significant growth of industry in the North with a concomitant widening of the gap between the two regions.

Southern writers have argued that the reinstitution of custom bars after 1887 had only imperceptible effects upon industry in the Mezzogiorno. They contend that the artisan workshops of the South still suffered from the competition of northern plants, that the latter were actively encouraged by the State (a debatable point), and that the new tariff barriers increased the costs for manufactured goods at a time when southern agriculture was already lagging behind that in the North. Perhaps more important was the lack of entrepeneurial spirit in the Mezzogiorno. In contrast to northern Italy, the prevailing southern attitude apparently deemed it to be "socially degrading" to accumulate wealth in industry or commerce. Then too, the pitifully low literacy levels in southern Italy were reflected in worker attitudes and skills, and the availability of social overhead capital was clearly inferior to that in the North. Finally, the "Meridionalists" have traditionally contended that public expenditures in the South, after 1861, were far lower than tax revenues received by the central government. This is refuted by the available evidence, one facet of which was the heavy expenditure by the state on railroad construction in the Mezzogiorno after Unification. According to Clough, southern Italy had only 6% of the nation's rail mileage in 1861 and that value had increased to 32% by 1875. He also indicates that the South never received less that its share of state expenditures for public works in comparison with its percentage contribution to state tax receipts.

After 1900, the relatively depressed economic conditions in southern Italy apparently discouraged further industrial investment there, either in the form of new plants or in the expenditure of funds for the expansion and modernization of industrial facilities which might have reduced production cost differentials between the two regions. The Liberal government did make some rather ineffectual attempts to reduce the socioeconomic gap between North and South, but agriculture and infrastructure were the chief beneficiaries. In 1904, however, customs and tax concessions were granted to firms who would locate industries in the Naples area. These concessions were extended to the rest of the South 2 years later. However, Naples was the only center which experienced significant industrial growth during that era.

The emphasis, up to this point, has been primarily on the lag in industrial development in Southern Italy in the late 19th and early 20th centuries, and indeed industrialization in that region is the major initial focus of this study. However, given the linkages between industry and agriculture, an abbreviated discussion of agricultural developments in the post-unification period is obviously in order.

The traditional heavy weighting of low productivity agriculture in the southern economy has always been one of the region's major handicaps. While the backward state of this sector cannot be explained solely in terms of the region's physical environment, clearly the inferior quality of the land in the South played a key role in the evolving dichotomy between the two areas. The limited availability of level land which drove grain cultivation into the hills accelerated soil erosion and even contributed to the spread of malaria in the lowlands of the Mezzogiorno. Then, too, the low amounts, high seasonality, and great variability of precipitation were also certainly negative elements.

In contrast, the North, with far more level land, more fertile soils, and a more reliable precipitation, had clear physical assets for the development of agriculture. These advantages, as we shall see, coupled with more favorable land tenure practices and cropping patterns, have meant that the North led southern Italy in crop productivity and per capita agricultural incomes over most of the past century. Its agriculture was, in many ways, healthier than that of the Mezzogiorno and much more comparable to that of Western Europe. Clearly, given its physical and economic constraints, the southern agricultural sector could not readily produce the surplus needed for investment in industry.

The French conquest of Italy, as a result of the Napoleonic wars, brought drastic changes to land tenure patterns in Italy, especially to those prevailing in the South. By 1806, the feudal system was virtually destroyed, and in theory, at least, the land was distributed to the peasants. It was the beginning of a new era in southern agriculture because for a time that region had a relative advantage over the North. This change in land ownership was upset by the repossession of land by the landowners and the development of the "latifondo," which were large estates inefficiently organized and often controlled by absentee landlords. Southern agriculture was undercapitalized, fields were fragmented, and extensive rather than intensive methods were utilized. All of these resulted in low productivity. The land was tilled by laborers hired according to daily work requirements or, at best, operated on a share-cropping basis, which was grossly unfair to the tenant farmers and afforded them no incentives for increasing agricultural productivity. The leading crop raised, even on former pasturelands in the hills, was grain. It was typically grown without fertilization or rotation (already established practices in the North), and these shortcomings contributed to soil exhaustion. In contrast, unlike grain, which was normally a subsistence crop, fruit, wine, and olive oil were competitive and continued to be produced for sale to external markets. Although, as we have seen, the South was behind in industrialization, its major disadvantage when the nation was unified was in agricultural production. Eckhaus[4] has estimated that as of Unification per capita agricultural income in northern Italy was 26% over that in the South. After 1861, the Mezzogiorno had a short-lived period of agricultural prosperity because the elimination of tariff barriers permitted the export of the region's surplus of fruit, wines, and olive oil. When custom bars were reimposed in the period from 1883 to 1887, southern farmers suffered because of the loss of foreign markets and the burden placed upon them by the high cost of manufactured goods from the North. As a result, there was a retreat from commercial to subsistence agriculture. The pressure of population on the land led, as we shall see, to mass migration. After 1887, there were no significant increases in agricultural production in the South, and northern agricultural output and per capita incomes from agriculture exceeded those in southern Italy.

Thus northern agriculture, in contrast to that in the South, could provide the surplus needed for investment in industry as well as a market for expanding industrial output. In addition, its high productivity per man permitted a ready diversion of labor from agricultural to industrial pursuits. Southern agriculture could not do the same because the region was unquestionably close to the margin of subsistence. The building of railroads in the region in the post-unification period, cited earlier, did not materially help its agricultural sector because there was no significant surplus to be transported. Thus, the argument of Eckhaus that "with differing initial conditions in 1861 in agriculture, transport and even industry, it is not surprising that the disparities between the two regions intensified"[5] cannot be refuted from the evidence of economic development in the post-unification era.

Even the modest subsidization of southern agriculture and industry by the Liberal government, prior to the first World War, through the media of tax credits, loans, and public works could not compensate for the South's physical and economic ills. Allen's recent analysis of early Italian development policy is an effective critique of those efforts.[6] He argues that although the program was well intentioned and probably in advance of efforts in other European countries, it was inadequate to overcome regional disparities in Italy and the inherent advantages of the North. The gap actually widened significantly after 1900 despite such efforts. He ascribes the failure to the limited funds allocated to the program, to the lack of a conceptual framework, the measures adopted were ad hoc and uncoordinated, and there was too great an emphasis on agricultural development and infrastructure with far too little on industrial subsidization. He also attributes the lack of success to the protectionist policy adopted in the 1880s, inflation, and the war. With regard to the last causal element, there is no question that the first World War spurred the rapid expansion of industry in the North in response to demands for military supplies. This, of course, intensified the North-South dichotomy in industrial development. In sum, at this stage in Italian economic development, there were already marked regional income inequalities with a depressed southern economy and a flourishing one in the North.

This leads us logically to the next step in our historical analysis. What were the demographic developments in Italy during the period following Unification, and, in particular, what were the effects of these events on

Table 1. Regional Variations in Natality and Mortality Levels in Italy Between 1881 and 1974[a]

Years	North			Mezzogiorno		
	Birth rates	Death rates	Natural increase	Birth rates	Death rates	Natural increase
1881–85	36.1	26.2	9.9	40.5	28.8	11.7
1886–90	36.2	25.7	10.5	39.4	29.4	10.0
1891–95	35.0	24.5	10.5	38.3	27.6	10.7
1896–00	33.3	21.8	11.5	35.8	25.1	10.7
1900–05	32.0	20.8	11.2	33.4	23.6	9.8
1906–10	31.7	20.0	11.7	33.8	22.9	10.9
1911–14	30.5	17.8	12.7	33.6	21.1	12.5
1915–18	20.4	24.3	-3.9	27.6	30.8	-3.2
1919–20	24.3	17.2	7.1	31.9	22.3	9.6
1921–25	27.0	16.0	11.0	34.5	18.6	15.9
1926–30	22.8	14.8	8.0	32.5	17.6	14.9
1931–35	20.6	13.0	7.6	29.6	16.1	13.5
1936–40	20.2	12.8	7.4	29.1	15.6	13.5
1941–45	16.9	13.7	3.2	24.9	16.2	8.7
1946–50	17.8	10.7	7.1	27.5	11.1	16.4
1951–55	14.4	9.9	4.5	23.5	9.2	14.3
1956–60[b]	14.5	9.9	4.6	23.1	8.8	14.3
1961–65[b]	16.5	10.3	6.2	22.5	8.6	13.9
1966–70[c]	15.7	10.3	5.4	21.0	8.4	12.6
1974–[d]	13.7	10.0	3.7	19.3	8.5	10.8

[a]Svimez, *Un Secolo di Statistiche Italiane, Nord e Sud: 1861–1961* (Rome: 1961), p. 79. Per thousand.

[b]Comitato dei Ministri per il Mezzogiorno, *Studi Monografici sul Mezzogiorno* (Rome, 1966), pp. 64, 68.

[c]Ibid., Rome, 1971, p. 52.

[d]Computed from *Annuario Statistico Italiano, 1975* (Rome, 1975), various pages.

population pressure in the Mezzogiorno? Between 1861 and 1913, as evidenced by Table 1, there were no major disparities in vital rates between the two Italies. Rates of natural increase were not high (10 to 11 per thousand) simply because high birth rates (33 to 40 per thousand) were paralleled by relatively high death rates (20 to 29 per thousand). The overall natural increase for the period was nearly 15 million people, but contemporaneously there was an estimated net outmigration of roughly 4 million inhabitants. Thus the effective increase in population was only 11 million. This massive outflow undoubtedly had major repercussions on national economic growth.

Our analysis of emigration patterns must commence with a major reservation. While there are abundant data on out-migration, statistics on return flows before 1904 must be estimated. Nevertheless, the absolute volume of emigration during this period was massive, considering the size of the base population during this era (26 million in 1861 and 37 million in 1913).[7] However, much of the out-migration was ephemeral, with permanant expatriates accounting for only a fraction of the total. Thus, during the same period, the total number of emigrants has been estimated at 13 million, while the net value was less than 30% of that number.

The major peak in the history of Italian emigration came during the period from 1876 to the first World War, when the enactment of bars to immigration abroad, followed by Fascist legislation against emigration in the 1930s, stemmed the outflow. This period logically falls into two time segments. From 1876 to 1900 the number of emigrants totaled over 5 million, for an average of about 210,000 per year. The volume increased markedly from 1901 to 1913, when the total reached 8 million, for an annual average of over 600,000 persons.

Although population pressure prior to 1913 was intense in both southern and northern Italy, on a relative scale it was clearly less so in the North because of its greater agricultural productivity and its comparatively higher levels of industrialization. Yet the numbers of emigrants from both regions were remarkably similar. The southerners came predominantly from Campania and Sicily, for these were the regions of greatest population pressure in the South. In the case of Campania, the exodus resulted from sheer population numbers relative to the available agricultural land; in the second instance, the outmigration resulted from Sicily's history of political instability, its land tenure system and its environmental handicaps. This group went mainly to the Western Hemisphere, particularly to the United States and Argentina. In contrast, the northern Italians migrated almost exclusively to countries in close proximity, especially France and Switzerland. The most important source region was Veneto in the eastern Po Plain, a badly drained region of comparatively low agricultural productivity.

The available evidence indicates that the largest proportion of the northerners were men, their movements were mainly temporary, and there were disproportionately heavy return flows. The converse was true of the southern emigration, for although, here too, there were large numbers of returnees, the migration was far more permanent. Entire family units commonly emigrated at the same time.

It is clear that emigration played a major role in slowing population growth, particularly in southern Italy, and reducing population pressure. Then too, in a human sense, the opportunity to emigrate was a boon to those who came from particularly depressed regions, but the raising of "emigrants for export" was obviously costly, for the expense of rearing them was borne by the parent regions. In addition, those emigrants were frequently the most enterprising people in the working age groups of the lower classes. Thus the drain on the labor force may have had some negative repercussions for the

South which were not offset by foreign remittances. Nevertheless, this massive outmigration clearly did reduce population pressure in what was already the backward region of Italy—the Mezzogiorno.

According to Vöchting,[8] when the Fascist regime came into power it did little to counterbalance past regional economic trends. In fact, during the 1930s government industrial development policies supported the further development of manufacturing in the North in order to achieve economic autarchy as quickly as possible and build the nation's military potential. The only exception was the investment by IRI (the state holding company created after the financial collapse of the early 30s) in expanding metallurgical and machinery production in Naples. By 1938, the South had only about 18% of the nation's manufacturing employment as compared with 37% of its population.

Fascist policies in the agricultural sector included attempts to drain marshlands as part of a general reclamation program but also as a step towards conquering malaria, irrigation, reforestation, and the control of soil erosion. However, the South apparently received no special priority in any of these programs other than those warranted by physical circumstances. In addition, there was, of course, the "battle" for grain self-sufficiency related to the virtual isolation of the Italian economy from world markets after 1929. In the Mezzogiorno, this drive led to the conversion of more pasture and woodland into grain cultivation, with rather disastrous consequences in terms of soil erosion. Nor did this program appreciably increase grain productivity in that region. Then too, the restrictions on internal migration (to be discussed later) increased population pressure on these marginal lands. In general, the agricultural measures that were adopted were uncoordinated. No attempt was made to provide funds for an undercapitalized agriculture, to rationalize land tenure patterns by land redistribution or land reform, nor were there significant attempts to improve agricultural methodology in the South. Nevertheless, in relative terms, agriculture and its associated infrastructure did receive higher priority than industrial development.

As noted earlier, until the first World War, there were only minimal regional differences in birth and death rate levels in Italy (see Table 1), and natural increments per thousand of population were remarkably uniform. Population pressure was intense in both regions, but, on a relative scale, greater in the South. Nevertheless, prior to the first World War, both areas experienced major outflows of population to other countries. After the war, with enactment of immigration bars, particularly in the United States, the flow of Italians abroad declined drastically from its previous peaks. Also, Fascist policy actively discouraged emigration except to Libya and the newly conquered territories in Africa.

It will be recalled that the period after the first World War was one of accelerated industrialization in northern Italy, particularly in Lombardia and Piemonte, reflected in increasing employment opportunities in those regions. It was also a time of sharply declining birth rates in the North. Increasing rural population pressure in the Mezzogiorno with continuing high fertility levels, coupled with declining possibilities for emigration overseas, should have resulted in a major exodus northward. However, the record, though fragmentary, does not support this thesis.

Unlike the data on emigration, which are readily available and of reasonable validity, statistics on internal migration prior to 1951 are far more limited in their availability and utility. Italian demographers appear to agree that the relative size of such internal movements was small compared to the massive volume in the '50s and '60s.

There are two sources for such materials: registration (termed anagraphical) records and the reports of the various population censuses. The former, covering inter-commune movements, have been collected since 1862, but these data were not consistent nor reliable until 1930. However, even the dimensions of the flows after that date are in doubt because of the nature of the anagraphical data which only record official transfers of residence and ignore what were, in fact, illegal movements. This complication results from laws promulgated during the Fascist era which were in essence anti-urban in character and designed to prevent the so-called depopulation of the countryside. A short review of this legislation is vital to an understanding of the nature and dimensions of internal migration in Italy prior to 1961, when the earlier legislation was modified. These restrictive laws, the first of which was contained in the law of December 24, 1928, stiffened in 1931 and finally made most rigorous in legislation passed in 1939, were designed to govern the

Table 2. Internal Migration in Italy, Based on Transfer of Residence Reports, by Commune, for the Years from 1931 to 1950[a]

Region	Outmigration	Inmigration	Net migration
Piemonte	2,319	2,445	+126
Lombardia	3,591	3,753	+162
Lazio	1,259	1,673	+414
Other Northern Italy	8,897	8,557	-340
NORTHERN ITALY	16,066	16,428	+362
Campania	1,436	1,431	- 5
Abruzzi and Molise	646	595	- 51
Puglia	1,045	1,013	- 32
Basilicata	173	158	- 15
Calabria	654	594	- 60
Sicily	1,667	1,566	-101
Sardinia	645	646	+ 1
SOUTHERN ITALY	6,266	6,003	-263
ITALY	22,332	22,431	+ 99[b]

[a]Compiled from data in the *Annali di Statistica,* 1965, op. cit., p. 673.
[b]Reporting errors.

conditions under which a person or family could transfer residence from one commune to another, to another nearby region, or, where appropriate, to another part of the country. One could not receive authorization to move from one commune to another, if the new commune had over 25,000 inhabitants, or in special instances if the destination was a significant industrial center (with a lower level of population), unless evidence could be provided of an assured job. In addition, illegal migrants could not enter their names on the unemployment roles of the commune to which they might have moved, so that clandestine migrants were, in many instances, forced to take jobs that were at best marginal in character. The strictness of interpretation of laws varied considerably from commune to commune. In some areas, illegal migrants were forcibly repatriated, fined and even imprisoned, while in others, like Turin, local industrial interests used their influence to ensure an abundant labor supply, so that virtually all who desired to move to that city were permitted to do so. These laws were apparently enforced with increasing stringency through the '30s; however, clandestine movements continued, but at a reduced pace. To all of these were added the dislocations of population resulting from the War. Tens of thousands, particularly in the South and Center, fled without permission or cancellation of official residence and remained as illegal migrants in their new homes.

Other problems in the reliability of the anagraphical records include time lags in the reporting of outmigration by local officials in the South and, in some instances, deliberate understatements of such movements. These have arisen because of the potential loss of political prestige by a commune and, most importantly, because of the fear of reductions in key financial subsidies for roads, schools, etc., which are linked to specific population levels.

Table 3. Estimated Outmigration from the Mezzogiorno to Northern Italy through 1951, Based on Residence and Place of Birth Data[a]

Southern region	Proportion of outmigration to the North (%)	Proportion of the resident population of the South in 1951 (%)
Abruzzi and Molise	15.9	9.5
Campania	20.1	24.6
Puglia	21.2	18.2
Basilicata	2.6	3.6
Calabria	11.2	11.5
Sicily	22.0	25.4
Sardinia	7.0	7.2
SOUTHERN ITALY	100.0	100.0

[a]*Annali di Statistica*, 1965 op. cit., pp. 666–667. Based on the 1951 Census Population.

Table 4. Estimated Internal Migration Data for the Mezzogiorno through 1951, Based on Residence and Place of Birth[a]

Place of birth	Place of residence						
	Abruzzi and Molise	Campania	Puglia	Basilicata	Calabria	Sicily	Sardinia
Abruzzi and Molise	1,609.0	14.3	9.6	0.7	1.1	1.9	1.2
Campania	12.2	4,146.8	32.1	8.7	10.4	10.9	4.6
Puglia	10.1	28.7	3,092.3	13.3	5.6	6.3	2.4
Basilicata	1.1	15.7	15.8	593.9	4.0	0.9	0.5
Calabria	1.5	19.1	9.2	3.0	1,988.8	16.8	1.3
Sicily	2.8	18.6	9.7	0.9	13.6	4,381.8	8.0
Sardinia	0.8	3.4	1.6	0.2	0.6	2.6	1,227.2
Southern Italy	1,637.5	4,246.6	3,170.3	620.7	2,024.1	4,421.2	1,245.2
Northern Italy and abroad	46.5	99.7	50.2	6.0	20.2	65.5	30.8
Resident population	1,684.0	4,346.3	3,220.5	626.7	2,044.3	4,486.7	1,276.0
Proportion born in other regions of southern Italy[b] (%)	1.7	2.3	2.4	4.4	1.7	0.9	1.4

[a] *Annali di Statistica*, 1965, op. cit., pp. 666–667. Based on the 1951 Census of Population.
[b] The average for southern Italy was 1.7%.

Finally, and perhaps most important for our purposes, the published anagraphical data from 1931 through 1954 do not indicate interregional flows. Despite all of these limitations, an array can be constructed for the period from 1931 to 1950, as is demonstrated in Table 2, which does provide some clues as to broad internal migration trends in that period.

From these data it appears that at least one-quarter of a million people left the South for northern Italy during this era. Their destinations can only be inferred from the anagraphical data, but it appears that the largest share moved to Lazio, presumably to take advantage of new employment opportunities which were developing as the nation's administrative structure grew and became more centralized in Rome. Less striking flows appear to have taken place to the expanding industrial centers of Milan (Lombardia) and Turin (Piemonte).

These statistics on transfers of residence can be supplemented by the published reports of the Population Census of 1951, which provide a regional matrix of places of birth and residence. As of the date of that enumeration, 975,000 residents of northern Italy were born in the South, or about 3.3% of the population of the North (the comparable figure for 1931 was 2.5%). The percentage was higher for Lazio (11.1), and undoubtedly the shares for Rome and some of the provinces and industrial centers of the Northwest would presumably have been even greater. The relationship between the shares of the individual southern regions in the overall migration to the North and their proportions of the population of the Mezzogiorno is demonstrated in Table 3.

Table 4, which indicates the migration patterns within the South, clearly demonstrates the minimal internal mobility of the region as well as the tendency for such movements to take place between adjoining regions, e.g., Campania-Basilicata-Puglia and Sicily-Calabria.

Finally, there is the question of the degree to which population pressure in the rural areas of the Mezzogiorno had been relieved by migration to the larger cities of the region. Here one must recall the anti-urban legislation of the Fascist period which was still in effect during the postwar era and presumably impeded the flight of the cities. Published statistics from the 1931 and 1951 censuses provide a basis for evaluating the dimensions of rural-urban flows. Natural population increases in the southern cities were significantly lower than those in the countryside, so that greater than average growth in the urban centers presumably would have been the result of internal migration. During the intervening period, the population of all communes in the Mezzogiorno with over 50,000 inhabitants coupled with that of all provincial capitals whose size fell below this level increased by nearly 1,000,000 or double the rate of increase for the remainder of the region, 29 vs. 15%.[9] However, it must be emphasized that the larger centers still only accounted for one-fourth of the population of the South in 1951, significantly less than was true in northern Italy where nearly one-third lived in such agglomerations. It is clear that there had been a flow of population to the southern cities, but neither this movement, the secondary migration abroad, nor the northward flows had reduced the Mezzogiorno's share of the Italian population, nor had it relieved population pressures in the region. Of course, the pressures would have been even more intense had these shifts not taken place.

Between 1936 and 1951, manufacturing employment in Italy increased by roughly 8%, but the increase in the North was twice that in southern Italy.[10] Vöchting has attributed this lag to the uneven regional incidence of war damage. The destruction suffered by industry in the South was estimated to have been 35% compared to 12% in the Po Valley. This was coupled with greater damage to the limited hydroelectric facilities in the South. The resulting shortages of electrical power seriously hampered industrial recovery in that region, whereas in the North less than 10% of such facilities were damaged. In addition, according to Molinari,[11] the compensation paid by the government for war damage was, apparently in practice, far more generous in the North. Once recovery began in Italy, after 1945, reconstruction progressed at a much slower pace in the Mezzogiorno than in the northern segments of the nation. Thus by 1951 southern industry showed only a modest absolute increase in employment over prewar levels and in relative terms had fallen even further behind that in the North.

NOTES

[1] Friedrich Vöchting, *Die Italienische Sudfrage* (Berlin: Duncker and Humblot, 1951); "Considerations on Industrialization of the Mezzogiorno," *Banca Nazionale del Lavoro Quarterly Review* (September 1958), pp. 349–355.

[2] Shepard B. Clough, *The Economic History of Modern Italy*, (New York: Columbia, 1964), pp. 163, 168.

[3] Richard S. Eckhaus, "The North-South Differential in Italian Economic Development," *Journal of Economic History*, Vol. XX, No. 3 (1961), pp. 311–318.

[4] Ibid, p. 300.

[5] Ibid, pp. 315–316.

[6] Allen, 1970, p. 44.

[7] *Un Secolo di Statistiche Italiane Nord e Sud, 1861-1961*, SVIMEZ (Rome: 1961), pp. 124–126.

[8] Friedrich Vöchting, "Industrialization or Pre-Industrialization of Southern Italy," *Banca Nazionale del Lavoro Quarterly Review* (April–June 1952), pp. 67–68.

[9] ISTAT, *Popolazione Residente e Presente dei Comuni ai Censimenti dal 1861 al 1961* (Rome, 1967), various pages.

[10] Rodgers, 1960, p. 47.

[11] Alessandro Molinari, "Southern Italy," *Banca Nazionale del Lavoro Quarterly Review*, No. 8 (1949), pp. 24–47.

Chapter 2

THE SOCIOECONOMIC PATTERN ON THE EVE OF THE DEVELOPMENT EFFORT

DEMOGRAPHIC CHARACTERISTICS

By 1951, southern Italy typified the socioeconomic characteristics of most underdeveloped regions within the developed world. Some comparative indicators of the relative positions of the Two Italies and representative provinces in 1951-52 are presented in Table 5. While such statistics cover only a limited sample of the available data for that period, they are illustrative of a broader range of socioeconomic contrasts between the two regions. It is also evident that it was not an understatement to have termed the Mezzogiorno a classic area of sub-national underdevelopment.

Demographic studies of southern Italy invariably emphasize its high population densities (averaging nearly 600 inhabitants per square mile in 1951), particularly when viewed on communal levels. However, simple arithmetic population densities in the Mezzogiorno were not markedly different from those of other sections of Mediterranean Europe, and when considered on a micro-level were far lower than those of many areas of northwestern Europe. Unlike the latter regions, however, well over half of the South's active population, nearly 4,000,000 workers, was directly dependent upon agriculture, compared to one-third in northern Italy and far lower proportions in most of northern Europe. Thus the density of farm population per unit of area was relatively high for a major segment of a "developed nation." It should also be stressed that given the region's difficult physical environment and backward economic structure (in terms of inefficient land tenure and cropping practices) which were reflected in low agricultural

Table 5. Socioeconomic Indicators, by Province and Region in Italy in 1951-52

Indicator	Milan	N. Italy	Naples	Agrigento	S. Italy
1. Net annual income (thous. of lire) per inhabitant (1952).[a]	350.2	201.7	141.7	68.1	104.4
2. Percent illiterate or lacking elementary certificate (1951).[b]	12.6	20.9	43.0	53.4	48.5
3. Bank and postal savings (thous. of lire) per inhabitant (1952).[a]	23.6	13.6	8.4	3.1	4.9
4. Electrical energy consumption, illumination, KWH per inhabitant (1952).[c]	111.1	56.3	49.8	14.4	23.9
5. Radio licenses per thous. inhabitants (1952).[c]	190.9	113.0	83.6	33.3	52.1
6. Index of "motorization" per thous. inhabitants (1952).[c]	172.8	103.0	36.9	12.9	29.3
7. Proportion of active population engaged in agriculture (1951).[c]	7.0	34.9	20.4	61.4	55.6
8. Proportion of active population engaged in manufacturing (1951).[d]	53.8	28.6	26.2	8.9	13.2

[a]Tagliacarne, op. cit., 1960, pp. 82–84, 112–14.

[b]*IX Censimento Generale della Popolazione*, (1951), Vol. VII, Dati Generale Riassuntivi, Rome 1956, various pages. These percentages refer to the share of the population over 5 years of age in November of 1951.

[c]G. Tagliacarne, "Calcolo del Reditto nelle Provincie e Regione d'Italia nel 1952, "*Moneta e Credito*" Second Trimester, 1953, pp. 174–67. The index of motorization refers to personal motor vehicles calculated on the following base: motor scooter–1, motorcycle–1.7, and car–3.2. In succeeding articles covering calculations for more recent years, these equivalents were changed by Professor Tagliacareo. The exchange rate at the time varied between 620 and 625 lire to the dollar.

[d]IX Censimento, op. cit., these values are not comparable to those from the industrial census because of the methods of enumeration. The province values for agriculture for Naples and Milan are not truly meaningful because these are highly urbanized areas.

productivity, the pressure of farm population in relation to cultivated land was exceedingly high.[1] My computations, using official 1951 statistics, indicate that the value added per worker in the North was 48% above that in southern Italy, while the comparable value for net income per worker was 44% above that in the Mezzogiorno.[2]

As of 1951, the South, with less than 39% of Italy's population, accounted for half of the nation's births, and, as a result of declining mortality levels, three-fourths of the natural increase in her population. Table 6 demonstrates

Table 6. Age Composition of the Population in Northern and Southern Italy in 1951[a]

Age grouping	Northern Italy (%)		Southern Italy (%)	
	male	female	male	female
0–15	23.4	22.1	33.2	29.8
16–55	59.3	58.4	53.8	54.7
56–65	8.6	10.0	6.3	7.8
over 65	8.7	9.5	6.7	7.7
Total (%)	100.0	100.0	100.0	100.0
Absolute (thous.)	14,450	14,904	8,809	9,353

[a] IX Censimento, Dati Riassuntivi, op. cit., various pages.

the age structure of the two regions in 1951. The larger share of the Mezzogiorno's population in the younger dependent cohort is readily apparent from the statistics. If the inefficient structure of southern agriculture and the extremely limited development of industry are taken into consideration, we can readily understand the push forces that have helped to produce the continuing depopulation of the countryside. These pressures had intensified over the previous half century as the population of the region increased by 42% despite a massive exodus abroad.

THE INDUSTRIAL PATTERN IN 1951

As of the 1951 Census, less than one person in eight of the active population of the South in 1951 was employed in manufacturing, while the comparable value for northern Italy was double that level. However, even these data exaggerate the role of manufacturing in the South, for, as will be demonstrated shortly, those industries which had developed in the region were predominantly of minute scale, with limited capital investments and low productivity.

Figure 3 illustrates the distribution pattern of manufacturing employment, by province, in Italy in 1951 when the Mezzogiorno had less than 600,000 industrial workers, about one-sixth of the Italian total, compared with nearly two-fifths of the nation's population. The dominance of the North is readily apparent on this map, as is the importance of Naples within the Mezzogiorno. Figure 4 shows the communal pattern in the same year. Note the high degree of spatial concentration of the symbols. Naples, alone, had about 62,000 employees (without the commune's industrial suburbs). Yet even this value was far less than that of competing centers in the North (one-third of the

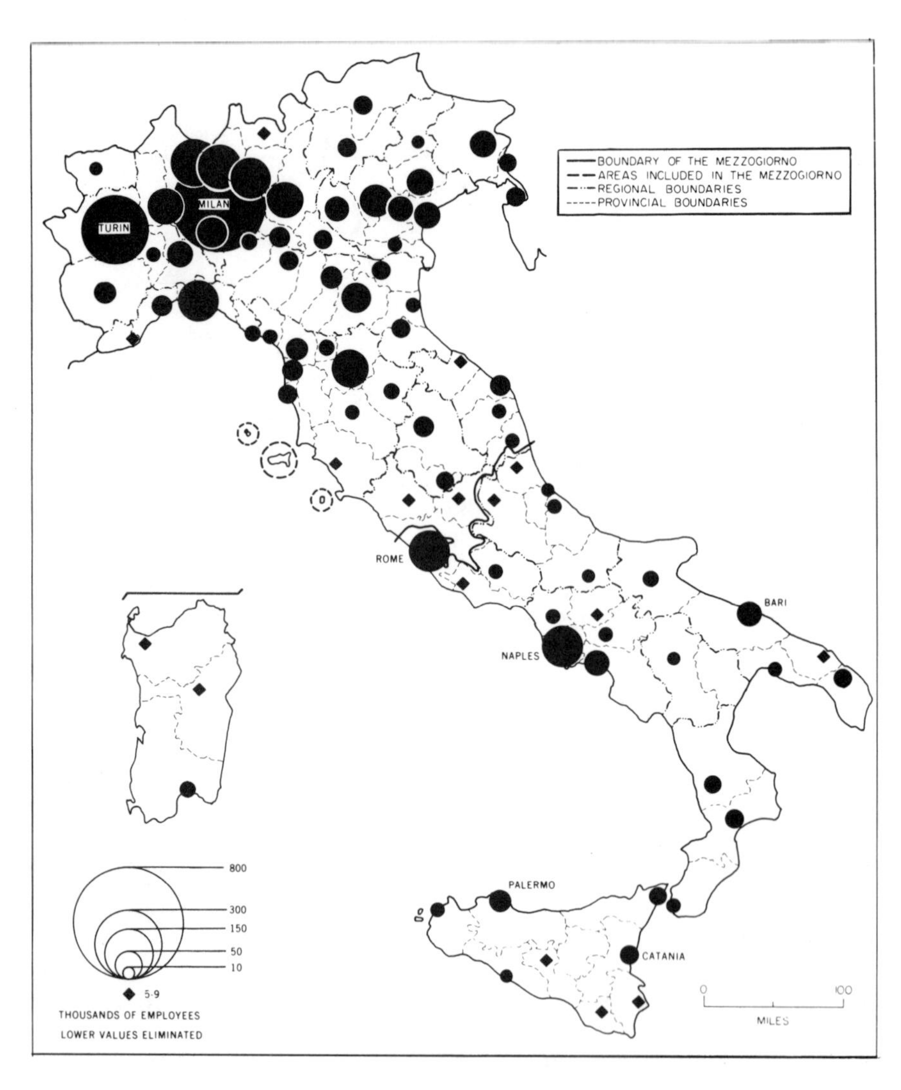

Fig. 3. Manufacturing employment, by province in 1951.

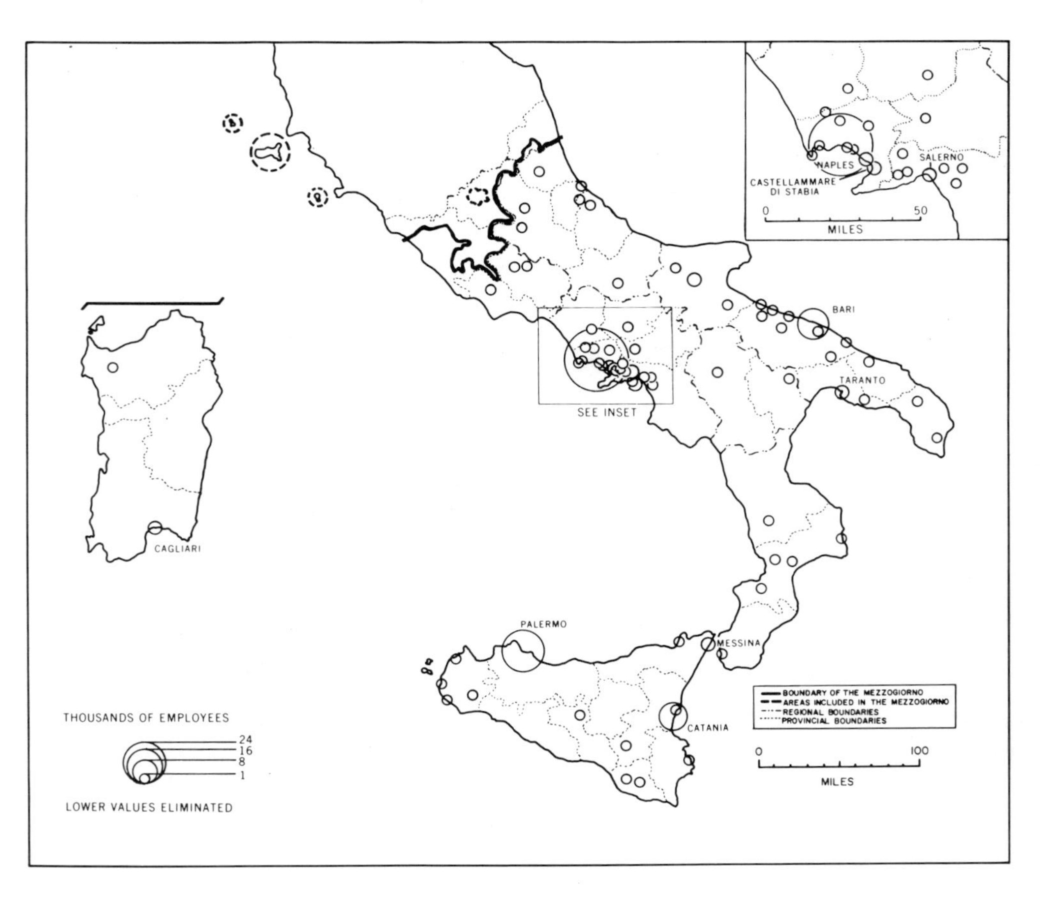

Fig. 4. Manufacturing employment, by commune, in southern Italy in 1951.

Table 7. Manufacturing Employment Characteristics, By Province and Region, In Italy, In 1951

Province and region	Proportion of working age population engaged in mfg.[a] (%)	Proportion of mfg. employment in firms[b]		Proportion of mfg. employment in firms lacking motor power[b] (%)	Proportion of mfg. employment in first stage industries[b] (%)
		below 11 employees (%)	more than 99 employees (%)		
Milan	31.3	16.3	57.0	8.3	28.0
Turin	28.3	15.3	65.7	7.6	22.0
NORTHERN ITALY	16.6	25.1	50.4	14.2	40.6
Naples	7.5	34.2	41.8	32.2	59.3
Bari	3.4	61.1	18.2	36.1	82.7
Palermo	4.7	62.2	18.6	47.7	78.9
SOUTHERN ITALY	4.6	63.9	19.2	48.7	82.9

[a] *IX Censimento Generale della Popolazione, 1951*, and *III Censimento Generale dell'Industria e del Commercio, 1951*, Vol. I, various pages.

[b] *III Censimento Generale dell'Industria e del'Commercio, 1951*, Vol. I, various pages. The working age population was defined as the resident population between 15 and 65 years of age. First stage sectors comprised food and tobacco processing, lumber products, clothing, leather and shoes, machine repair shops, printing, and the stone, clay and glass industries. The division was necessarily arbitrary. Artisan employment not included.

number of Turin and only one-fifth of the employment in Milan). Aside from Naples, the only nodes of even modest proportions in the Mezzogiorno were Bari in Puglia and Palermo and Catania in Sicily.

Table 7 illustrates several key facets of the structure of southern industry compared to those of the North. It covers such elements as the relationship of manufacturing employment to population (of working age), size of firm, motive power, and type of industry. All of these provincial and regional values clearly support the notions of the general backwardness of manufacturing in the South prior to the development effort. Also, we should add some of the well-documented handicaps of the South as of that date. These include its lack of a significant resource base; a labor force, which though numerous and expanding, had an excessively high rate of illiteracy and low technical skills; deficiencies in local enterprenurial initiative; the limited availability of investment capital; and the absence of a well-developed infrastructure.

NOTES

[1] Although agricultural densities in southern Italy might not appear to be high compared to the North (65 per km^2 vs. 61 for northern Italy), the quality of land was of key importance.

[2] *Annuario di Contabilità Nazionale*, Vol. III, 1973, Tomo II (Rome, 1973), various pages.

Chapter 3

THE INDUSTRIAL DEVELOPMENT PROGRAM

In 1950, the Italian government, despite considerable opposition from northern industrialists and politicians, instituted a program designed to lead to the solution of the economic and social problems of the South. The agency that was created for the implementation of this policy was the Cassa per il Mezzogiorno, which was under the supervision of the Committee of Ministers for the South.

Initially the plan called for a 10-year program of investment in southern Italy, but the funds, which were extremely limited, were to be channeled predominantly in nonindustrial sectors such as agriculture (77%) and infrastructure. Although the need for industrial development was recognized, the planners, working within the framework of existing development theory, called as an initial step for the pre-industrialization of the Mezzogiorno. They envisaged the South mainly as a developing market for northern industrial products. However, within a relatively short time it was realized that the provision of limited funds restricted to these branches was hardly sufficient to cure the ills of the South. The solution would only be sought through immediate and direct industrial investment in that region, and that end could only be achieved by permitting it to outdraw the attractive power of northern industrial centers. It should be recognized at this stage that no one envisaged the coming turn of events in the late '50s and '60s, i.e., the northern Italian miracle nor the massive emigration from the Mezzogiorno to northern Italy and abroad. That miracle was undoubtedly, in part, a result of the so-called "hemmorrhage" of the South.[1]

However, the creation of modern, large-scale manufacturing plants in southern Italy proved to be extremely difficult because the industrial giants of the North were clearly reluctant to invest in the region. In other words, to them "the industrial climate" was not attractive; there was also the obvious element of inertia. Thus there came the necessity of direct government expenditures to stimulate the location of new and modern private industrial establishments in the South and also the need to force government-controlled firms to invest in the region. The initial mechanism for this drastic policy shift was the passage of the 1957 legislation for the Mezzogiorno (No. 634). This, in turn, was followed in 1965 by a new law (No. 717) which resulted in major modifications of previous policies. Although there have been significant changes in the development legislation for the South since 1970, these permutations will not be discussed until the concluding section of this study. The logic for this temporal division is based upon my main focus which, as noted earlier, is an analysis of the industrial changes that occurred between 1951 and 1971 and an interpretation of their socioeconomic impact.

Perhaps the most important element of the new legislation was the shift of control over decision-making for development in the South to CIPE (The Inter-Ministerial Committee for National Economic Planning). Thus, planning for that region was now to be coordinated with that on the national level, and the Cassa became the implementation instrument for decisions taken by CIPE in consultation with the various ministries.

While not *all* of the remaining features need be discussed here (for they are amply treated in the literature), the major facets do require some elaboration, particularly their evolution and ramifications. The structure of the development program will be discussed under four headings: financial incentives; location policy; implementation of the incentive program (time, sector, and location); and direct state intervention.

FINANCIAL INCENTIVES

These stimuli, which were extended by the 1965 legislation, first called for tax exemption or reduction of a variety of levies imposed at national or local levels for industrial plants locating in the Mezzogiorno. Such charges included income and turnover taxes, excise duties on power consumption, registration fees, and local imposts. Reductions of customs duties, which had been one of the earlier incentives, were eliminated in 1965. Allen has argued that, in combination, these could conceivably amount to a saving as high as 8% of industrial income.[2]

Perhaps far more important than tax inducements has been the policy of direct and indirect stimulation through the media of grants and loans. These incentives will be discussed as of the late '60s, commenting, where appropriate, on changes from the 1957 legislation. This loan and grant program provided incentives that were highly competitive with those available in other parts of the European community. Thus it was possible to obtain loans and grants totaling as much as 85% of fixed investments, a theoretical level which was presumably rarely received by any single applicant. The loan program is the older of these two types of stimuli. It first began with the creation of three special banking institutes (ISVEIMER for the southern mainland, IRFIS for Sicily, and CIS for Sardinia). Other banks were later authorized to offer loans under the same terms. These credits were supposedly available primarily for small- and medium-sized enterprises (under 6 billion lire), but data provided by the Cassa indicate that many far larger establishments took advantage of both the loan and grant programs. These included petro-chemical works at Gela (ENI) and Priolo (Montedison) in Sicily, and Brindisi (Montedison) in Puglia as well as the huge integrated steel mill at Taranto (IRI) in the same region. Most recently, huge loans and grants were allocated for the construction of the Alfa-Sud automobile plant (IRI) at Pomigliano D'Arco near Naples.

The loans, themselves, are highly favorable because of their low interest terms and the 15-year repayment period. Interest rates vary from 3 to 6% depending upon the size of the investment, with the highest level for undertakings valued at over 12 billion lire. Loans could reach a maximum ceiling of 70%, but for amounts over that lire value the recipient could only receive 50% of fixed investment costs. An explicit arithmetic weighting scheme was utilized which was based on three key criteria (with varying levels within each category). These were size of investment (20%), sector (25%), and location (25%). With regard to size, preference was given to larger investments (the inverse of the grant priorities). The sector weights favor what are considered to be high growth potential industries such as secondary chemicals, machinery and metal products, all of which could now profit from "material" proximity. Location is the third element in the triad. Here the highest points supported the creation of plants locating in growth centers termed "areas" and "nuclei" and especially in the industrial zones or "agglomerati" of such districts. This locational facet will be treated in greater detail in a succeeding section.

Direct grants, also termed contributions, are a more recent innovation in the development program. They were approved in 1957 but were not implemented until 2 years later. This feature calls for contributions for buildings and equipment with a maximum level of 20% up to 6 billion lire coupled with an additional 10% on an additional 6 billion. Investments exceeding 12 billion lire could receive a maximum grant of 12% of their total expenditures. The criteria used for the determination of the size of such contributions are in principle similar to those used for loans: size of investment, sector, and location. They favored modern, high-growth potential, captial intensive industries. As we shall see later, the *current* legislation now supports labor-intensive plants. The only major variance between this form of assistance and the loan program, other than the funds allocated for both and the difference in percentages of support, is that, in contrast to the loan program, the grants favor smaller firms. The rationale for this divergence is that the smaller plants are less able to cope with long-term debts. There was also an obvious omission in the grant program in that it failed to aid very small craft-type establishments. This has since been corrected!

The concessions on transport costs, which were suspended in 1972-3, have apparently had only a limited impact on industrial location decisions. True, such costs are still of considerable importance in some heavy industries, but most scholars agree that they are only of secondary consequence in many modern high value-added types of manufacturing. However, the comparatively new policy on social security rebates does require some elaboration. We could have argued, at least in the initial stages of the development program, that one major locational advantage possessed by the South was its lower labor costs; but by contractual agreement, which was implemented in 1971, regional differences in hourly wage costs were theoretically eliminated. This accord clearly negated some of the attractiveness of the Mezzogiorno for new industries, particularly those in the labor-intensive branches. However, as a partial offset for this change, the government instituted a reduction of social security costs to employers in the South beginning in 1968 and lasting at least until 1980. Such social changes are important, for they constitute an important share of the labor cost burden in Italy. It was hoped that these reductions (averaging 28.5% of the base renumeration)[3] may offset the abolition of regional wage differentials. The results are still difficult to evaluate, but it would appear that this decrease does counteract the contractual changes in wage rates. To illustrate, prior to 1968, wages in the Mezzogiorno were roughly 10 to 20% below those in Milan and Turin (depending upon sector, skill and location).[4]

The final element in our analysis of these financial incentives ("agevolazioni") is the obvious question: Were they sufficiently attractive to counterbalance the inherent cost advantages, external economies, inertia, etc. of the North? This is a question which has clearly intrigued both foreign and Italian economists over the years beginning with the Ackley-Dini study in 1959.[5] Their admittedly crude, non-spatial analysis showed cost advantages for hypothetical firms in the Mezzogiorno (which took full advantage of the incentive program) ranging from 4 to 16%. They argued that the financial incentives were primarily designed to reduce capital costs, so that in those industries where such costs were a major share of total costs the attractiveness of the South was far greater, with the reverse holding true for industries which were labor or material-intensive. As we shall see later, it is precisely the capital-intensive industries which did receive an overwhelming share of the loans and grants. Allen has estimated the average incentive level for the Mezzogiorno at 11%.[6] My calculation of the differences in value added per employee in modern manufacturing was 8.4% in favor of the North in 1971. Therefore, the differential, based on these two sets of values, is not that great, particularly when such imponderables as external economies and conveniences, inertia and traditional northern attitudes toward investment in the South are taken into consideration.

Italian studies by ILSES (1968)[7] and ISVET (1973)[8] have probed this problem in far greater depth, and their recommendations appear to have influenced recent legislation for the Mezzogiorno. Both investigations treated the effects of the financial incentives in terms of *sector* and *location* (within the South), using essentially the same methodology. However, the ISVET study concentrated primarily on contrasts between the Italian incentives and those available in other segments of the EEC; therefore, my focus will be on the earlier ILSES report.

Its objective was to measure the level of advantage (or disadvantage) for a private firm locating in the South in comparison with a similar unit in the so-called industrial triangle of the Northwest. It was assumed that the firm's decision would be rational, in an economic sense; i.e., to minimize costs and maximize profits. The sectors studied covered a wide range of modern industries with differing technologies and dimensions. In all instances, the case studies included both new plants and the modernization or expansion of existing facilities. Costs were derived from industrial experience gained through a detailed interview process. These included transport costs, the availability and cost of investment capital, labor costs, and external economies, with the last often qualitative necessitating the use of non-parametric methodology. The major conclusions of the ILSES report follow:

(1) The incentives were not sufficiently strong to overcome negative elements in the industrial climate of the South.

(2) The incentive system (1968) provides no meaningful coordinated locational policy, for it was minimally differentiated from one area to another within the Mezzogiorno.

(3) The system strongly favored capital-intensive industries.

(4) In contrast, despite the program's strong dependence upon low labor costs and an automatically elastic labor supply, the incentives do *not* support labor-intensive industries.

(5) The incentives do not offset high transport costs. This is an element which is still crucial for those industries in which such costs are a major share of total costs and which serve distant markets. In particular, the system does not favor industries designed to serve foreign markets.

LOCATION POLICY

Another aspect of the development program has been the problem of industrial location. In its early stages, there was no coherent spatial design. If thought was ever given to such a locational policy, i.e., the critia for the selection of areas to be supported, it was never implemented. Given the initial focus on "pre-industrialization" with very limited credits for industry and the quasi-exclusive reliance on the market mechanism, the comparatively few plants that were sited in the region tended to gravitate to the few established nodes. Thus Naples was a primary focus with secondary growth in urban centers like Bari, Palermo, and Catania-Siracusa. In effect, then, the disparity between the few preexistent industrial regions of the Mezzogiorno and the less developed areas widened.

In the mid-fifties, however, several key government officials–Professor Pasquale Saraceno (the chief architect of the southern development effort), members of the research staff of the Cassa, and the Committee of Ministers for the South in collaboration with scholars associated with SVIMEZ (The Association for the Development of Industry in the Mezzogiorno)–began to promote the design of a location policy which would support the restriction of investment to a limited number of viable regions within southern Italy. In essence, these planners favored a growth center policy at a stage when only the United Kingdom had begun to move in that direction. However, their efforts were diluted by the influence of local politicians who lobbied successfully for a program of dispersion to aid their own constituencies. The ultimate result was an amalgam of these two philosophies in the legislation of 1957. To a degree, the growth center strategy was strengthened by the provision of Law No. 717 passed in 1965. As noted earlier, both the loan and grant incentives used criteria based upon industrial sector, size of investment, and *location*. It is that locational element which requires summarization here.

There are two facets to this policy. The first was agglomerative, for it supported the creation of "areas of industrial development." These were described as zones that were extensive in area incorporating a number of communes; they would normally have a single focal center, frequently a port. There should be a reasonably well-developed infrastructure, a large literate labor force, and the zone should already have a significant number and variety of industrial plants. These areas must prepare a regional plan analyzing their present economic situation and detailing proposals for improvement of their infrastructure and for industrial development. It was felt that with careful planning, improvement of road and port facilities, etc., these zones should act as magnets for further industrial growth. This growth would be supported by heavy financial contributions by the state, including all of the infrastructure improvements plus significant direct support for the building and equipping of plants. The "areas" must have a population of at least 200,000 and the population of the principal commune must be at least half that size.

The second type of zone is the so-called "industrial nucleus" whose population should normally not exceed 75,000. These smaller regions should have a relatively restricted number of soundly based industries either in being or in process of realization and a more limited infrastructure. The notion of industrial "nuclei" is clearly diffusive in character and has strong political overtones.

It is clear that the locational goal, at least as specified by development officials, was to secure the implantation of new plants within these "areas" and "nuclei" and particularly inside what is termed the "agglomerati" or industrial zones so as to minimize the expenditure of scarce investment resources on infrastructure improvements. The gradation of support favored developments in these more favorable industrial districts.

The idea of planned industrial areas, particularly as it applied to the "areas of industrial development," appears sound, and it was in conformance with recommendations long supported in the growth pole literature. On the other

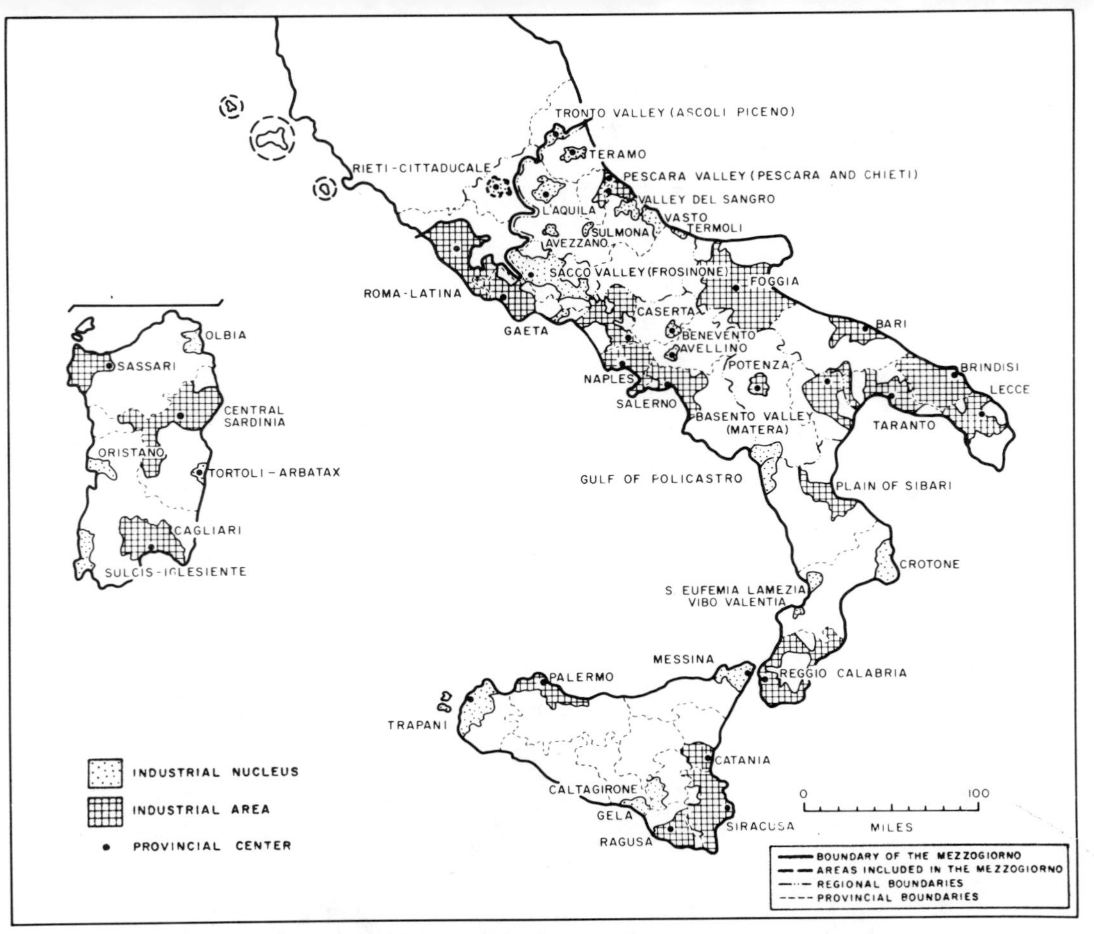

Fig. 5. Industrial development regions in the Mezzogiorno.

hand, the notion of "industrial nuclei" is subject to serious question. If support for such areas were kept to a minimum level, there might have been grounds for optimism. However, this has not proved to be the case. There are now (1978) 48 development zones, and half of these are classified as "nuclei" (see Fig. 5).[9] Their area now covers more than one-fifth of the territory of the Mezzogiorno, practically all of its level areas, and, as of 1971, nearly three-fifths of the population of southern Italy. Thus it appears that because of political pressures there has been an inordinate multiplication of these industrial zones, with the raising of the status of many "nuclei" to the level of "areas" (this is important for these zones, because they now become eligible, at least theoretically, for greater state expenditures on infrastructure). There has clearly been an impairment of the effectiveness of what was first termed a "growth center" policy. By 1975, there had been excessive delays (even for Italy) in the implementation of this location program, and comparatively few of the newer industrial zones had been provided with the necessary infrastructure nor were they fully operational. Some probably exist mainly on "paper." The delays may possibly be intentional because of funding problems, but I should also note that the various stages in the approval process are inordinately cumbersome and time-consuming.

Members of the Italian planning community have argued with me that the proliferation of areas and nuclei was a result of the political bargaining for votes in the South, and that there was never any real intention of ever providing meaningful support to many of these zones. It should also be recalled that the ILSES report (1968) found that the gradation of support on the basis of location was not a meaningful incentive for spatially differentiated investment within the South. Planners also point to the fact that in the *Plan for the Coordination of Investment, 1966–70,*[10] there was a discussion of "aree di sviluppo globale."* Five areas were so designated: the first in southern Lazio and Campania; the second in the southeast, particularly in Bari, Brindisi and Taranto (Puglia) and Matera (Basilicata); the third in the Catania-Siracusa region of eastern Sicily; the fourth around Palermo; and the last focusing on Cagliari in Sardinia. Given the still comparatively limited resources available, a policy of *concentrated investment* in areas such as these appears to be the most rational alternative open to planners of industrial development in the Mezzogiorno, particularly if the main concern is for sustained regional (in a macro-sense) and national economic growth. Unfortunately, the "equity" or "justice" theses typically receive priority in decision making in areas with marked regional income inequalities, perhaps on humanitarian grounds, but more often because of potential political returns. This dispersal of scarce resources is often wasteful, may result in excessively high production costs and, in the long run, may retard national economic growth. In southern Italy, political pressures, such as the relatively recent riots in Calabria, soaring inflation, and rapidly rising unemployment levels, probably make rational decisions unfeasible. As the Communists were successful in the

*Areas of "global" development.

recent elections, they might introduce a system of industrial licensing such as has long been practiced in the United Kingdom. These controls might prohibit the construction of new plants in the already highly congested urban areas of the North. This policy has long been advocated by proponents of southern development but unfortunately never enacted. Finally, it should be noted that the past decade has witnessed the passage of legislation to aid the "depressed areas" of the Center-North. While only limited funds so far have been directed to these regions, an expansion of such investments has long been supported by "northerners." The potential consequences of an expansion of such expenditures, given limited overall resources, are a constant concern for the "Meridionalisti" (southern advocates).

IMPLEMENTATION OF THE LOAN AND GRANT PROGRAM

Over the period between 1951 and 1974, 6 trillion lire were allocated through the loan program or about $10 billion (at the old exchange rate of 625 lire to the dollar) and 1 trillion lire as direct grants.[11] Although the loan program began in 1951 and the grants (contributions) commenced in 1959, the bulk of the funds were, in fact, disbursed within the past decade. Thus three-fourths of the loans and 94% of the grants were approved in the period following the passage of the 1965 legislation. The past few years have witnessed an even more rapid growth of these disbursements. Roughly two-thirds of the loans have been granted for the building of completely new establishments, with the remainder allocated for the modernization and expansion of existing facilities (often plants whose construction was originally financed by the incentive program). As might be anticipated, the relative share for expansion has grown significantly since 1965. On the average, loans accounted for about half of the overall anticipated fixed investment; that level is roughly concordant with the provisions, detailed earlier, of the loan program. Table 8 demonstrates the sectoral composition of the loan and grant program. These data reflect the point stressed previously that the bulk of the investments (both loans and grants) has been disbursed to capital rather than labor-intensive industries. It should also be emphasized that the aggregated data conceal the fact that a major share of the funds has gone to support a few large petro-chemical plants producing solely "basic" products, employing mainly engineers and highly qualified technicians, and to a single huge integrated iron and steel works at Taranto. None of these facilities, to date, has produced notable multiplier effects leading to the creation of linked secondary industries. The only major exception to this pattern is the funding of the Alfa-Sud automobile plant (near Naples). Although greater attention is now being paid to such industries, particularly those that involve limited investments per worker employed, the evidence even for the 1971–74 period still did not demonstrate any significant effects of this policy reversal.

Table 9 demonstrates the provincial pattern of loans and grants through 1973–74. These summary statistics are supplemented by commune data which

Table 8. Manufacturing Loans and Grants Disbursed to Firms in the Mezzogiorno, by Sector, Through 1974[a]

Sector	Loans		Grants	
	(trillions of lire)	(%)	(trillions) of lire)	(%)
Food and tobacco	468.3	7.6	136.2	12.3
Textiles	212.7	3.5	43.1	3.9
Clothing	39.1	0.6	12.2	1.1
Leather and products	25.0	0.4	7.9	0.7
Metallurgy	766.6	12.4	224.4	20.3
Machinery	577.6	9.4	182.8	16.6
Stone, clay and glass	505.7	8.2	120.0	10.9
Chemicals	2,492.3	40.4	260.0	23.6
Artificial fibres	525.6	8.5	11.0	1.0
Plastics	65.1	1.1	21.2	1.9
Rubber	99.2	1.6	12.0	1.1
Wood products	100.9	1.6	26.9	2.4
Paper	218.2	3.5	27.4	2.5
Printing	31.1	0.5	8.6	0.8
Varied	45.2	0.7	10.4	0.9
All sectors	6,172.6	100.0	1,104.1	100.0

[a]*Bilancio 1974*, Vol. II, Appendice Statistica, Cassa Per il Mezzogiorno (Rome, 1975).

are illustrated on Figures 6 and 7.[12] I should stress that the coverage on these maps is incomplete, for only those communes with values over 5000 million lire were plotted. In the case of the loan statistics, the 130 administrative units recorded accounted for 92% of the total disbursements. In contrast, the data for grants (contributions), which were plotted on a comparable scale, and covered 27 communes, comprised only about 57% of these direct subsidies. Despite the considerable locational dispersion that occurred in the previous decade, by 1974 the pattern of investments was still highly concentrated in a limited number of regions. Note that there are four dominant zones on the loan map, accounting for roughly three-fourths of the total disbursements. These include, first, the region from Latina and Frosinone on the northwestern fringe of the Mezzogiorno (to these should be added the communes south of Rome) through Caserta and Naples to Salerno; second, the provinces of Bari, Brindisi, and Taranto in Puglia and Matera in Basilicata; third, Catania and Siracusa in eastern Sicily (to which might be added Palermo in the northern part of that island); and finally, fourth, a number of comparatively new centers of industrial development in Sardinia. I should stress the high degree of correlation of the tabular and cartographic concentrations with the "aree di sviluppo globale" noted earlier.

Table 9. Manufacturing Loans and Grants Dispersed to Firms in the Mezzogiorno for Expansion of Facilities or New Plants Through 1973–74[a]

Region	Province[b]	Manufacturing loans through 12/31/73 (millions of lire)	Manufacturing loans through 12/31/73 (%)	Manufacturing grants through 9/30/74 (millions of lire)	Manufacturing grants through 9/30/74 (%)
Southern Lazio	Frosinone	230,917	3.31	47,811	4.97
	Latina	166,814	2.39	33,388	3.47
Abruzzi e Molise	Campobasso	43,774	0.63	5,833	0.61
	Chieti	74,524	1.07	20,452	2.13
	L'Aquila	60,801	0.87	8,961	0.93
	Pescara	51,554	0.74	10,618	1.10
	Teramo	33,422	0.48	13,071	1.36
Campania	Avellino	26,104	0.37	9,796	1.02
	Benevento	23,449	0.34	3,714	0.39
	Caserta	183,521	2.63	33,343	3.47
	Naples	709,112	10.16	128,567	13.37
	Salerno	199,628	2.86	38,970	4.05
Puglia	Bari	198,797	2.85	48,192	5.01
	Brindisi	182,987	2.62	28,368	2.95
	Foggia	116,371	1.67	14,303	1.49
	Lecce	25,674	0.37	9,343	0.97
	Taranto	1,060,221	15.19	165,951	17.26
Basilicata	Matera	126,238	1.81	12,851	1.34
	Potenza	53,384	0.77	10,554	1.10
Calabria	Catanzaro	287,853	4.12	7,770	0.81
	Cosenza	59,158	0.85	9,002	0.94
	Reggio Calabria	94,709	1.36	8,027	0.83
Sicily	Agrigento	32,519	0.47	4,963	0.32
	Caltanisseta	105,224	1.51	17,851	1.86
	Catania	130,019	1.86	17,626	1.83
	Enna	23,066	0.33	1,879	0.20
	Messina	106,400	1.52	13,849	1.44
	Palermo	513,799	7.36	17,857	1.86
	Ragusa	37,666	0.54	4,841	0.50
	Siracusa	420,168	6.02	26,070	2.71
	Trapani	15,606	0.22	6,472	0.67
Sardinia	Cagliari	606,895	8.70	65,560	6.82
	Nuoro	574,363	8.23	7,324	0.76
	Sassari	402,854	5.77	108,411	11.27
Southern Italy		6,977,591	99.98	961,588	100.00

[a]*Source: Cassa per il Mezzogiorno*, special tabulation, courtesy of Professor Pescatore, its President.

[b]In this tabulation certain communes of the provinces of Livorno, Rome, Rieti, and Ascoli Piceno, which are legally part of the Mezzogiorno and received loans and grants, were excluded.

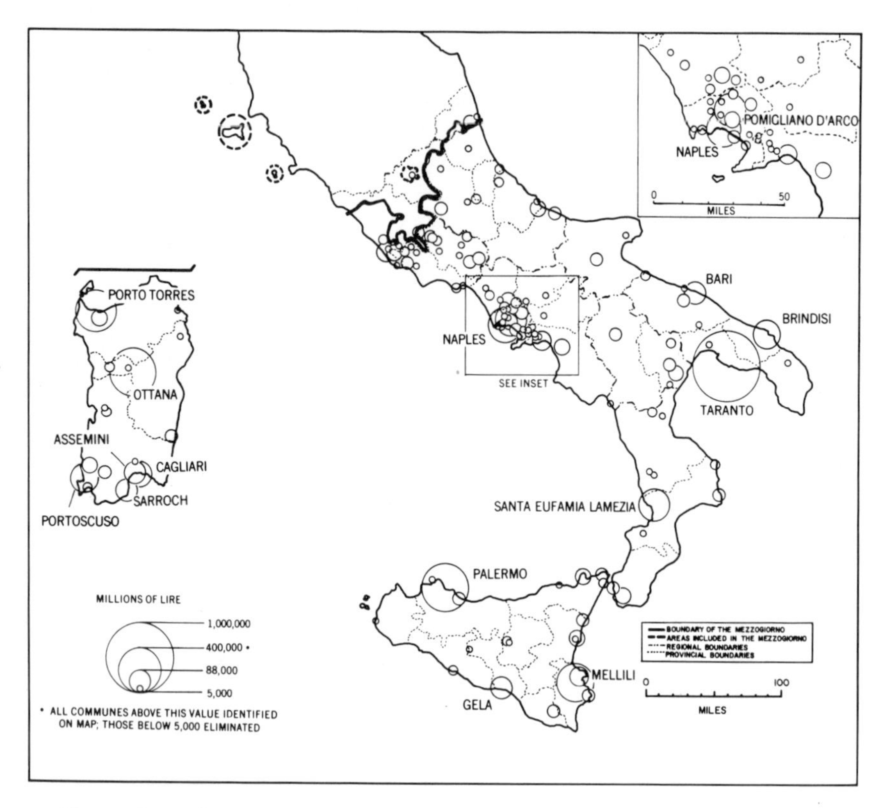

Fig. 6. Manufacturing loans, by commune, in southern Italy through 1973.

Although statistics on anticipated employment to be derived from such disbursements were also available, these were considered to be exaggerated and unreliable. Thus centers with highly capital-intensive industries, such as Brindisi in Puglia, Gela and Mellili in Sicily, and Porto Torres and Cagliari-Sarroch in Sardinia with their massive petro-chemical complexes, are all notable on these maps. Yet the numbers of workers supported by these plants is extraordinarily low considering the total investment involved. We should also add Taranto to this list of communes with capital-intensive industries. Its huge steel mill stands out strikingly on the loan and grant maps. In contrast, centers like Palermo, Naples, and Pomigliano D'Arco (with its new Alfa Sud Plant) have gained industries that in a relative sense, at least, are labor-intensive.

Given the gradation of financial support favoring investment in the areas and nuclei of industrial development, we could anticipate that these zones would have received the lion's share of the subsidies. This proved to be the

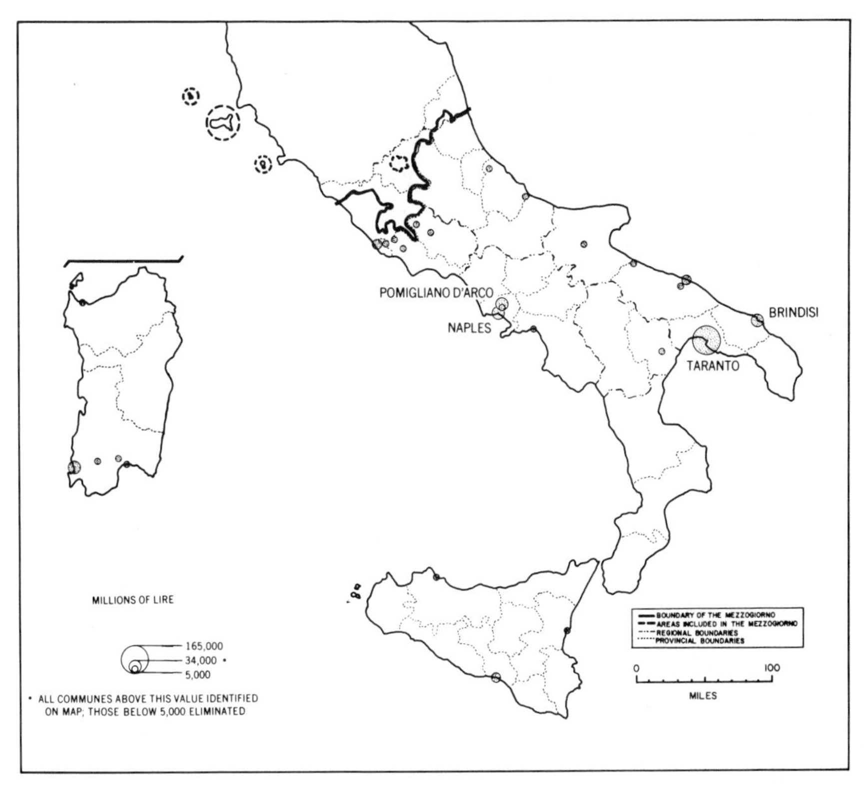

Fig. 7. Manufacturing contributions, by commune, in southern Italy through 9/30/74.

case when I summarized the values for the districts. Tabulating those communes with loans totaling over 2,000 million lire showed that at least 84% of *all* loans were disbursed to firms located in the development zones. A similar count for grants (in this instance, the lowest level was 1000 million lire) produced a value of 77%. While I should emphasize the fact that these are necessarily minimum estimates, for many communes within the areas and nuclei received loans or grants whose total was under the values listed above, these percentages clearly demonstrate the dominance of the industrial development zones in the allocation of these funds. If, however, we accept the ILSES argument, posed earlier, that the locational incentives were of minimal significance, then the only defensible explanation would link the investments to the inherent array of advantages possessed by these areas rather than to any existing locational subsidies.

DIRECT STATE INTERVENTION

Although the financial incentives were initially designed to encourage private investment in the South, they have been used by the state sector as well. I should, however, stress that this practice is by no means illegal under the terms of the development legislation. The exact share absorbed by the agencies could not be accurately estimated from the data made available to me by the Cassa because of the strictures of Italian banking laws. However, the summary data by commune coupled with first-hand knowledge of the industries that have been established in these districts make it abundantly clear that a very large proportion of the loans and grants went to state-controlled firms. It is the role of these firms in the development program that is next on our agenda.

Largely as an inheritance of the Fascist period, a major share of modern Italian industry outside of the "traditional" sector is in state hands. Currently the most important of these state-controlled firms are IRI (Istituto per la Ricostruzione Industriale) and its subsidiaries, ENI (Ente Nazionale Idrocarburi) and EFIM-BREDA. These three account for practically all of the *direct* industrial investments by the State in the South. Their main manufacturing activities are the production of iron and steel, machinery, cement, refined petroleum products and petro-chemicals. The legislation of the late '50s and '60s required that at least 60% of the new industrial investments of these firms and 40% of their total industrial investments be located in the Mezzogiorno.

According to a recent report of the Ministry of State Participation, 64% of the industrial investments of all state-controlled industrial firms in 1972 was located in the Mezzogiorno.[13] As Table 10 illustrates, the share of such investments has risen strikingly since 1957, with roughly half of all such funds concentrated in southern Italy during the 17-year period covered by these statistics. Thus the directives of the legislation of 1957 and 1965 have, at least in the industrial sector, been observed by the State-controlled firms. I should add, parenthetically, that without the investments by those state entities, the results of the industrial development program for the 1951–1971 period would have been a virtual disaster.

Table 11 illustrates the sectoral subdivisions of this investment effort for the same period. Clearly the "basic" industries accounted for the bulk of the investments of the State-controlled firms. While not all of these firms were equally profitable during this era, the State-controlled enterprises do owe responsibility to their shareholders as well as to the State, and in fact, a number of uneconomic plants have been closed over the years. We will return to the role of this group of firms in the course of a discussion of changes in the magnitude and location of manufacturing in southern Italy from 1951 to 1971.

One final point should be appended to my discussion of the role of the State in the industrial development of the Mezzogiorno; here I refer to the question of locational controls of ordinary government expenditures. The

Table 10. Industrial Investments by State Firms from 1957 to 1972 in Relation to Total Industrial Investments in that Region (Billions of Lire at Current Prices)

Year	Total ind. investments by State firms in Italy[a]	Ind. investments by State firms in the South[a]	%	Total ind. investments in the South[b]	% of total ind. investments by State firms in the South
1957	221.0	44.1	20.0	183.7	24.0
1958	189.7	56.7	29.9	189.6	29.9
1959	193.9	57.1	29.4	205.4	27.8
1960	211.0	87.1	41.3	267.9	32.5
1961	319.0	124.3	40.0	343.8	36.2
1962	501.0	215.0	42.9	481.6	44.6
1963	552.6	124.2	51.4	670.4	42.4
1964	532.8	284.1	53.3	650.8	43.7
1965	412.1	226.8	55.0	487.8	46.5
1966	340.9	166.8	48.9	466.0	35.6
1967	344.5	149.6	43.4	549.4	27.2
1968	409.7	161.9	39.5	557.9	29.0
1969	521.2	234.7	45.0	700.7	33.5
1970	871.8	457.8	52.5	1,140.7	40.1
1971	1,224.3	738.0	60.3	1,585.5	46.6
1972	1,338.9	869.0	64.9	1,852.4	46.9
1957–72	8,184.4	3,997.2	50.8	10,333.6	40.2

[a] *Annual Reports* of the Ministero delle Participazione Statale. Industrial, in this instance, includes the production of energy and the distribution of petroleum products.

[b] Conti Economici Territoriali, *Bolletino Mensile di Statistica*, Supplemento Straordinario, Istituto Centrale di Statistica, No. 9 (Rome, Settembre 1970) and *Annuario Statistico Italiano 1973*, Istituto Centrale di Statistica (Rome, 1974). South does not include southern Lazio.

legislation of the '50s included a provision requiring that government ministries and agencies place at least one-fifth of their purchase orders with southern industrial firms. My informants in Rome have indicated that their requirement was frequently evaded because the law provided no mechanism to ensure compliance. By 1965, that required level had been raised to 30% with the additional provision that it now was also applicable to State holding companies like IRI and ENI. In addition, CIPE was now made responsible for its enforcement. While no data are available to evaluate the current levels of compliance, the indirect evidence would make it appear that the quota is being enforced with greater rigor than in the past. Thus southern firms now have an assured market which is obviously of vital importance in their chance

Table 11. Industrial Investments, by Sector, of State-Controlled Firms in the Mezzogiorno from 1957 to 1972 (Billions of Lire at Current Prices)[a]

Sector	Investment	%
Manufacturing		
Metallurgy	2,019.1	48.6
Machinery and shipbuilding	522.3	12.6
Petro-chemicals	452.5	10.9
Cement	85.1	2.0
Other[b]	262.0	6.3
Energy		
Petroleum refining, transportation, and distribution[b]	547.7	13.2
Electrical and nuclear energy	270.6	6.5
Industrial investment	4,157.2	100.1

[a] *Annual Reports* of the Ministero delle Participazione Statale.

[b] The other category mainly includes the production of textiles, foods, and paper, but some secondary extractive industry employment is also in this total. Note that some employment in manufacturing is involved in petroleum refining, but this sector is primarily nonmanufacturing in character.

of success. Of course, their ability to tap that government market depends on a host of economic and political variables, often beyond the control of individual firms other than the State firms or the private "giants."

NOTES

[1] Rodgers, 1970, pp. 111–136.

[2] Allen, 1970, p. 55.

[3] Ibid, p. 98.

[4] S. K. Holland, "Regional Underdevelopment in a Developed Economy. The Italian Case," *Regional Studies*, Vol. 5 (1971), p. 82.

[5] G. Ackley and L. Dini, "Tax and Credit Aids to Industrial Development in South Italy," *Banca Nazionale del Lavoro Quarterly Review* (December 1959), pp. 340–367.

[6] Allen, 1970, p. 98.

[7] ILSES, *Effetti degli Incentivi Diretti nelle Convenienze all Insediamento delle Industrie* (Milan, 1968), 84 pp. plus tables.

[8] *Le Misure di Incentivazione a Favore delle Attivita Industriale nei Principali Paese C.E.E.: Valutazione degli Effetti sui Bilanci di Impresa*, ISVET, Vol. I, 45 pp. and Vol. II, 531 pp. (especially Vol. II, Part III) (Rome, 1973).

[9] Information provided by IASM (Istituto per L'Assistenza allo Sviluppo del Mezzogiorno) and the Cassa per Il Mezzogiorno.

[10] Comitato dei Ministri per il Mezzogiorno, *Piano di Coordinamento degli Interventi Pubblici nel Mezzogiorno* (Roma, 1968), pp. 25–27.

[11]Cassa per il Mezzogiorno, *Bilancio 1974*, Vol. II, *Appendice Statistica* (Rome, 1975), various pages.

[12]The data plotted on Figures 6 and 7 came from a special compilation provided through the courtesy of Prof. Gabriele Pescatore, then President of the Cassa per il Mezzogiorno.

[13]See Rodgers, 1976, p. 64 and the various annual reports of the Ministero delle Participazioni Statali.

Chapter 4

AGRICULTURE AND INFRASTRUCTURE PROGRAMS

AGRICULTURE

I have dwelt at some length on the problems of agriculture in southern Italy during the post-unification era. In particular, the heavy dependence of such a large fraction of its population upon the land (in some provinces that level exceeded three-quarters of the active population in 1951) has been noted. This reliance upon agriculture was exacerbated by both the physical handicaps of the region and the inefficient use of its farmland. The latter was reflected in outmoded land tenure patterns, fragmentation of land holdings, poor cropping practices, and low productivity crop and livestock systems.

As we have seen, in the initial stages of the development program the focus was on pre-industrialization with by far the highest precedence given to agriculture. Clearly, the Italian authorities envisaged a twofold solution to population pressure in the South. One priority was an overall improvement in agricultural productivity, and the second was a reduction in the size of the agricultural labor force (3,800,000 in 1951). Over time, the need to find employment outlets for displaced southern farmers took precedence. This was evidenced in the increased allocation of funds to the industrial sector and secondarily to the improvement of the transport network. It was hoped that the economic development of the urban centers of the South would provide such employment possibilities, but such opportunities did not develop at a rate or scale that was sufficient to absorb the exodus from southern agriculture. Ultimately, of course, it was the expanding job markets of northern Italy coupled with those of western Europe that provided the solution.

As stated above, one of the key priorities was the improvement of productivity in this sector which had been traditionally noted for its high levels of unemployment and underemployment. Agriculture in the South was also known for its excessive costs reflecting low output per man and per hectare. Thus it was clear that it needed to be reorganized and its operations rationalized and improved. These changes were envisaged as having two thrusts: The first was changes of direct importance to the farmers themselves, and the second was designed to stimulate agricultural productivity so as to influence the overall economy of southern Italy. It was hoped that these changes would ultimately result in an expanding market for goods and services in the agricultural regions of the Mezzogiorno. Parallel to this, of course, would be the anticipated urbanization of the South with a simultaneous expansion of markets for agricultural produce. All of these would culminate in a major growth of northern and international markets, particularly for products such as fruits, vegetables, and wine.

The planners were determined to generate a lasting transformation of southern agriculture through the mechanisms of land improvement and land reform.[1] The creation, in 1950, of the Cassa per il Mezzogiorno provided the vehicle for planning and funding these programs. Three laws were passed in the same year (Legge Sila, Legge Stralcio, and Legge Siciliana), all designed to provide the instruments for the regional implementation of these policies.

It should be recalled that a very large portion of the cultivated land of the South (55%) was located on slopes exceeding 15% and that these areas were typically without effective drainage. The cultivation of such areas had accelerated soil erosion and gullying. Then, too, precipitation was low, irregular, and rendered ineffective by evaporation and run-off. Given those physical handicaps, the land improvement program called for aqueduct and dam construction for the expansion of irrigation and flood control, the afforestation of denuded hill and mountain areas, the stabilization of slopes, and the building of drainage networks both in the hills and marshes.

It was noted earlier that there was a bi-modal land tenure system with large estates often owned by absentee landlords who cared little for increased agricultural productivity. Their holdings were paralleled by those controlled by the "borghesia" (middle-class land owners) who rented their farms on a share-crop basis to peasants on terms that were grossly unfair to their tenants. These holdings were tiny, with the great bulk (80%) less than 5 hectares in size and these properties were typically fragmented into miniscule fields which made mechanization and even irrigation difficult at best. Clearly, a 5-hectare (roughly 12 acres) farm in a fertile region could be quite prosperous, particularly if the area were irrigated. However, a similar operation in an infertile or hilly area meant a marginal existence. Thus the plans called for a land reform program which provided for expropriation (with compensation) of the large land holdings. These properties were to be parcelled out in units presumed to be adequate in size to support individual families given land quality and accessibility. The settlement pattern was to be dispersed rather than nucleated with the building of new homes, barns, wells and access roads "in the fields." In addition, service centers were to be created along with hospitals and schools (the latter to cope with the enormous illiteracy problem of these regions).

Simultaneously, the program promoted major changes in crop and livestock patterns. This meant shifts from subsistence to commercial agriculture and from extensive to intensive cultivation. In essence, this would entail a reduction of the role of cereal grains coupled with a rise in the importance of fruits, vegetables, and grapes. Concomitantly there would be a return of marginal cultivated land to pasture or forest. It also meant a transformation of livestock raising from its previous emphasis on sheep and goats to beef and dairy cattle (reared on planted hay crops and purchased feed grains). With the proposed commercialization of agriculture would come the need for rural electrification, a reliable water supply system, improved marketing facilities, refrigeration, plants for the initial processing of agricultural products, and above all a marked expansion of the transport network of the region.

TRANSPORT FACILITIES

If forced to select the single most important economic deficiency of the Mezzogiorno as of the inauguration of the development program, I would choose its primitive transportation network. The lag in the development of both rail and road system impeded the flows of people and goods, and in turn acted as a major deterrent to economic growth.

The road network was unbelievably primitive considering that the Mezzogiorno was a segment of a "developed" nation. Thus one writer described the system in this fashion: "Most of the roads date from Roman times and have changed only for the worse during the past 2,000 years."[2] While this comment is somewhat exaggerated, with few exceptions the roads did tend to be narrow, twisting, pocketed, and often impassable after convectional downpours. Of course, part of the problem was the nature of the terrain which made construction so difficult and expensive. However, of equal or greater importance was the failure of the State to invest in the development of a highway network in the South prior to the 1950s. Many areas were virtually isolated and could only be reached by dirt track traversable only by cycle, mule, or foot. Nevertheless, in a quantitative sense, the Mezzogiorno did possess 43,000 km of roads in 1951, but its densities in relation to both area and population were roughly half of those in northern Italy.[3] The main difference lay in the far larger mileage of "extracommunal" roads (outside of cities) in the North. However, these values shed no light on the quality of the road network in southern Italy, which was decidedly inferior to that in the North. A final element of the road system is the question of pattern, viewed in a geographic sense. Here a map of the network, itself, is not very instructive, for the questions of width, quality, and capacity cannot be interpreted based on the Italian designations of State, provincial and communal roads. Far better would be a map of commodity flows on these arteries; such a map was prepared by EAM for 1954.[4] It shows not only the contrasts between northern and southern Italy, cited earlier, with far heavier flows in the Center and North, but it also demonstrates a high level of concentration within the South itself. The heaviest flows were found in the northwestern coastal region (Latina, Caserta, Naples, and Salerno).

Greater than average flows were also evident on the northeast coast (Teramo, Pescara, and Chieti); in Puglia (Foggia, Bari, and Taranto); and in two Sicilian coastal strips (Palermo–Trapani and Messina–Siracusa). The lack of transverse connectivity on the mainland reinforces the notion of isolation of the various populated areas.

Turning to the rail network, here too the South, despite rail construction during the post-unification era, lagging markedly behind northern Italy. The data on kilometers of track are misleading, for the Mezzogiorno did have about 38% of the nation's mileage.[5] Thus, density ratios in relation to area and population would appear to show the absence of regional disparities. However, if we examine the nature of the rail facilities of southern Italy compared to those of the North, a new pattern emerges. Only 7% of the rail lines in the Mezzogiorno were double tracked in 1951 (compared to 37% in northern Italy). In addition, 17% of the southern rail system were electrified compared to 46% in the North. While automatic block signalling and centralized traffic control (on single track lines) were not the norm in either region, these systems which add significantly to the capabilities of a rail system were beginning to make significant headway in northern Italy. Thus, in the overview, the capacity of the southern rail network was far less than that in the North. While our study region had roughly its proportionate share of the nation's rail mileage, it accounted for only one-quarter of the commodity traffic[6] (measured in ton-kilometers).

Like the road pattern, the flows of goods on the railroads was highly concentrated in a limited number of areas. In 1954, according to statistics published by the Ferrovie dello Stato (the State-controlled rail lines),[7] the basic commodity movements on the mainland followed the Tyrrhenian and Adriatic coasts. In Sicily, as was true of road traffic, the main flows were found on the north and east coasts. Freight movements, by rail, were minimal in Sardinia (but here the railroads are largely "in concession"). Yet, as might have been expected from the existing economic structure of the South in 1951, one area was outstanding and that was the zone from Rome to Salerno (including Latina, Caserta, and Naples) in the northwest.

Finally it must be stressed that there was no meaningful unification of the transport network as a whole, despite the fact that commodity movements often involved several transfers, particularly in the hill and mountain areas of the Appenines. These commonly took the form of movements from rail to truck and even to wagon or mule. Such transfers meant a high level of loss through spoilage, damage or pilferage; and of course they meant excessively high transport costs which impeded economic development.

As indicated earlier, the improvement of the infrastructure of the South always had a high priority in the development program. In the case of the road system, the plans called for a massive investment in the building of autostradas and superstradas. This was to be coupled with a quantitative improvement of the existing roads and the building of critical linkages. In the rail sector, the funds were to be utilized for electrification, double tracking, the addition of automatic signaling equipment, the improvement of the preexisting rail system (both trackage and equipment), and the building of new lines. Thus the plans for the improvement of the transport system were extremely ambitious, but they were viewed as critical for the economic development of the South.

NOTES

[1] See Russell King, *The Questione Meridionale in Southern Italy*, Department of Geography, University of Durham, Research Paper Series, No. 11 (Durham, 1971).

[2] Gustav Schachter, *The Italian South* (New York: Random House, 1965), p. 146.

[3] Derived from Un Secolo, 1961, op. cit., p. 487.

[4] Map shown to me by EAM (Ente Autotrasporti Merci) in Rome but not available for reproduction.

[5] Derived from Un Secolo, 1961, op. cit.

[6] Roberto Cagliozzi, *L'Ammodernamento delle Ferrovie ed il Ruolo del Trasporto Ferroviario nel Mezzogiorno*, SVIMEZ (Milan-Rome: Giuffrè, 1975), p. 24.

[7] Internal (classified) statistics compiled by the Ferrovie dello Stato (the State Railway System).

Chapter 5

THE IMPACT OF THE INDUSTRIAL DEVELOPMENT EFFORT

What then have been the effects of these investments on the changing industrial employment pattern of the South? My assumption has been that, given industrial incentives of the dimensions and the attractiveness of those available in the Mezzogiorno, all companies seeking to build plants in Southern Italy or expand existing facilities would have taken advantage of such inducements. In fact, as noted earlier, a very large share of the loans were disbursed to establishments built by state holding companies like IRI, ENI, and EFIM-BREDA as well as such industrial giants as Fiat, Pirelli, Olivetti, Monte-Edison, etc. These firms can be assumed to have had access to other funding sources but presumably none so attractive as the government-supported, long-term, low-interest loans and direct contributions.

As is illustrated by Table 12, between 1951 and 1971 industrial employment in the South grew by 200,000 to a level of nearly 774,000 workers, but the growth rate during this two-decade interval was notably less than that of the North despite massive development efforts.

These data, coupled with the decline in the share of the South in the nation's overall industrial employment, would appear to support the arguments of the southerners who maintain that not enough has been done for their impoverished region, at least in the way of industrialization. The same statistics may also provide additional ammunition for those northerners who have contended that subsidy of the Mezzogiorno may have retarded overall national economic growth (an argument that is scarcely testable here). What, however, is revealed by the same table is the fact that the overall

Table 12. Changes in Manufacturing Employment in Italy from 1951 to 1971

Variable	Year	North	South	Italy
Industrial employment (in thou.)	1951	2,925	574	3,499
	1961[a]	3,844	649	4,493
	1971	4,513	774	5,287
Absolute change (in thou.)	1951–61	919	75	994
	1961–71	669	125	794
	1951–71	1,588	200	1,788
Relative change (%)	1951–61	31.4	13.1	28.4
	1961–71	17.4	19.3	17.7
	1951–71	54.3	34.9	51.1
Share of industrial employment (%)	1951	83.6	16.4	100.0
	1971	85.4	14.6	100.0

[a]Data for 1961 derived from the *IV Censimento Generale dell'Industria e del'Commercio, 1961*, Vol. 1 (Rome, 1962), various pages.

1951-1971 permutations cloak two distinct periods of change which are surprisingly disparate in their values. Thus, if we separate the study interval into the two inter-censal periods (as demonstrated by the same table), the decade from 1961 to 1971 was one in which the Mezzogiorno, on a relative basis, grew more rapidly than did its northern counterpart. Nevertheless, the South's share had declined appreciably during the overall era despite the hopes and perhaps the dreams of proponents of industrial development in the South.

On the more positive side, a comparison of the data in Table 13 with those of Table 7 in Chapter 2 provides clear evidence of a shift in the structure of southern industry. Although one could hardly term the area "a highly industrialized region," there is strong evidence to support the notion of the increasing scale and modernization of southern plants as well as increased productivity resulting from changes in entrepreneurial initiative, improved literacy levels, and certainly a growth in technical skills on the part of youthful segments of the Southern labor force. The local industrial climate is clearly changing, but whether such changes are proceeding at a sufficiently rapid pace to counteract the dramatic effects of the current Italian recession is debatable.

SECTORAL ANALYSIS

Another facet of the changing manufacturing pattern, over time, in the Mezzogiorno is the growth of the individual industries. These changes from 1951 to 1971 are demonstrated in Table 14. It is clear from this sectoral

Table 13. Manufacturing Employment Characteristics, by Province and Region in 1971

Province and region	Proportion of working age population engaged in mfg.[a] (%)	Proportion of mfg. employment in firms[b]		Proportion of mfg. employment in first stage industries[b] (%)
		below 10 employees (%)	more than 99 employees (%)	
Milan	28.8	14.3	55.4	31.4
Turin	29.4	9.5	72.4	17.1
Northern Italy	17.1	20.6	47.1	52.3
Naples	8.8	25.6	48.5	52.7
Bari	8.4	39.5	32.0	73.5
Palermo	4.8	45.2	30.3	66.8
Southern Italy	6.2	39.6	33.8	68.3

[a] *XI Censimento Generale della Popolazione, 1971,* Vol. I (Rome, 1972) various pages and *V Censimento Generale dell'Industria e del'Commercio, 1971*, Vol. I, Tomo I (Rome, 1972), various pages. Because of data problems, the working age population was defined in this instance, as the *resident* (rather than the *present*) population between 15 and 65 years of age.

[b] *V Censimento Generale dell'Industria*, op. cit., various pages. First stage sectors (as in 1951) comprised food and tobacco processing, lumber products, textiles, clothing, leather and shoes, machine repair shops, and the stone, clay and glass industries.

array that these compositional shifts are more encouraging than the overall absolute and relative change values. Thus, there were major declines in such traditional industries as foods, shoes, and wood products, while simultaneously there were notable increases in sectors like chemicals, metallurgy, machinery, and metal fabrication. These, of course, could have been anticipated from the loan and grant data, for it should be recalled that the bulk of the disbursements went to capital-intensive industries.

LOCATIONAL ANALYSIS

Table 15 demonstrates absolute and relative changes in the manufacturing employment pattern, by province, in the South from 1951 to 1971. The absolute changes are plotted on Figure 8. Judging from these data, there were clearly two major growth regions—the northwest (Latina, Frosinone, Caserta, and Naples) and the southeastern segment of the mainland (particularly Bari and Taranto in Puglia). There was also significant growth in areas such as eastern and southeastern Sicily and parts of Sardinia, but, as stressed earlier, the financial incentive program in these areas had supported capital-intensive

Table 14. Growth of Manufacturing Employment, by Sector, from 1951 to 1971[a]

Sector	Mfg. employment 1951 (thou.)	Mfg. employment 1971 (thou.)	Absolute change 1951–71 (thou.)	Relative change 1951–71 (%)
Foods and tobacco	168.7	116.2	-52.5	-31.1
Textiles	26.3	41.9	15.6	59.3
Clothing	61.3	81.7	20.4	33.3
Leather	5.0	6.5	1.5	30.0
Shoes	56.3	27.3	-29.0	-51.5
Lumber	56.6	53.2	-3.4	-6.0
Furniture	24.2	16.2	-8.0	-33.1
Stone, clay and glass	38.6	74.5	35.9	93.0
Machine repair	55.4	99.1	43.7	78.9
Metallurgy	10.1	33.8	23.7	224.7
Nonelectrical machinery	11.1	24.7	13.6	122.5
Electrical machinery	1.8	34.9	33.1	1,838.9
Transportation equipment	15.3	35.4	20.1	131.4
Fabricated metals	3.0	27.9	24.9	830.0
Chemicals	18.4	49.9	31.5	171.2
Rubber	0.8	10.1	9.3	1,162.5
Paper	8.2	11.9	3.7	45.1
Printing	7.8	12.7	4.9	62.8
Varied	4.6	16.0	11.4	247.8
Southern Italy	573.5	773.9	200.4	34.9

[a]Derived from the 1951 and 1971 Censuses of Industry.

industries, such as petrochemicals, with the result that the expansion of the labor force there was far less than could have been forecast from the loan and grant maps. It should also be emphasized that there were a number of provinces where employment had actually declined. These were located along the Appenine spine, particularly in the region from Avellino and Benevento to Reggio Calabria. Lesser declines were experienced in western Sicily.

The per capita (working age population) manufacturing employment values for 1951, 1961, and 1971 are demonstrated in Table 16. Since in this instance I was interested in internal spatial variation *within* the Mezzogiorno, the southern Italian average was used as a base level to compute those index

Table 15. Change in Manufacturing Employment by Province in the Mezzogiorno from 1951 to 1971[a]

Region	Province	Mfg. employment 1951 (thou.)	Mfg. employment 1971 (thou.)	Absolute change 1951–71 (thou.)	Relative change 1951–71 (%)
Southern Lazio	Frosinone	13.7	27.0	13.3	96.9
	Latina	6.9	31.6	24.7	213.3
Abruzzi e Molise	Campobasso	10.4	8.3	-2.1	-20.4
	Chieti	13.3	19.9	6.6	49.4
	L'Aquila	8.9	11.9	3.0	33.7
	Pescara	10.3	13.5	3.2	31.1
	Teramo	8.5	15.6	7.1	84.4
Campania	Avellino	12.1	11.5	-0.6	-5.0
	Benevento	8.8	6.7	-2.1	-23.8
	Caserta	12.9	29.0	16.1	124.2
	Naples	98.4	148.9	50.5	51.2
	Salerno	41.9	40.6	-1.3	-3.2
Puglia	Bari	41.3	70.4	29.1	70.7
	Brindisi	9.3	15.1	5.8	61.4
	Foggia	17.9	22.0	4.1	22.4
	Lecce	25.0	24.5	-0.5	-1.9
	Taranto	12.1	30.0	17.9	148.1
Basilicata	Matera	4.6	8.2	3.6	78.9
	Potenza	11.2	10.2	-1.0	-9.3
Calabria	Catanzaro	20.0	15.2	-4.8	-24.2
	Cosenza	18.5	14.5	-4.0	-21.7
	Reggio Calabria	14.5	10.3	-4.2	-29.0
Sicily	Agrigento	10.2	9.7	-0.5	-5.0
	Caltanisseta	5.6	8.8	3.2	56.6
	Catania	23.9	29.0	5.1	21.5
	Enna	3.7	4.1	0.4	10.9
	Messina	19.1	19.3	0.2	1.4
	Palermo	31.0	34.6	3.6	11.8
	Ragusa	6.7	7.4	0.7	10.3
	Siracusa	8.1	16.2	8.1	101.2
	Trapani	12.7	12.2	-0.5	-4.0
Sardinia	Cagliari	17.0	26.6	9.6	56.8
	Nuoro	5.4	5.8	0.4	8.3
	Sassari	9.6	15.2	5.6	58.0
Southern Italy		573.5	773.8	200.3	34.9
Northern Italy		2,924.7	4,512.9	1,588.2	54.3

[a]Derived from the 1951 and 1971 Censuses of Industry.

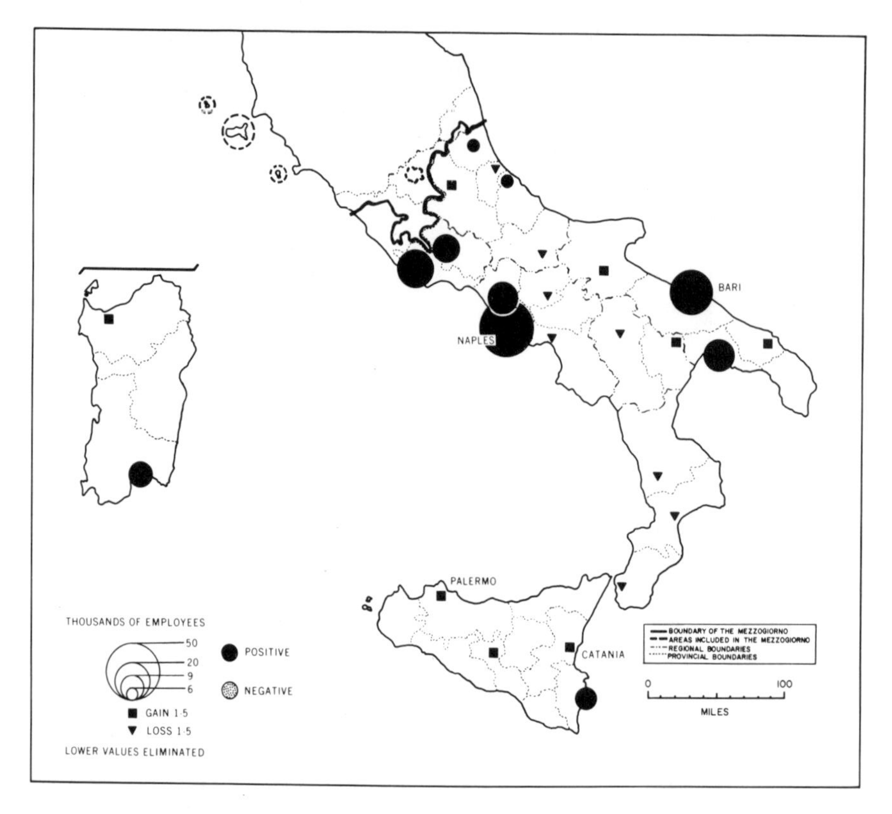

Fig. 8. Absolute change in manufacturing employment from 1951 to 1971, in southern Italy.

values. While the basic matrix resembles that found in Table 15, the weighted employment data do demonstrate patterns that were obscured by the absolute values. Note how many provinces in the extreme South had values far below 100 even as late as 1971. Only Siracusa, in eastern Sicily, exceeded that level. Another pattern emerging from this tabulation is the belt of high values along the northern tier of provinces. Their growth was clearly a phenomenon of the past two decades, as were the increases recorded in the adjoining provinces of Bari, Brindisi, Taranto, and Matera. Then too, the change in per capita levels of employment is instructive because it reflects, in part, the net outmigration from the Mezzogiorno from 1951 to 1971.

Table 17 summarizes the geographical shifts in provincial employment data by using coefficients of variation for the three years. As is evident, the changes in these coefficients are not striking, but the trend was evidently towards polarization rather than dispersion.

Table 16. Per Capita (Working Age Population) Manufacturing Employment in Southern Italy for 1951, 1961 and 1971[a]

Region	Province	Southern Italy = 100		
		1951	1961	1971
Southern Lazio	Frosinone	92	105	158
	Latina	76	135	206
Abruzzi e Molise	Campobasso	79	68	63
	Chieti	103	93	139
	L'Aquila	75	84	98
	Pescara	130	142	121
	Teramo	95	121	146
Campania	Avellino	77	71	66
	Benevento	82	73	57
	Caserta	69	78	108
	Naples	155	169	142
	Salerno	160	147	107
Puglia	Bari	112	125	134
	Brindisi	95	89	106
	Foggia	90	95	87
	Lecce	128	95	89
	Taranto	91	80	148
Basilicata	Matera	82	72	107
	Potenza	81	61	63
Calabria	Cantanzaro	94	76	57
	Cosenza	88	73	55
	Reggio Calabria	75	64	47
Sicily	Agrigento	69	61	54
	Caltanisseta	61	46	80
	Catania	97	84	79
	Enna	50	38	52
	Messina	90	85	73
	Palermo	97	100	77
	Ragusa	88	86	73
	Siracusa	80	107	112
	Trapani	95	100	74
Sardinia	Cagliari	84	82	87
	Nuoro	69	57	55
	Sassari	88	79	97
Southern Italy		100	100	100

[a]Calculated from provincial data in the 1951, 1961, and 1971 Censuses of Population and Industry. Since no "present" population data, by age group, were available for 1971, age proportions, by province, for *1961* were used for all three periods. The working age population was defined, arbitrarily, as all between the ages of 16 and 65.

Table 17. Coefficients of Variation of Manufacturing Employment, on a Provincial Level, for Southern Italy[a]

CV	1951	1961	1971
Employment share[b]	0.97	1.20	1.07
Per capita employment[c]	0.26	0.33	0.40

[a] $CV = 1/\overline{X}\sqrt{1/N \sum_{i=1}^{N} (X_i - \overline{X})^2}$ where N = number of provinces (34); X_i = value for ith province, and $\overline{X}$ = Southern Italian mean. The same formulation was used for the index numbers in the per capita data.

[b] The employment values were the *shares* of the manufacturing employment of the South, by province, for the three census years. These CVs are skewed by the disproportionate role of Naples (1951–17.7%, 1961–20.7%, and 1971–19.2%).

[c] As noted earlier, working age population proportions for 1961 were used for all 3 years.

Figure 9 demonstrates the manufacturing employment pattern, on a commune level for 1971. It should be compared with Figure 3 in Chapter 2 the map of manufacturing employment by commune in 1951. The changes during the 20-year interval are illustrated on Figure 10. Clearly the bulk of the employment growth was restricted to the few centers previously noted: Latina and Frosinone with their proximity to Rome, Caserta, Naples, Pomigliano D'Arco, and their industrial suburbs, the Bari region, Brindisi and Taranto, and finally eastern and southeastern Sicily, particularly Mellili (near Siracusa), Catania, and Gela.

SHIFT-SHARE ANALYSIS

Given the apparent differential in employment shifts in the two inter-censal periods (1951-61 and 1961-71), demonstrated on Table 12 above, and the availability of sector data by province for the three census years, I used shift-share procedures to appraise those locational changes. This technique has been widely used in the regional planning literature, although it has been subjected to considerable criticism, particularly with respect to its utility as a forecasting tool.[1] Since, in this study, I have used this device primarily as a descriptive mechanism for examining past locational changes rather than for prediction, this criticism appeared inappropriate. I also felt that Southern Italy might provide a useful laboratory for testing the utility of this method.

It should be noted first that the technique ultimately adopted follows the

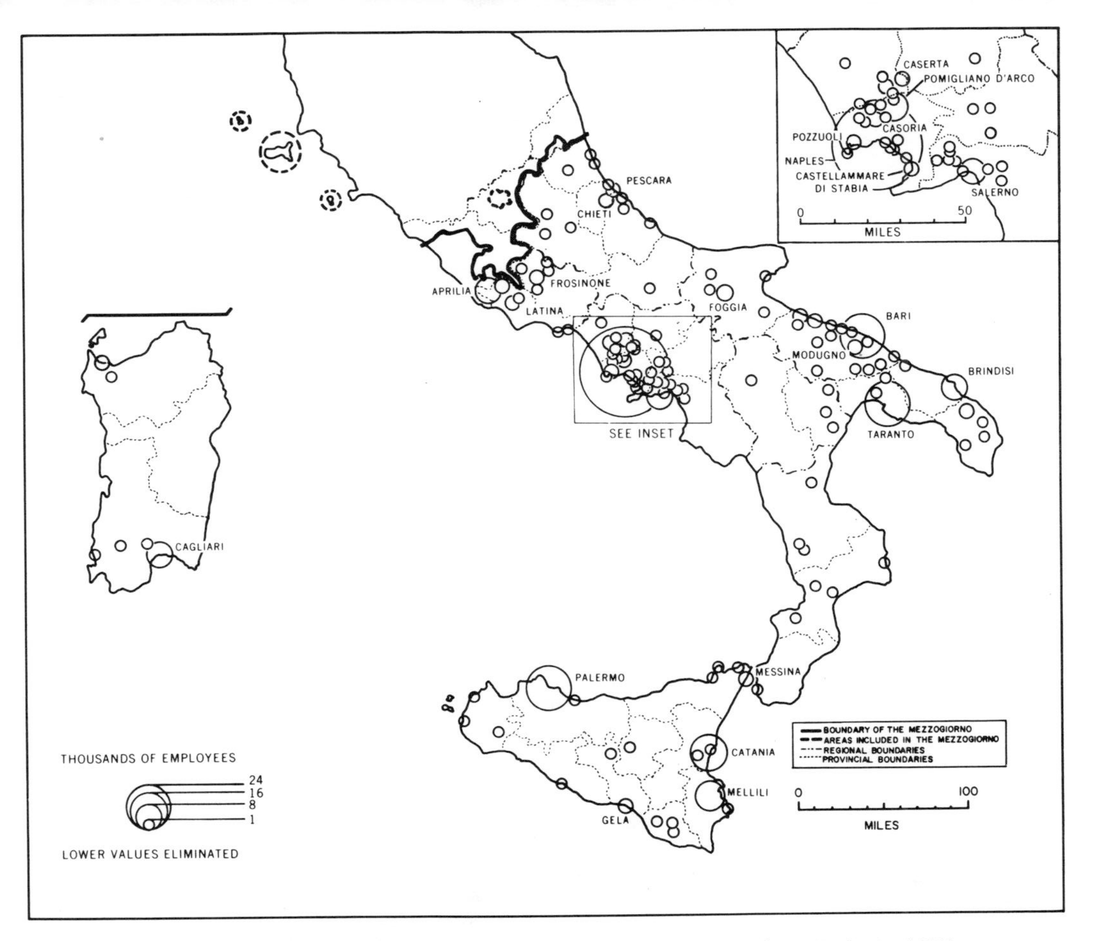

Fig. 9. Manufacturing employment, by commune, in southern Italy in 1971.

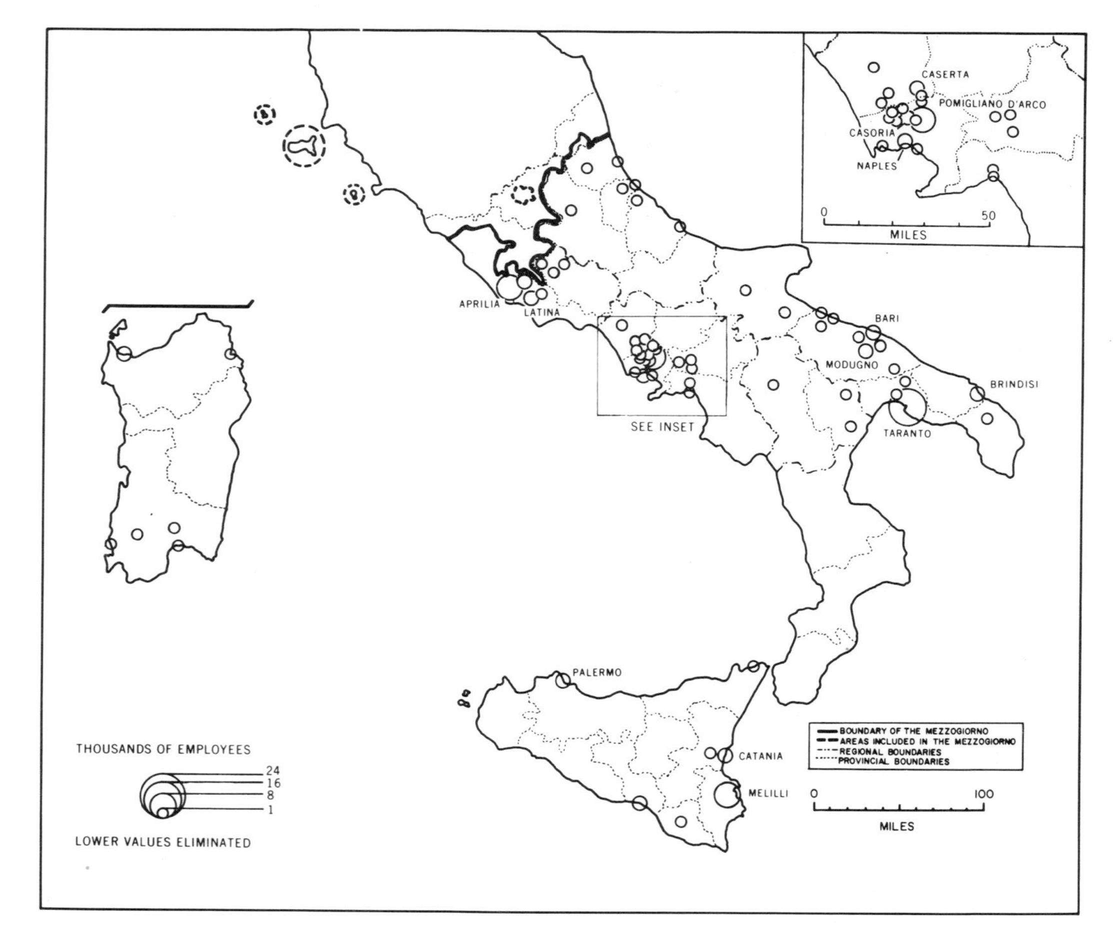

Fig. 10. Absolute changes in manufacturing employment, by commune, in the Mezzogiorno, from 1951 to 1971.

"American" approach rather than that more comonly used in the United Kingdom.[2] In its simplest form, shift-share analysis uses a national rate of industrial growth as a base level, and this rate is applied to the actual employment values in the individual regions in the base year to arrive at an "expected" value for employment in the terminal year. From this point, positive and negative shifts are calculated using actual employment levels at the end of the study period. In this instance, since my concern was with differential internal growth in the Mezzogiorno, it was decided to use the values for the South as a base level rather than those for the nation. As noted earlier, the available data (after revisions for the purposes of comparability over time) were for 19 manufacturing sectors in 34 provinces for the 3 census years. These permitted the calculation of what are commonly termed industry mix and competitive shifts.

The industrial mix effect for each sector (e.g., foods) is computed by multiplying the provincial employment in that sector by the difference between *the growth rate for that activity* and *the growth rate for all manufacturing* for Southern Italy. Mix effects arise from the fact that nationwide or, in our case in Southern Italy, some manufacturing sectors expand more rapidly than others. Therefore, those provinces that tend to specialize in slow growth industries experience net outward shifts, and those that specialize in rapid growth industries experience net inward shifts.

In contrast, the competitive component is self evident. It is calculated using *the difference between the local growth rate for that activity* and *the growth rate for that activity in Southern Italy* multiplied by the provincial employment in that sector. Negative shifts here are traceable to the failure of specific industries to keep up with the employment gains of their broad regional counterparts at the provincial level, or, in a reverse view, the local industry may have grown at a faster rate than that of the Mezzogiorno as a whole. The mix and competitive shifts are then cumulated into values which are called combined effects or–as we will use the term–net shifts. The resulting summary of my calculations is demonstrated in Table 18.

NET SHIFTS BETWEEN 1951 AND 1971

My analysis, at this point, will be confined to the net shifts (combined effects) during the study era, particularly the contrast between locational shifts in the inter-censal intervals. Figure 11 demonstrates the pattern of net shifts that occurred between 1951 and 1971. It must be emphasized again that these gains or losses are only relative values; they were computed as departures from the growth, by sector, of the Mezzogiorno treated as a single entity and as a closed system–which, of course, it is not. Note that aside from the two major growth areas that were apparent on the absolute change map, there are now a number of provinces with relatively large negative values. These include Salerno (Campania), Lecce (Puglia) and the three provinces of Calabria. Lesser negative shifts are also evident in northern Sicily.

Table 18. Shift-Share Analysis, by Component and Province, for Manufacturing Employment in Southern Italy from 1951 to 1971[a]

Year / Province	1951–61 (thou.)			1961–71 (thou.)			1951–71 (thou.)		
	Mix	Competitive	T.	Mix	Competitive	T.	Mix	Competitive	T.
Frosinone	0.3	0.4	0.7	0.2	7.4	7.6	-0.2	8.7	8.5
Latina	0.3	6.8	7.1	3.4	10.3	13.7	-0.4	22.7	22.3
Campobasso	-0.5	-2.2	-2.7	-1.1	-1.3	-2.4	-2.4	-3.3	-5.7
Chieta	-0.6	-1.9	-2.5	-1.2	6.1	4.9	-3.0	4.9	1.9
L'Aquila	-0.6	0.3	-0.3	-1.2	1.3	0.1	-2.4	2.3	-0.1
Pescara	1.3	-0.3	1.0	0.6	-2.2	-1.6	2.3	-2.7	-0.4
Teramo	-0.1	1.8	1.7	-0.9	3.0	2.1	-1.2	5.4	4.2
Avellino	-0.7	-1.2	-1.9	-2.1	-0.5	-2.6	-3.2	-1.7	-4.9
Benevento	-0.6	-1.2	-1.8	-0.4	-2.5	-2.9	-2.4	-2.7	-5.1
Caserta	-0.7	3.2	2.5	-0.2	8.7	8.5	-2.8	14.3	11.5
Naples	14.3	-0.9	13.4	29.6	-29.8	-0.2	64.9	-48.8	16.1
Salerno	-2.9	1.4	-1.5	-8.7	-5.5	-14.2	-9.6	-6.4	-16.0
Bari	1.1	5.0	6.1	-1.2	8.6	7.4	2.0	12.8	14.8
Brindisi	-0.5	0.5	0.0	-0.9	3.5	2.6	-1.2	3.7	2.5
Foggia	-0.7	1.5	0.8	-2.7	-0.5	-3.2	-3.2	1.0	-2.2
Lecce	-3.6	-2.5	-6.1	-4.5	2.7	-1.8	-10.4	1.2	-9.2
Taranto	0.1	-1.0	-0.9	0.0	14.9	14.9	1.9	11.8	13.7
Matera	-0.3	0.0	-0.3	-0.9	3.2	2.3	-1.3	3.3	2.0
Potenza	-1.0	-2.3	-3.3	-1.4	0.4	-1.0	-3.5	-1.5	-5.0
Cantanzaro	-0.8	-3.3	-4.1	-1.2	-5.5	-6.7	-3.2	-8.7	-11.9
Cosenza	-1.7	-2.1	-3.8	-2.9	-2.8	-5.7	-6.1	-4.3	-10.4
Reggio di Calabria	-0.4	-3.1	-3.5	-1.5	-3.5	-5.0	-2.8	-6.5	-9.3
Agrigento	-0.9	-0.6	-1.5	-1.4	-0.9	-2.3	-3.1	-0.9	-4.0
Caltanissetta	-0.4	-1.3	-1.7	-0.5	3.8	3.3	-1.6	2.8	1.2
Catania	0.2	-1.7	-1.5	-0.3	-1.0	-1.3	-0.1	-3.1	-3.2
Enna	-0.2	-0.9	-1.1	-0.5	1.0	0.5	-0.9	0.0	-0.9
Messina	0.3	-1.7	-1.4	-1.1	-3.6	-4.7	-0.7	-5.6	-6.3
Palermo	1.5	1.4	2.9	7.5	-18.2	-10.7	5.0	-12.2	-7.2
Ragusa	-0.6	0.5	-0.1	-1.0	-0.5	-1.5	-2.0	0.3	-1.7
Siracusa	-0.3	4.0	3.7	2.1	-1.2	0.9	-1.3	6.6	-5.3
Trapani	-0.9	1.2	0.3	-3.1	-2.3	-5.4	-3.6	-1.3	-4.9
Cagliari	0.7	0.6	1.3	0.1	2.1	2.2	0.4	3.3	3.7
Nuoro	-0.4	-0.3	-0.7	-0.9	0.3	-0.6	-1.2	-0.3	-1.5
Sassari	-0.5	-0.2	-0.7	-1.6	4.6	3.0	-2.1	4.4	2.3

[a]Data calculated from the *V* Cens. Ind., op. cit., Vol. I, various pages.

T. = combined.

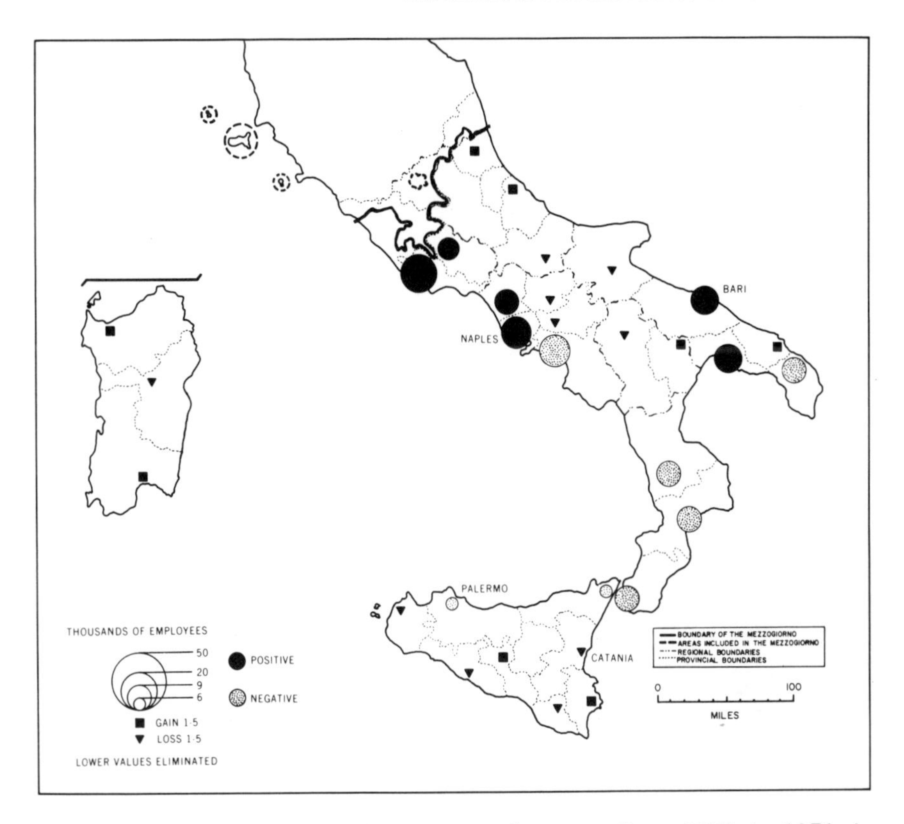

Fig. 11. Net shift in manufacturing employment, from 1951 to 1971, in southern Italy.

These variances are clearly important, but I will focus here on the shifts during the two inter-censal periods.

The 1950s have been termed "the miracle years" of the North by some Italian writers. In contrast, this period was one of limited industrial growth and heavy outmigration for the Mezzogiorno. This lag is partially attributable to the fact that the loan program only began to "take off" after 1959 and that direct grants had just begun. The consequence of these two contrasting developments was a significant *increase* in the disparity in the share of manufacturing employment between the two regions. Thus the southern proportion was roughly 16% in 1951, and that share had dropped to 14% by 1961. Figure 12 is the net shift map for this decade. Note the few areas with increases such as Naples and to a lesser degree Latina and Bari. Fifteen provinces experienced minor losses during that era. Naples had witnessed the growth of a wide variety of labor-intensive industries plus the reconstruction and modernization of a number of state owned enterprises.[3] In addition, Naples

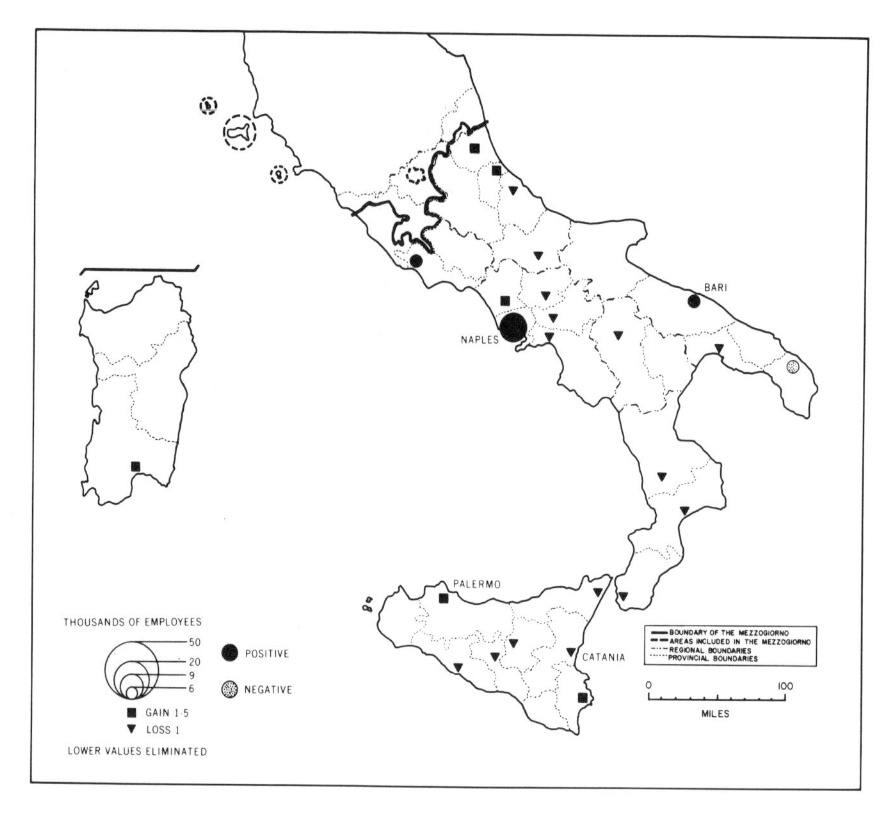

Fig. 12. Net shift in manufacturing employment, from 1951 to 1961, in southern Italy.

was the headquarters of the industrial credit institution for the Italian mainland (Isveimer); thus it was a early "innovator" in the request for government subsidized loans. Proximity to Rome had stimulated industrial developments in Latina,[4] while Bari had long possessed the self-propulsive mechanism for further industrial development.

During the following decade, the loan program finally blossomed, and new plants were created particularly by State enterprises, who had been enjoined by law, to devote 40% of their total investment capital and 60% of their new industrial investments in the South. These expenditures were complemented by those made by private companies who sometimes voluntarily and in other cases by public "suasion" had developed subsidiaries or branches in Southern Italy. It is true that many of these investments were in industries that were capital rather than labor-intensive such as the petrochemical plants cited earlier. Nevertheless, the net shift map for the 1961–71 period (Fig. 13) shows a large number of provinces with significant positive shifts in employment, including

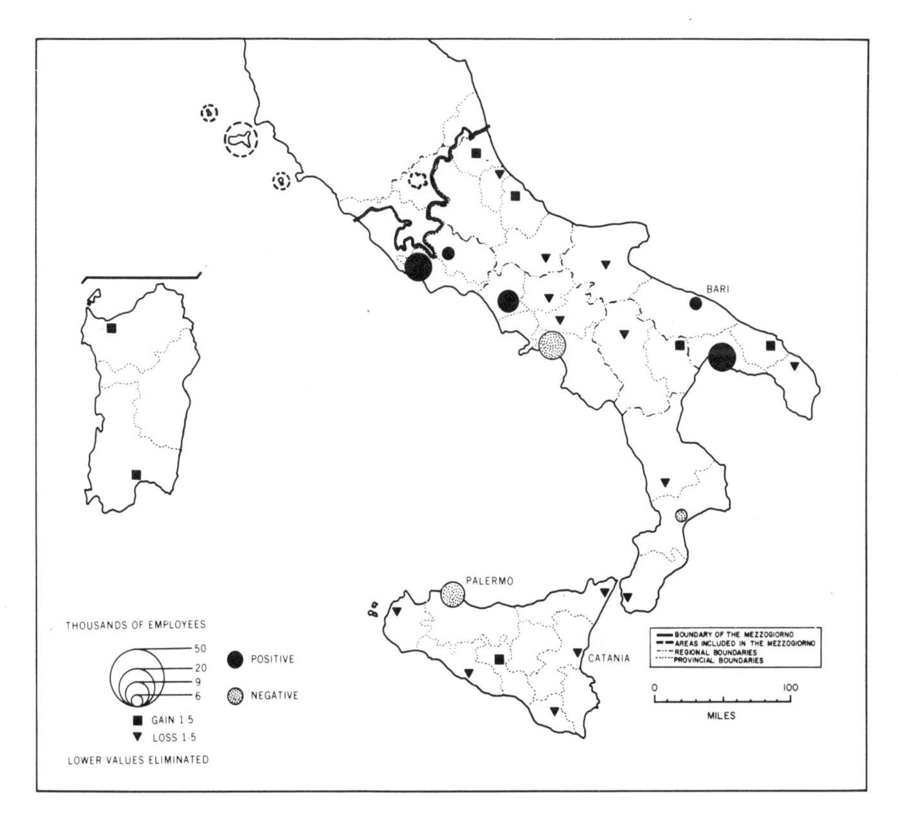

Fig. 13. Net shift in manufacturing employment, from 1961 to 1971, in southern Italy.

several new centers such as Taranto (with its huge new steel mill) and Frosinone. However, the negative values for Salerno, Palermo, and Catanzaro should also be noted. In all, half of the provinces of the Mezzogiorno showed negative shifts of over 1,000 employees. Clearly, growth during this era, when the South on a relative basis developed more rapidly than Northern Italy, was confined to a restricted number of areas. It is notable too that Naples, which had grown quite rapidly in the previous decade, actually experienced, on a relative basis, a minor loss during the 1960s.

Figure 14 is an attempt to synthesize the information depicted on both of the previous maps. The result is clearly qualitative in nature, yet instructive. It demonstrates again the fact that industrial growth in Southern Italy has been highly polarized.

With respect to the components of growth shown in Table 18 an analysis of these elements demonstrates a clear trend for the "healthy" group. The provinces that were classified as such on Figure 14 almost invariably had their

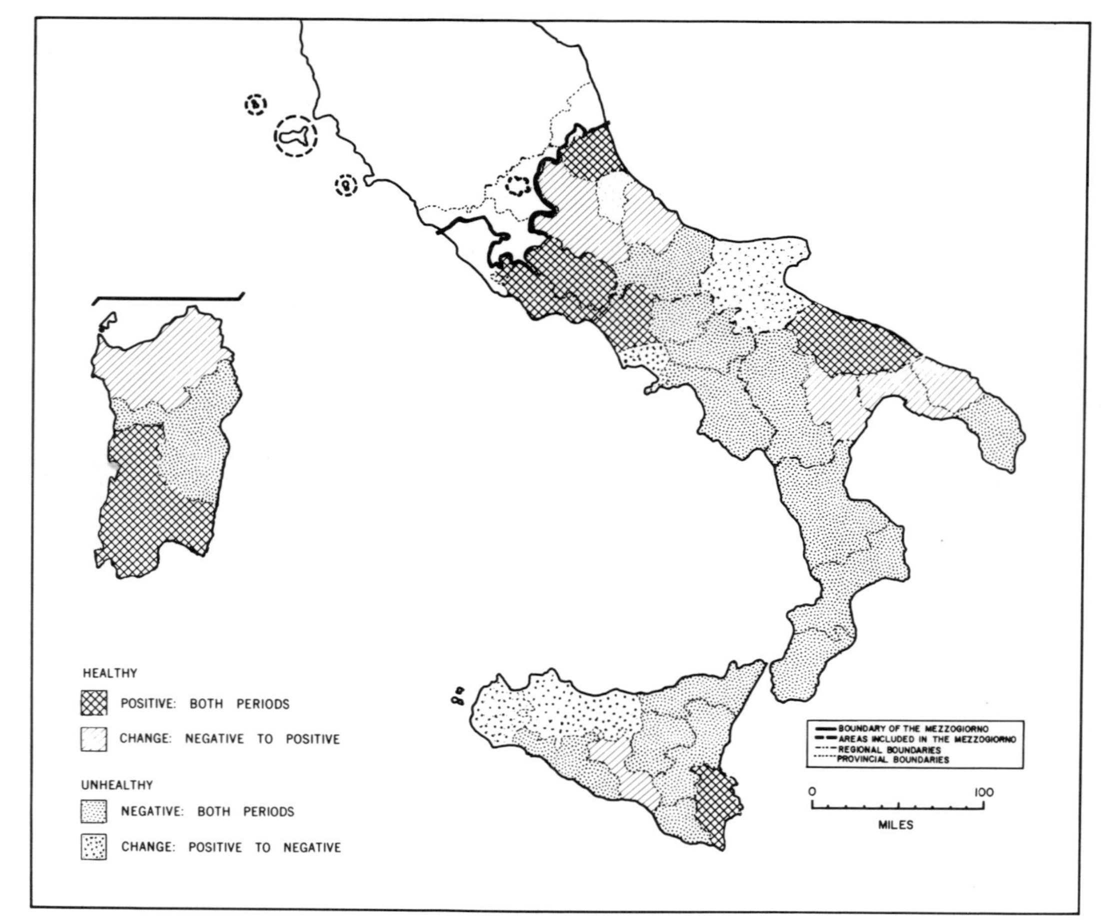

Fig. 14. Comparative shift-share analysis: 1951 to 1961 vs. 1961 to 1971 in southern Italy.

growth concentrated in the competitive shift category. In contrast, the "unhealthy" provinces were far more difficult to classify. For some, the negative values could be attributed to an undesirable mix, while in others the negative values were apparently due to the failure of local sectors to keep up with growth at the broader regional level.

A final word about shift analysis is applied to regional inequalities in industrial growth in the Mezzogiorno. It is my view, based on this experience, that the shift-share technique does permit a disaggregation of data and a comparison of provincial growth with that of the South, as a whole. In addition, it makes possible a study of "components" of change. Thus it is possible to discuss those shifts attributable to "structure" as opposed to those tied mainly to success or failure in a "competitive" sense. My studies, however, have shown that absolute and relative change values produce more plausible results when one tries to explain regional variation in industrial growth. Thus my experience with economic data for the Mezzogiorno and the USSR proved to be at variance with those of Fuchs in this study of the growth of manufacturing in the United States.[5] The evidence from my research would appear to support the notion of the weakness of shift-share analysis as a *predictive* tool. Though I believe that the method is a good mechanism for the analysis of past locational changes, it does not appear to provide any "answers." As Isard[6] noted a number of years ago, this technique, among others, is a scaffolding for further analysis and interpretation. It does not take into account changes in other major variables, i.e., regional realignments of population, income payments, etc. Shift analysis may reveal certain statistical tendencies but it does not "explain" nor identify the economic forces which interact to produce such tendencies. This method, as he has indicated, is an aid in ordering and classifying empirical data, and it may help in deciding which avenues of research are likely to be most fruitful.

THE ROLE OF STATE-CONTROLLED ENTERPRISES

It was stressed earlier that the growth of manufacturing employment in the Mezzogiorno during the period between 1951 and 1971 was disappointingly small—on the order of 200,000 workers. Yet without the direct intervention of State-controlled enterprises, even this growth might never have been achieved. Between 1953 and 1971, employment in manufacturing in State firms in the Mezzogiorno grew by almost 72,000, or roughly 36% of the increase in manufacturing employment in Southern Italy. It can be estimated that the 1951-71 value would be closer to 40%. Thus, the role of these direct state industrial investments in this underdeveloped region was of crucial importance. Admittedly, the decision on their part to "move into the South" was not solely altruistic; it was obviously the result of both political pressure and the development legislation requiring major investments by State firms in the Mezzogiorno.

With regard to political pressure, a good example was the long struggle between economists and politicians in the IRI bureaucracy on the decision to build a new integrated steel mill at Taranto. That plant has been so successful that it is currently being expanded into one of the largest in Western Europe. The decision to build an additional integrated steel mill at Gioia Taura (near Reggio Calabria) has now been cancelled despite social and political pressures. The decision in favor of building the Alfa-Sud automobile works at Pomigliano d'Arco (now in operation) is another case in point, complicated by the economic problems that have befallen this industry both in Italy and abroad. One final point should be made about the government ventures, whether they be individual undertakings or those jointly with private corporations. These State-controlled firms have it within their power to set up groups of plants simultaneously to take advantage of external economies. Such opportunities are often unavailable in the private sector. Then, too, the decision on the part of the State to open a plant in the Mezzogiorno such as Alfa-Sud has forced a private firm such as Fiat to set up competitive facilities in Southern Italy to retain a share of the "southern" market. In sum, although officials of the private sector often decry the uneconomic nature of some state plants, without their initiative the growth of modern industry in the Mezzogiorno would have been greatly retarded.

THE ROLE OF THE "GROWTH CENTERS" IN INDUSTRIAL DEVELOPMENT

Table 19, which demonstrates the changes in manufacturing employment in the development zones of Southern Italy during the study era, makes it abundantly clear that industrial growth was confined almost exclusively to these "areas" and "nuclei." In fact, their relative increase (in percent) was *greater* than the average for the North. It is also apparent that the remaining regions of the Mezzogiorno actually experienced a decline in manufacturing employment.

The growth of employment in the individual development zones is detailed in Table 20.

It was emphasized earlier in the discussion of location policy that the goal of the development planners was to secure the implantation of new establishments within the "areas" and "nuclei" and particularly inside their "agglomerati" or industrial zones, so as to minimize the expenditure of scarce investment resources on infrastructure improvements and to take advantage of external economies. However, although the gradation of support in the legislation for the South favored such areas, the ILSES and ISVET reports (cited previously) found no *significant* financial advantage accruing to firms locating in such districts compared to sites in other areas of the Mezzogiorno. It was also stressed that there had been an inordinate growth in the number of such development areas during the period since 1957, so that a meaningful "growth center" policy was negated. However, data in Table 20 provides

Table 19. Growth of Manufacturing Employment in the "Areas" and "Nuclei" of the Mezzogiorno between 1951 and 1971[a]

	Manufacturing employment			
Regions	1951 (thou.)	1971 (thou.)	Absolute increase (thou.)	Relative increase (%)
Nuclei	39.2	64.3	25.1	64.0
Areas	260.9	480.0	219.1	84.0
Areas and nuclei of the South[b]	300.1	544.3	244.2	81.4
Other regions of the South	273.4	229.5	-43.9	-16.1
Northern Italy	2,924.7	4,512.9	1,588.2	54.3

[a]Calculated from the Censuses of Industry for 1951 and 1971. The communal limits of the "areas" and "nuclei" were given in materials provided by the Committee of Ministers for the Mezzogiorno, CIPE, and SVIMEZ. These data cover only 42 of the 48 "areas" and "nuclei" either approved or in process of approval as of 1974.

[b]These "areas" and "nuclei" accounted for 52% of the manufacturing employment of the South in 1951 and 70% in 1971.

conflicting evidence. Despite the absence of true locational incentives, market forces have produced a concentration of investment in a remarkably limited number of advantaged industrial zones. Sixty percent of the increase in manufacturing employment occurred in six "areas" comprising Latina, Caserta, Naples, Bari, Brindisi, and Taranto. If we add to this group five other "areas" with increases over 5,000 the share rises to 75%. Clearly, the great majority of the employment growth took place in a very limited number of regions. Thus, by a self-selection process the State and private firms have chosen to *concentrate* rather than disperse their resources. One of our major goals in a succeeding chapter will be to ascertain the reasons for the selection of these areas.

NOTES

[1]The most commonly cited early papers are Lowell Ashby, "The Geographical Redistribution of Employment: An Examination of the Elements of Change," *Survey of Current Business*, Vol. 44 (1964), pp. 13–20, and Edgar Dunn, "A Statistical and Analytical Technique for Regional Analysis," *Papers and Proceedings of the Regional Science Association*, Vol. 6 (1960), pp. 98–112. Considerable criticism has been levied at

Table 20. Changes in Manufacturing Employment in the "Areas" (A) and Nuclei (N) of the Mezzogiorno from 1951 to 1971[a]

Region	Manufacturing employment			Region	Manufacturing employment		
	1951	1971	Absolute change		1951	1971	Absolute change
Ascoli Piceno (N)	1,821	4,187	2,366	Lamezia Terme (N)	841	1,379	538
Rieti (N)	2,824	3,786	962	Crotone (N)	2,734	3,741	1,007
Latina (A)	4,542	41,952	37,410	Policastro (N)	403	1,511	1,108
Frosinone (A)	2,113	8,367	6,264	Sibari (A)	1,232	1,676	444
Gaeta-Formia (N)	2,973	4,913	1,940	Reggio di Calabria (A)	4,802	4,694	-108
Pescara (A)	11,247	20,128	8,881	Vibo Valentia (N)	1,017	1,491	474
Avezzano (N)	2,209	2,795	586	Catania (A)	13,970	20,789	6,819
Teramo (N)	1,383	3,050	1,667	Palermo (A)	24,369	29,013	4,644
Vasto (N)	1,841	4,619	2,778	Siracusa (A)	6,716	15,342	8,626
Caserta (A)	10,743	25,666	14,923	Caltagirone (N)	1,217	902	-315
Naples (A)	95,547	145,805	50,258	Gela (N)	939	5,206	4,267
Salerno (A)	16,177	19,908	3,731	Messina (N)	10,678	13,240	2,562
Avellino (N)	1,803	3,560	1,757	Ragusa (A)	2,665	4,461	1,796
Benevento (N)	2,263	2,478	215	Trapani (A)	8,209	8,088	-121
Bari (A)	19,783	37,013	17,230	Central Sardinia (A)	2,863	3,224	361
Brindisi (A)	4,167	15,062	10,895	Cagliari (A)	7,438	13,938	6,500
Taranto (A)	8,225	25,350	17,125	Olbia (N)	1,445	2,636	1,191
Foggia (A)	4,018	7,103	3,085	Oristano (N)	747	1,354	607
Lecce (A)	13,396	14,067	671	Sassari (A)	3,698	8,557	4,859
Potenza (A)	985	3,388	2,403	Sulcis-Iglesiente (N)	1,912	2,645	733
Basento (A)	3,384	6,451	3,067	Tortoli-Arbatax (N)	147	830	683

[a]These data, derived from the Censuses of Industry for 1951 and 1971 using the communal limits from sources in Table 19, cover 42 of the 48 "areas" and "nuclei" either approved or in process of approval as of 1974.

the utility of this tool, particularly at its value in studying projected changes in the composition of employment; see David Houston, "Shift-Share Analysis of Regional Growth: A Critique," *Southern Economic Journal*, Vol. 33 (1967), pp. 577–81. This criticism has not gone unchallenged; see Lowell Ashby, "The Shift and Share Analysis, A Reply," *Southern Economic Journal*, Vol. 34 (1968), pp. 423–25, and F. J. B. Stilwell, "Further Thoughts on the Shift and Share Approach," *Regional Studies*, Vol. 4 (1970), pp. 451–58.

[2]F. J. B. Stilwell, "Regional Growth and Structural Adaptation," *Urban Studies*, Vol. 6 (1969), pp. 162–178.

[3]Rodgers, 1966, p. 26.

[4]A. Mori, "Il Limite della Zona di Intervento, della Cassa del Mezzogiorno come Fattore d'Attrazione e di Localizzazione Industriale," *Rivista Geografica Italiana*, Vol. 72 (1965), pp. 19–41, and Ernesto Mazzetti, *Il Nord del Mezzogiorno*, (Milan: 1966), 233 pp.

[5]Victor R. Fuchs, *Changes in the Location of Manufacturing in the United States Since 1929* (New Haven: Yale, 1962).

[6]Walter Isard, *Methods of Regional Analysis* (Cambridge, M.I.T., 1960), pp. 266–281.

Chapter 6

THE IMPACT OF THE AGRICULTURE AND TRANSPORT PROGRAMS

Despite the growing emphasis on industrial investment by the planning and funding agencies, shifts in other sectors are also crucial for our understanding of southern economic development. Hence changes in agriculture and transport are important not only because of their direct linkages with industrialization but also because of their more general impact upon the economic evolution of the Mezzogiorno.

AGRICULTURE

In 1951, as noted previously, there were roughly 3,800,000 workers engaged in agriculture in Southern Italy or about 57% of its active population. Two decades later, that number had declined to less than 1,800,000, and the share of that component had been reduced to less than 30% of the total.[1] Clearly, there had been a massive decline of the numbers engaged in agriculture. If we could compute the *total* population directly dependent upon agriculture, the tally would be even more impressive. Were such multitudes "unleashed" on the urban centers of the South, the resulting population pressure would have been enormous, for the rise in employment opportunities in these cities in no way matched job requirements, nor were the housing and social services of these centers capable of coping with such demands. Though we have already noted the minimal increase in industrial jobs (200,000), there remains the question whether the rise of employment in tertiary functions (including construction) might have made up the differential. Here the answer is necessarily mixed. True there was a significant increase in those engaged in these sectors (936,000), but the available evidence indicates that the bulk of such opportunities came in marginal activities, typically those with low and unreliable incomes. Then too, our data ignore the natural increase in the active agricultural population of the South arising from its relatively high birth rates and low death rates (compared with the North). The solution, as we shall see, lay mainly in outward migration to take advantage of expanding employment opportunities in northern Italy and abroad. Thus we have the main bases for the massive outflows from the Mezzogiorno from 1952 to 1972. Of course, an important share of those displaced from agriculture did move first to southern cities. Cella believes that this group comprised about two-thirds of the migrants, while the remainder went directly to the North and abroad.[2] However, he also feels that most of those migrants who initially remained in the Mezzogiorno ultimately departed. Thus he has described what, in effect, was a two-stage process of migration resulting from disappointment with the lack of alternative employment opportunities within the South itself.

What then had caused the exodus from agriculture in that region, and what changes had occurred in the structure of its agriculture? With regard to the first question, the exodus might be attributed solely to excessive population pressure on the land, particularly that common in vast areas of low environmental quality. Such pressures were clearly most intense in the isolated hilly and mountainous interior. The argument here then was that the land could not support the existing agricultural population at an equitable standard of living. However, the push and the pull forces were far more complex and such a simplistic answer does not suffice, for there was also an exodus, admittedly more limited in scope, from the plains as well. Here too, there appears to have been a two-fold movement, for a notable development during this era was the "return" from the hills, which had been refuges from brigandage and malaria, to the coastal plains. Yet these lowlands, which had received major allocations of investment funds for drainage and irrigation, ultimately experienced some out-migration because of problems arising from the land reform program.

Formica summarizes the push forces in this fashion: agricultural income levels in southern Italy were extremely low when compared with those of the North, but also when contrasted with incomes earned by those working in southern cities, and there was a great instability of agricultural employment, particularly for the dependent workers or bracciante, sometimes amounting to less than 100 days per year.[3] To all this could be added the basic deficiencies in the infrastructure of the rural areas such as: inadequate housing, primitive sanitation, insufficient rural electrification, poor services particularly the provision of schools and hospitals, and miserable transport facilities.

Thus there was indeed a flight from misery and even hunger. However, to all this could be added a general feeling of isolation from the active life of urban centers (to which they were exposed through radio and television), and a desire on the part of the peasants to seek a new way of life, drastically different from that of their parents and their forebears.

Formica, while stressing that the exodus was valuable and necessary in the broadest possible sense, argues that it was far too highly age and sex selective. It left a population which he describes as senile (really "aged") and feminine in composition. Thus we have the typical descriptions, in the literature, of many villages in the hills and mountains with either abandoned houses, or homes populated mainly by older men, women and infants. One set of statistics he cites illustrates some of the demographic changes in the composition of the agricultural population. In 1951, roughly 23% of the active labor force engaged in agriculture was under 20 years of age; that proportion had declined to 6% two decades later.[4] Thus the exodus has had both positive and negative effects. In some areas there are actually seasonal labor shortages which have resulted in the importation of North African migrant workers. I will return to the question of the rural exodus in my subsequent discussion of demographic changes in the Mezzogiorno since 1951.

Turning to the question of the changing structure of southern agriculture, I noted previously that the development programs, as implemented by the funding and planning agencies, have received far lower priority in recent years than those for industry. Nevertheless, the original plans envisaged vast improvements in the environmental conditions under which agriculture functioned in the South. Significant funds were to be spent for drainage and irrigation programs in the more favored segments of that region. The plans also called for changes in crop and livestock patterns, mechanization, increases in agricultural productivity and reform in land tenure patterns.

The program for improvements in the environmental basis for agriculture encompassed a wide range of projects generally termed "bonifica." These included irrigation, drainage of swampland, soil conservation and stabilization, and afforestation of denuded hill and mountain regions. Irrigation clearly received the highest priority. Here the goal was not only an increase in the irrigated area, but a change from traditional methods (in part from wells) to reservoirs, canals and modern pumping equipment. The expansion in the irrigated area was envisaged as a growth from about 250,000 hectares in 1951 (roughtly 3% of the agricultural land) to 600,000 hectares by 1980. While the available statistical data vary enormously, the Census of Agriculture of 1970 reported an irrigated area then of 587,000 hectares (or 7% of the agricultural land in that year).[5] Thus the goal may have been achieved a decade earlier than originally conceived. However, other sources report the irrigable area (that possessing an irrigation network) as varying from 371,000 to 400,000 hectares, but more importantly, it was argued that in the late 1960s only 35% of that area, on the average, was actually irrigated in any one year. Other sources indicate that plans are currently underway for a vast increase in the irrigated area of the Mezzogiorno, but such reports must be treated with extreme caution.[6]

Paralleling the decline in the population engaged in agriculture, there was a reduction, on a smaller scale, of the land devoted to this activity. From 1953 to 1974 that decrease was over 1,500,000 hectares or about 13% of the total farm land. This decline has been most notable in two rather diverse types of regions; first, in hilly and mountainous provinces such as L'Aquila, Benevento, and Cosenza and, second, in industrialized areas like Latina, Caserta, and Naples. In the latter cases, the reduction may, in part, be attributable to encroachment by urban land use.

More impressive, perhaps, has been the relative and absolute decline in the area devoted to food grains. It was noted earlier that the campaign for food self-sufficiency, which began in the Mussolini era, had forced grain production into marginal hilly and mountainous regions. The case of wheat, the most important grain, is particularly noteworthy of the post-war decline. Here the average decrease was nearly *one-fourth*, but because of a growth in productivity, the overall output had increased by over 17% during the same interval. The decline in grain acreage was paralleled (on a smaller level) by an increase in the area devoted to planted feed crops designed to support growing herds of beef and dairy cattle. The total number of cattle in the Mezzogiorno increased by at least one-fifth from 1953 to 1970, and by the latter date there were almost 500,000 dairy cows in the South. This quantitative increase coupled with modest improvements in feed supply resulted in a two-and-a-half-fold growth of beef and milk production in the South, exceeding the expansion in population.[7] Nevertheless, per capita consumption of these products still lagged far behind that in northern Italy.

Although the agricultural problems of the Mezzogiorno are unquestionably related to the physical handicaps of that region, to attribute the backwardness of this sector solely or mainly to environmental constraints would be grossly misleading. That arrested development also reflected land tenure practices that had evolved over the centuries. By this, I mean the long-term insecurity and low income levels experienced by both the bracciante (day laborers) and the contadini (particularly those contadini or peasants who rented their holdings on short duration leases). Other unsatisfactory elements in the land tenure pattern were the inadequate size of the majority of the farm holdings and their fragmentation.

Thus, in addition to the easing of population pressure, which clearly did take place after 1951 (but not as part of any conscious planning process) and land improvements, cited earlier, there was an obvious need for land reform. This problem has been surveyed in depth by King.[8]

The land reform program, which began in the early 1950s, arose from economic, political, and social pressures. It should be recalled that the development legislation of 1950 considered land reform to be a major priority. The program that was funded by the "Sila," "Stralcio," and "Sicily" laws of the 1950s was administered by the Ente di Riforma (under the supervision of the Ministry of Agriculture). It called for the expropriation of land from the large estates, typically owned by absentee landlords; the "preparation" of that land for intensive farming; and the necessary provision of buildings, roads, and services for the peasants who would ultimately operate these new, small dispersed farmsteads.

According to King, in all some 700,000 hectares were expropriated and the beneficiaries were 113,000 bracciante families. However, only about 70% of that land (and possibly the same share of the total number of families) was located in the Mezzogiorno.[9] Thus, in fact, I am describing a program that encompassed only about 5% of the agricultural area of southern Italy. But as King and others have stressed, despite its limited scope the trend was extraordinarily valuable, and the nature of the land reform process overshadows its actual dimension. The land that was expropriated (apparently with reasonable compensation) ranged from marginal hill and mountain country to flat plains which with drainage and irrigation could be potentially excellent for agricultural development. Where cultivation had been practiced before, the use was extensive rather than intensive in character, for the typical preexisting pattern had been a sheep-wheat combination. Low returns per hectare had been the rule. Thus before intensive farming could be undertaken in the relatively level areas, major land improvements were necessary as was the creation of a viable infrastructure. The program has been described as extraordinarily expensive, particularly when viewed in terms of funds expended either per hectare or per family, and the available evidence supports this contention.

There was, as stressed earlier, a notable shift of population from the isolated hill and mountain country to the once deserted plains, reversing a 2,000-year-old trend. This shift was most notable and most effective in southern Lazio, Campania, and Puglia and far less so in Calabria and Sicily. Environmental improvements coupled with the development of new roads and the initiation of a reasonable social infrastructure meant that these areas had the potential for the development of intensive-irrigated agriculture, focusing on vegetables, fruits (deciduous and citrus), table grapes, forage grains and grasses, and industrial crops like sugar beets. These areas with their evolving commercial agriculture could now be tied to the expanding economic foci of the South.

Lest all be considered positive in the land reform program, other than its limited scope, it should be stressed that the planners often paid far too little attention to social and psychological variables. The relative isolation fostered by the new dispersed settlement was alien to the typical southern farm family. Even with the development of cooperatives, severe social adjustment problems often ensued. Even in the economic sphere, which was the major thrust of the program, the focus, until 1966 at least, was more on "assistance" rather than "intervention." More recently there has been a move to direct grants for agriculture, paralling, on a smaller scale, the trend in industrial development. Then too, the holdings provided for the bracciante families were often too small for economic viability. There is evidence, noted previously, of out-migration even from some of the land reform areas in the coastal plains. Similarly, while marginal agricultural land in the hills that was not part of the land reform program was often, in theory at least, "abandoned," the owners in fact retained their land titles. All too often they moved to the cities and rented their fields, thus preventing the consolidation of fragmented holdings or their conversion to forest land. In addition, of course, many areas of the Mezzogiorno were completely unaffected by the land reform program and conceivably never will be, unless it is revived.

Given this background, what was the economic impact of the various agricultural programs and the rural exodus? It has already been noted that there has been a sharp decline in the sheer size of the agricultural labor force and the rural population. This, however, was diluted significantly by the age and sex selectivity of the out-migration. What has happened to the role of agriculture in the southern economy in contrast to other economic activities, and what has happened to its productivity, particularly when compared to that in the North? I am not "doggedly" insisting here on economic parity with northern Italy, but that *developed* region does provide a base level for evaluative purposes. In addition, despite all rational arguments to the contrary, the "southerner" still views that parity goal as a primary economic, social and political objective. His perception cannot be ignored.

It is clear from the data presented in Table 21 that along with the decline in agricultural labor force and the marginal rise in industrial employment, the share of agriculture, as a portion of total income, had declined significantly from 1951 to 1971 in both northern and southern Italy (despite a notable absolute growth). However, the decrease was even more striking in the North, where it duplicated trends in other developed regions of western Europe and Anglo-America. Formica also argues that the decline in the role of agriculture in northern Italy was relatively healthier, because its agriculture was more competitive, more rational, and the ties between urban and rural areas, he believes, were actually strengthened. In contrast, he postulates that the over-swollen southern cities, with their unemployed or underemployed workers, were parasitic, draining income from the countryside. In addition, in

Table 21. Net Income, by Sector, for the Two Italies, in 1951 and 1971 (at 1963 constant prices)[a]

Sector	Southern Italy		Northern Italy	
	1951 (%)	1971 (%)	1951 (%)	1971 (%)
Agriculture	29.9	20.8	16.6	8.6
Industry	18.9	28.5	32.0	44.7
Tertiary activities	32.0	36.9	37.9	38.0
Public administration	19.2	13.8	13.5	8.7
Total	100.0	100.0	100.0	100.0
In billion lire	3,730	9,605	10,731	30,348

[a]*Annuario di Contabilità Nazionale*, 1973, Tomo II, Rome, 1973, pp. 15–18. Northern Italy includes southern Lazio.

his view, southern agriculture was far less competitive than that in the North either in national or European markets.

Dell' Angelo, in a recent historical review of southern agricultural development,[10] notes that the gross agricultural production in the South grew faster between 1951 and 1974 than in northern Italy but that the production per worker was only three-quarters of that in the North and had not changed perceptably during that interval. Thus his evaluation is only marginally positive.

In sum, this brief review of the impact of the various agricultural programs has demonstrated that significant changes had occurred in southern agriculture during the study period. There had, in a relative sense, certainly been improvements both in the structure and productivity of this sector. However, the unforseen and certainly unplanned exodus from the land has been a mixed blessing. Clearly much remains to be accomplished in the resolution of agricultural problems in the Mezzogiorno.

TRANSPORT DEVELOPMENT

I noted in an earlier chapter that the single most important deficiency of the Mezzogiorno prior to the development program was its backward transport system. By this I mean deficiencies in road and rail networks as well as in its limited port capacities. Here I intend to analyze structural alterations in these facilities during the study era and changes in traffic volume and flow patterns and their implications.

The literature on the development of transport in the South is quite sparse and is mainly confined to studies of SVIMEZ (Associazione per lo Sviluppo dell' Industria nell' Mezzogiorno).[11] Regretfully, these monographs are based on data for the years preceeding 1970. They have been supplemented, where possible, by more recent statistics published by ISTAT (Istituto Centrale di Statistica) but more meaningful analyses must await the release of more recent data by ANAS (Aziende Nazionale Autonoma delle Strade) and the Ferrovie dello Stato (Ministero dei Transporti). My emphasis will be principally on road transport, the sector which has exhibited the most significant changes in recent years.

Road Transport

I noted previously that as of 1951 the southern road network whether in density related to population or area was half that of the North. The relative and absolute improvements in that network are illustrated in Table 22.

Data from the Ministry of Public Works, cited by Cecchini, show that in the early years of the program the South was allocated about half of the overall funds devoted to road construction. That value declined to 28% in 1960, but a decade later it had risen again to 47%.[12] Thus, in the overview, southern Italy received far more than its per capita share of that national investment in

Table 22. Length of Road Network, Areal and Population Density as of 1951, 1961, 1971 and 1975[a]

Southern Italy				
Road (km)	1951	1961	1971	1975[b]
Autostradas	–	66	1,135	1,751
State	10,524	15,287	19,633	20,087
Provincial	20,209	33,503	38,137	40,449
Communal[c]	12,164	5,167	37,498	36,817
Total	42,897	54,023	96,403	99,104
Density				
Area (per 100 km^2)	34.9	43.9	78.4	80.5
Population (per thou.)	2.4	2.9	5.1	5.0

Northern Italy				
Road (km)	1951	1961	1971	1975[b]
Autostradas	–	1,121	3,207	3,578
State	11,409	13,966	23,714	23,914
Provincial	21,909	44,245	55,005	58,934
Communal	95,004	79,616	108,167	105,339
Total	128,322	138,948	190,093	191,765
Density				
Area (per 100 km^2)	72.1	78.0	106.7	107.6
Population (per thou.)	4.3	4.3	5.4	5.2

[a]Cecchini, *Trasporto Stradale*, p. 115. Southern Italy does not include southern Lazio.
[b]Annuario Statistico Italiano 1976, Rome 1976, various pages.
[c]Extra urban.

this sector. The results are clearly demonstrated by data in Table 22, for the gap between North and South had narrowed drastically, at least in overall road densities. However, closer examination of these statistics reveals a lag in the building of extra-urban communal roads, although some improvement is evident there as well. Note the spectacular growth in the development of an autostrada network in the South. This is a very recent phenomenon, which coupled with the elimination of tolls on the Salerno-Reggio di Calabria and several of the Sicily autostradas is clearly designed to foster major increases in passenger and truck traffic in the Mezzogiorno.

Geographical changes in the route pattern within the region are illustrated by Figure 15. It shows only the major additions to the road network between 1951 and 1976. The development of the autostrada network was dramatic.

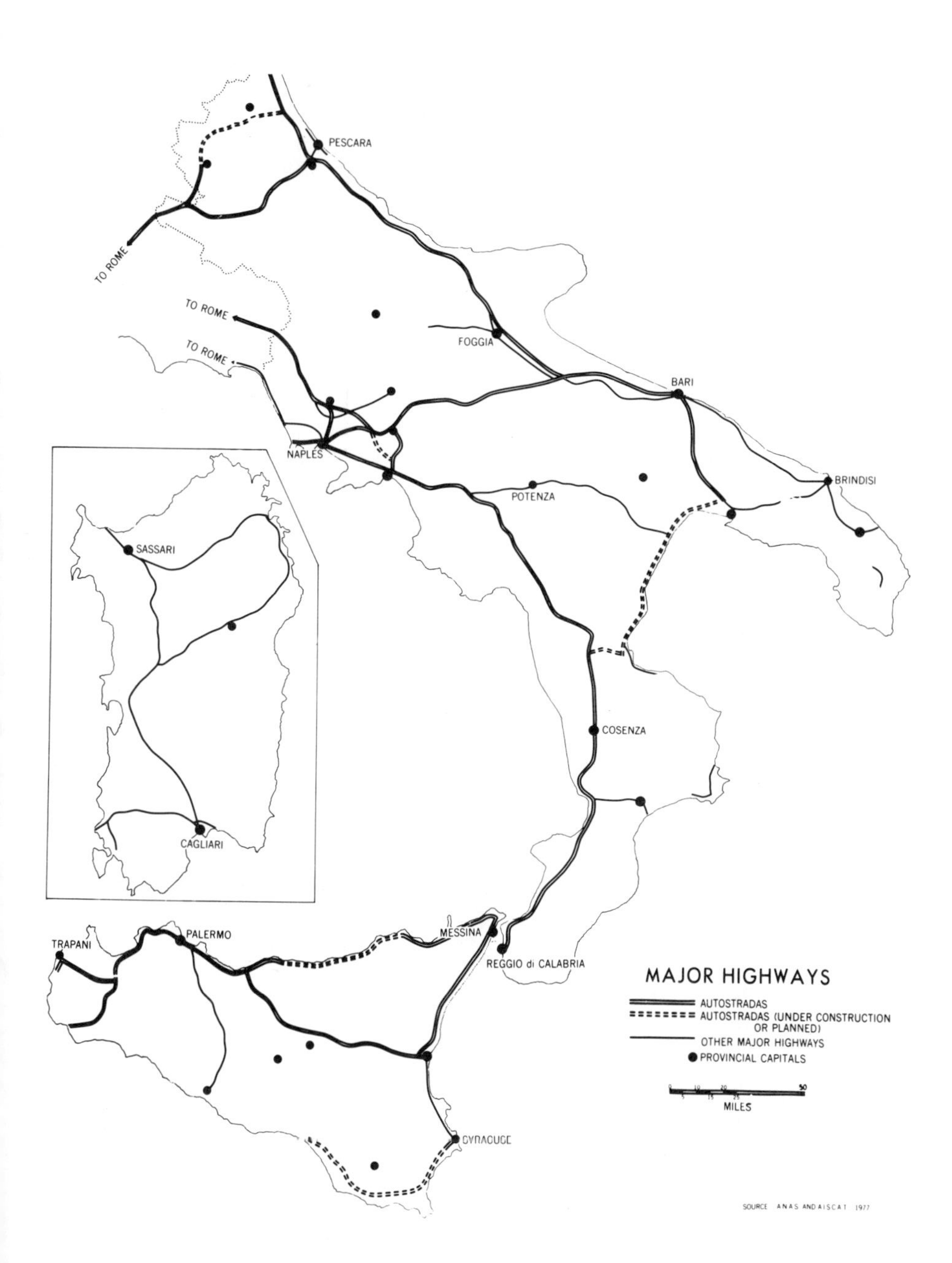

Fig. 15. Major changes in the highway pattern in southern Italy, from 1951 to 1976.

Sardinia, for reasons unknown, is the only region that has not participated in that program, although it too has witnessed the construction of a limited network of "superstradas." Note too, the new transverse links across the Appenines and through the interior highlands of Sicily.

While the map does not show the entire road system, it does demonstrate that as of 1976 some provincial capitals were still not tied to the major inter-urban network. Less evident (because of generalization) is the fact that some major roads bypass key urban centers, and these nodes must be served by communal access roads which are often indadequate for existing traffic. This has led to severe traffic congestion, particularly around major metropolitan centers. There is a pressing need for the construction of additional high capacity access roads and peripheral super highways (which would permit the bypassing of large urban centers). Such a tangential road has recently been completed around Naples.

In passing, it should also be noted that among the problems of road transportation in the South are the physical characteristics of the roads themselves. All too often, the roads have been narrow, circuitous and poorly paved. This is amply documented by Cecchini.[13] Considerable funding, during the development period, has been devoted to the upgrading of the southern road network, but much remains to be accomplished.

Turning from the road network to the vehicle stock and the traffic pattern of the South, the data gaps are, regretfully, major handicaps in appraising the transport development program given the pace of road construction in the 1970s. A detailed road census was taken in 1975, but its results have still not been released. I am indebted to Cecchini's 1975 study which utilized various road censuses taken by ANAS from 1955 through 1970. His pioneer work will be supplemented, where necessary and feasible, by more recent materials.

My discussion of the vehicle stock of the South focuses solely on the changing inventory of trucks in that region, during the 1955 to 1970 period, for commodity flows are a central concern of this chapter. Here the published data can be misleading. In 1955 southern Italy possessed 18% of the nation's trucks, and by 1970 that value had risen to 27%, for an absolute increment of over 400,000 vehicles and a four-fold relative increase. That growth rate was twice the rise in the North. The ISTAT data for 1955 and 1975 for "autocarri" or trucks show approximately the same relative change.[14] However, neither the ANAS nor the ISTAT values are adequate surrogates for changing capacity, for the increase registered in the Mezzogiorno came primarily in its inventory of light trucks. These vehicles are mainly designed for use in agriculture and for intra-urban transport. In contrast, the growth in tractor-trailers from 1955 to 1970 was less than 100,000 vehicles. Thus the share of the South changed only minimally in that crucial category, which is so reflective of a maturing industrial economy.

With respect to commodity flows, the ANAS data are more elaborate. There are two measures of such movements: ton-kilometers and vehicle-equivalent kilometers; the latter is available on a more detailed road category and regional basis. While these two measures cannot be used interchangeably,

Table 23. Changes in Truck Traffic Flow (all carriers) from 1955 to 1970[a]

Year	North	South	Southern share (%)	North	South	Southern share (%)
	(Vehicle-equiv. kms. millions)			(Ton-kms. millions)		
1955	10,451	4,178	28.6	12,578	4,570	26.4
1970	43,813	20,283	31.7	62,490	23,590	27.4
Change (%)	319.2	385.5		396.8	416.2	

[a]Adapted from Cecchini, pp. 19, 24-25, 34.

in Italy, they are relatively highly correlated as demonstrated in Table 23.

Thus truck traffic in the South grew rapidly during this era when measured by *both* indicators. In fact, it rose at a faster rate than in northern Italy with the result that the South's relative share of Italy's commodity traffic by this transport mode showed a modest improvement. However, the percentage shares as of 1970 for this region were still far below its proportion of the nation's territory and population.

Despite what was clearly a significant increase in commodity movements, Cecchini argues that, in fact, there was a negative facet of this growth which deserves summarization. He reasons that if one takes the relationship between annual increases in commodity movements, measured in ton-kilometers, and annual increases in income for the South, traffic growth outpaced the increases in net income. In particular, during the 1965–1970 period, the rate was more than twice that of "Northern" Italy (*excluding* in this instance, the Center).[15] Thus, for every additional unit of income in the Mezzogiorno there was double the "demand" for transport than was the case in the "North." In other words, the quantities of commodities transported with increments in income have increased over time in the South, and in contrast they have been significantly reduced in the more advanced segments of the nation. This argument, of course, is not new; it is a standard thesis of much of the western literature on development. Previous studies have shown that in developed countries commodity traffic tends to grow more slowly than in the less developed world. The trend in the Mezzogiorno then is another example of the same pattern.

Moving now to commodity flow patterns, here the only available detailed data are those for vehicle-kilometers based on the 1970 road census. Even though that survey covered only state roads and the autostradas, it did account for an estimated 87% of the total movements. Cecchini's data, derived from that census, show the high degree of spatial concentration of the southern network of autostradas and state roads as well as the focusing of

Table 24. Average Daily Traffic Flow on Autostradas and State Roads in Southern Italy in 1970[a]

Line of area	Commodity			Passenger	
	Traffic (% of total vehicle-kms)	Traffic (vehicle-equiv. kms/km	Index	Traffic (vehicle-equiv. kms/km	Index
Adriatic line	52.5	8,123	122	7,344	90
Naples area	37.3	6,949	105	11,653	143
Tyrrhennian line	49.5	6,046	91	6,169	76
Eastern Sicily line	30.3	5,110	77	11,737	144
Western Sicily line	40.6	5,719	86	3,357	103
Cagliari area	28.2	3,566	54	9,059	112
Sub-total	43.5	6,726	101	8,709	107
Rest of South	58.0	6,152	93	4,455	55
Southern Italy	45.0	6,647	100	8,124	100

[a]Adapted and modified from data in Cecchini, op. cit., p. 53. His statistics were derived from the Censimento Stradale of 1970 (ANAS). No traffic counts were published for provincial and communal roads.

passenger and commodity traffic in the same areas. Thus the east coast (Adriatic), the west coast (Tyrrhenian), and the Naples region together accounted for 68% of the major roads in the South in 1970 and 72% of their traffic.[16] What is more critical for my analysis is the intensity of use of the various links and the relative importance of commodity versus passenger traffic on them. These data are demonstrated in Table 24.

Note first the variation in the share of commodity traffic out of total traffic by line or area. Given a mean of 45%, higher than average values were registered for the eastern and western coastal zones and "the rest of southern Italy." In contrast, passenger movements dominated the traffic of the Naples, Sicily and Cagliari (Sardinia) networks. Turning to the index values, which measure the relative intensity of commodity and passenger traffic, it is clear that the Naples area was the only region that ranked above the average (100) in both categories. With respect to commodity flows, the Adriatic line also exceeded the mean. Aside from Naples, only eastern and western Sicily and the Cagliari region had intense passenger movements relative again to the average for the South.

While the number of vehicles *alone* is only a partial measure of traffic intensity, when computed and mapped on a detailed geographical basis it is useful in analyzing flow patterns. Figure 16 is a schematic representation of the 1970 commodity movements by line for the South based on the ANAS

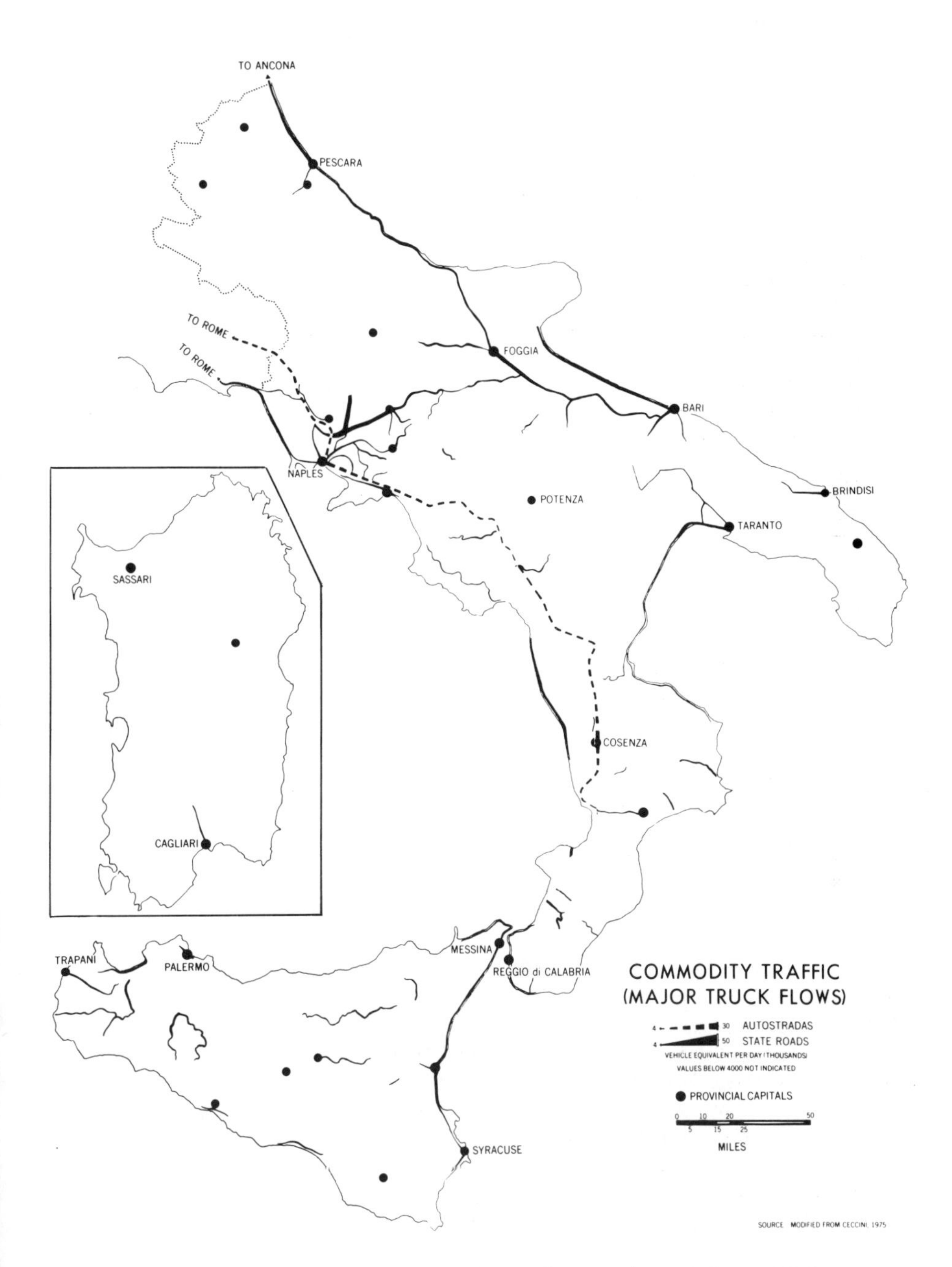

Fig. 16. Commodity movements, by road, in southern Italy, in 1970.

Table 25. Truck Traffic on Italian Autostradas during the April–June Quarter of 1976[a]

Route	Length (km)	Vehicle-equiv. (daily in thou.)	Vehicle-km in millions	Vehicle-km per km in thou.	Index
Rome-L'Aquila	115.1	0.7	3.8	33.0	8
Torano-Pescara incl. Torano-Avezzano	17.5	0.2	0.2	11.4	3
Pescara-Torre de 'Passeri	27.3	0.3	0.6	22.0	5
Rome-Naples	202.0	15.7	120.3	595.5	146
Caserta-Nola-Salerno	55.3	2.7	7.4	133.8	33
Naples Peripheral	20.2	4.9	4.9	242.6	59
Naples-Canosa	172.3	4.3	23.6	137.0	34
Pescara-Canosa incl. Pescara-Lanciano	49.7	4.5	11.5	231.4	57
Lanciano-Canosa	189.6	4.2	25.6	135.0	33
Naples-Salerno	51.6	9.3	22.2	430.2	105
Messina-Catania	76.8	3.6	20.1	261.7	64
Messina-Palermo incl. Messina-Patti	68.3	3.1	6.4	93.7	23
Buonfornello-Cefalu	12.4	0.1	0.2	16.1	4
Southern Italian autostradas	1,058.1	53.6	246.8	233.2	57
Northern Italian autostradas	3,517.4	291.2	1,622,2	461.2	133
Italian autostradas	4,575.5	344.8	1,869.0	408.5	100

[a] *AISCAT Informazione*, Vol. XI, Rome, April–June 1976, pp. 9–13. The autostradas were built by private contractors with government funds.

samples compiled by Cecchini. I have set as my lower limit an arbitrary value of 4,000 vehicles per day. Thus the map deliberately shows a series of disconnected flow lines with the missing segments falling below that minimum level. In essence, it portrays cartographically the key movements summarized in Table 24. Note the heavy traffic in the Naples region and the considerable movements on or near the Tyrrhenian and the Adriatic Sea as well as the lesser flows in eastern and northwestern Sicily. Only a single east-west flow appears across the Appenines and that is the road connecting Naples with Bari. This deficiency has since been corrected by the construction of autostradas from Rome to the Adriatic as well as the key Naples-Canosa-Bari autostrada.

The statistics used in the preparation of this map and the accompanying table are by now somewhat dated. Reviewing the record of road construction since 1970 (Table 22), it appears that if we confine our attention to the major highways (i.e., state roads and autostradas) the mileage increase in state roads has been minimal, even though there has been considerable upgrading of their capacity by widening, straightening and resurfacing. Given, however, the rapid pace of autostrada construction in the South during the 1970s and the availability of traffic data for these routes, an assessment is in order; I should note here that such statistics are only available for the roads built by private contractors and charging tolls. Thirty percent of the super-highway mileage built in the South was constructed directly by the State (IRI) and is toll free (presumably to stimulate transport and economic development). The traffic of these roads does not appear in published materials. Such routes include the Salerno-Reggio di Calabria, Catania-Palermo, Palermo-Punta Raisi-Mazara del Vallo autostradas for a total of 762 km. Data for these roads must await that publication of the 1975 road census.

By 1976, there were 1820 km of these superhighways in operation in the Mezzogiorno. This was clearly a major increase over the 1970 level when the overall length of the functioning autostradas was less than 1100 km. However, when data for the southern autostradas are examined in detail (as in Table 25), only the Rome-Naples segment exceeded the northern traffic intensity average. I should caution that these data must be viewed with some care, for many of these roads or tracts were so recently opened to traffic. They are, nevertheless, instructive. Using the mean for southern Italy as a base level, aside from the Rome-Naples superhighways, the only relatively intense areas of truck traffic appear to be the Naples-Salerno, the Naples Peripheral, the Pescara-Lanciano segment of the Adriatic line, and the Messina-Catania route in eastern Sicily. Were data available for the Salerno-Reggio di Calabria road, I suspect that segments of that line too might exceed the average for the Mezzogiorno.

Rail Transport

Unlike the case of highway transportation, where there have unquestionably been major structural and traffic increments during the study era, railroad development has proceeded at a far slower pace. Despite an improvement in the quality of rail facilities in the South and a modest growth in traffic, the share of this carrier in land transportation in Southern Italy has drastically declined over time paralleling trends in other regions of the developed and less-developed world. However, it must be stressed that the rail system did not receive the same priority in funding that had been the case in the road development program. In 1974, for example, the railroads received less than one-fifth of the moneys for work "initiated, underway, or completed" that were allocated to the highway sector.[17]

My treatment of railroad transport will be abbreviated due to problems with data. Much of the necessary statistical material was either unavailable or not aggregated at spatial levels that would permit use of analytical procedures employed in my discussion of road transport.

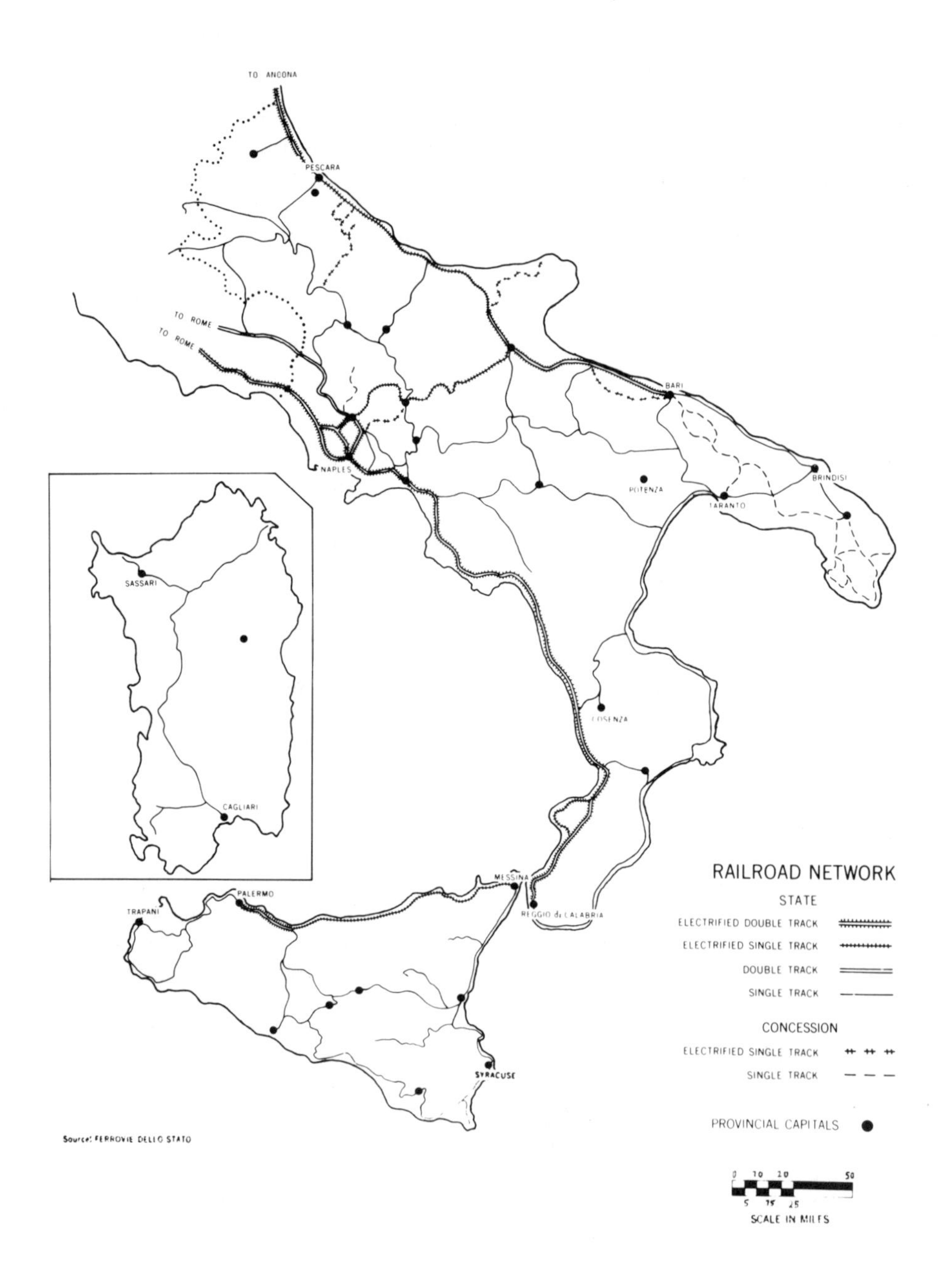

Fig. 17. The rail network in southern Italy in 1973.

I argued earlier that while the overall share of trackage in southern Italy in 1951 was not markedly below levels expected from its share of population and area, the network was backward and deficient in every other respect, such as: double track as a share of total trackage, the heaviness and quality of the track, levels of electrification, signaling equipment and the character and capacity of its freight and passenger stations. Thus my first comments relate to the changing structure of the southern railroad network.

It should be noted initially that the railroad system is essentially state owned and operated. In fact, 88% of its mileage is in the category called "Ferrovie dello Stato." However, the remainder are railroads "in concession." In general, the so-called "private" lines are often narrow gauge, rarely electrified; double tracking is unusual, and overall capacity is minimal (far below its share of the overall trackage of the South).

Given the fact that the rail trackage of southern Italy has expanded only minimally since 1951, was its capacity increased? The answer is clearly affirmative. This has been the thrust of the investment program. When viewed in isolation the improvements are notable; however, the system still lags far behind that of the North based on the criteria cited above. In 1951 roughly 7% of the southern rail lines were double tracked; by 1970 that share had risen to 15% (35% in the North). Similarly, at the start of the development program 17% of the southern network was electrified; by 1970 that value had risen to 26% (60% in the North). Although comparable statistics cannot be cited for the other criteria, there apparently had been significant improvements there as well. Nevertheless, the gap between the networks of the two Italies had not been narrowed. The rail network as of 1970, is demonstrated on Figure 17. While such elements as double tracking and electrification are identified, many other criteria which are correlated with capacity were not available in spatial detail or could be mapped.

Shifting to the question of commodity flows on the southern rail system, the general pattern for the region was demonstrated in Table 26. There had been a growth in this category of rail traffic in the 1955 to 1970 period, but it had been minor and far below that of the North. Thus in effect the southern *share* of the nation's commodity traffic by railroad, as measured in ton-kilometers, had actually declined.

What was the pattern of flows within the Mezzogiorno and had these movements shifted during the 15-year interval? The answers to these questions could clearly be anticipated from Figure 17 and the evolving geography of socio-economic development in southern Italy. Figure 18 illustrates schematically the broad elements of the commodity flow patterns c. 1973. This map is a synthesis of those produced by Ferrovie dello Stato and Celant.[18] The reader should be cautioned that this rail map cannot be contrasted with that by road for the scales are drastically different; therefore, any direct quantitative comparison would be misleading. However, the similarities in *pattern* are obvious.

The major flows are found in the two strips along the Adriatic and Tyrrhenian coasts. What is not clearly evidenced is the fact that within the

Table 26. Commodity Traffic by Carrier in Southern Italy 1955 to 1970

Carrier	1955		1970		Change (%)
	Ton-kms (millions)	Share (%)	Ton-kms (millions)	Share (%)	
Rail	3,409	42.9	4,731	16.7	38.8
Road	4,539	57.1	23,590	83.3	419.7
Total	7,948	100.0	28,321	100.0	256.3

[a]Cecchini, op. cit., p. 34.
[b]Cagliozzi, Roberto, *L'Ammodernamento delle Ferrovie ed il Ruolo del Trasporto Ferroviario nel Mezzogiorno*, SVIMEZ, Rome, 1975, p. 9.

western "band" the chief movements were concentrated in the zone from Rome to Naples and in the Naples region itself. Similarly the key flows on the east coast were found in Puglia (Bari-Brindisi-Taranto). The minimal trans-Appenine currents could have been expected from the facilities map. Here the only noteworthy movement was that from Naples to Bari. In Sicily the most important streams were those from Palermo to Messina on the north coast and from Messina to Syracuse on the eastern margin. As in the case of road traffic, rail movements in Sardinia were extremely limited. I suspect that recent economic developments on that island will foster considerable increases in both rail and highway flows of bulk commodities.

Returning to the question of the changing roles of rail and road commodity transport in the South, can the relative decline in rail transport be readily explained? It is well known and amply documented in both the press and scholarly literature that railroads in all of the developed nations, at least, are in serious financial difficulty. They cannot survive without heavy government subsidy. To illustrate in the Italian case, Cagliozzi cites the fact that the rail system may have lost nearly one billion dollars in 1973.[19] He also noted, in his study of changing traffic patterns from 1960 to 1969, that even on the best-endowed rail links, the construction of a parallel autostrada almost inevitably leads to a decline in rail commodity traffic. In part, the competitive failure of the railroads is due to higher costs, but, as in western Europe and this country, unquestionably, the more important factors can be summarized as "service advantages" which apparently cannot be overcome by improvements in rail facilities, lowering of rates, reduction of transfer time, containerization, piggy-back services etc.

Maritime Transport

The ports of the South were not treated previously simply because, with the exception of Naples, they had played such a minimal role in the economic

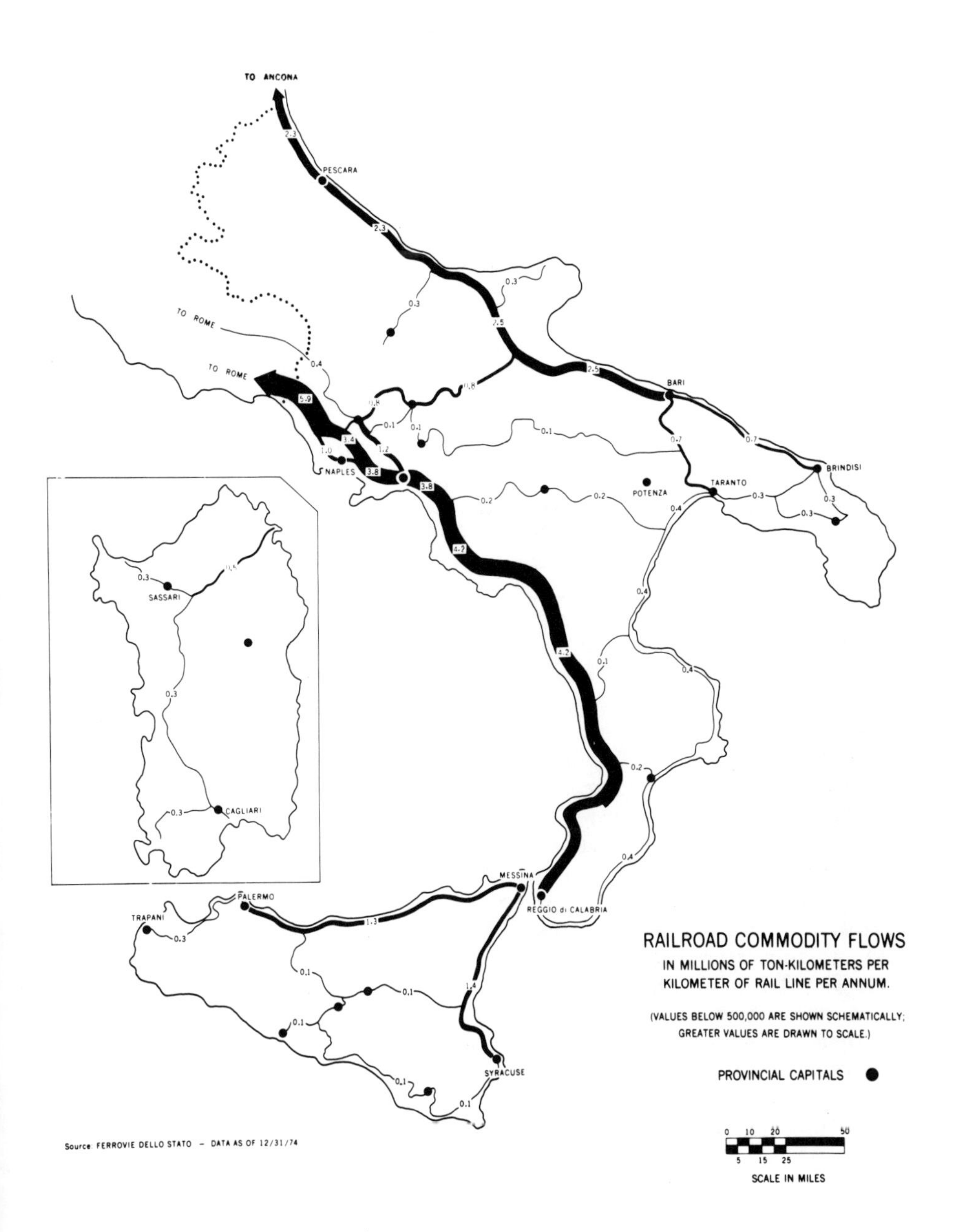

Fig. 18. Commodity flows on "State" railways in 1973.

Table 27. Commodity Traffic, by Port, in Italy in 1974[a]

Port (province)	International (thou. tons)		Coastal (thou. tons)		Total (thou. tons)	
	Inbound	Outbound	Inbound	Outbound	Inbound	Outbound
Gaeta (Latina)	2,271	363	95	308	2,366	671
Bagnoli (Naples)	4,029	60	321	488	4,350	548
Naples	7,109	1,353	1,661	785	8,770	2,138
Bari	2,381	40	761	850	3,142	890
Brindisi	1,264	210	1,640	527	2,904	737
Taranto	18,745	1,543	1,268	4,416	20,013	5,959
Palermo	190	44	1,617	661	1,807	705
Milazzo (Syracuse)	8,769	4,543	168	2,546	8,937	7,089
Augusta (Syracuse)	22,556	7,337	1,593	10,851	24,149	18,188
Gela (Caltanisetta)	3,065	352	682	2,229	3,747	2,581
Cagliari	318	282	1,445	1,088	1,763	1,370
Porto Foxi (Cagliari)	11,663	4,948	189	5,174	11,822	10,122
Porto Torres (Sassari)	3,817	1,578	1,362	1,279	5,979	2,857
Main South Italian ports	86,177	22,655	12,802	31,202	99,749	53,855
Genova	46,036	2,710	5,545	4,838	51,581	7,548
La Spezia	7,827	711	1,572	2,452	9,399	3,163
Savona	12,033	855	2,390	331	14,423	1,186
Venice	11,557	1,404	10,198	1,172	21,755	2,576
Trieste	30,304	1,319	1,258	808	31,562	2,127
Livorno	5,926	1,104	2,709	1,333	8,635	2,437
Piombino	2,833	102	1,082	834	3,915	936
Main North Italian ports	116,516	8,205	24,754	11,768	141,270	19,973
Italy (all ports)	225,313	35,796	55,117	55,995	280,431	91,791

[a]*Annuario Statistico Italiano 1976*, Rome, 1976, pp. 263–264.

development of that region since Unification. As of 1951, these ports accounted for less than 20% of the nation's maritime traffic with roughly three-quarters of that volume concentrated in Naples and secondarily Bari.[20] The remaining movements were scattered among a series of poorly equipped harbors. Taranto, the former naval center, was perhaps potentially more viable as was Palermo. I should add, however, that in terms of natural circumstances there were many suitable harbors along the extensive coastline of the Mezzogiorno that had not been developed. Then, too, the construction of transport connections across the Appenines has engendered heavy expenditures. Thus the propensity for maritime communication in the region tying it to the rest of the nation as well as to foreign sources of materials and markets has stimulated the creation of new ports and the expansion of existing harbor facilities.

By 1974, maritime traffic in the South had expanded enormously both in the international and coastal sectors. It had growth fourfold from 1950 levels.[21] More importantly an entire array of new ports had developed as is demonstrated in Table 27.

The South, by this date, accounted for 40% of the country's maritime traffic, and a number of the new ports that had been created now handled far larger flows than Naples. However, the data in this table are, in a sense, misleading, for Naples is a highly diversified port handling both bulk and general cargo, while most of the new harbors are tied to a single commodity group—oil and oil products. That high level of specialization is true of Augusta, Milazzo, and Gela in Sicily; Porto Foxi and Porto Torres in Sardinia; and Gaeta, Brindisi, and Bari on the mainland. In contrast, Bagnoli and Taranto are both major integrated metallurgical centers. Their traffic is dominated by inbound raw materials for their steel mills and outbound semi-fabricated metal products destined for national and export markets. Taranto is also an oil refining center, so that movements of petroleum and petroleum products are also important components of its traffic. In fact, Naples, Palermo, and Cagliari are the only major general cargo ports in the Mezzogiorno.

Celant has argued persuasively that the new maritime centers of southern Italy were designed to support industries that were implanted there as a result of the development program.[22] Unlike Genoa, Livorno, Venice and Trieste in the North, these southern ports do not serve hinterlands other than their immediate coastal industrial complexes. The only major exception to this rule is, of course, Naples. Thus, as Cagliozzi suggested, the maritime development of the South during the past several decades has not been integrated with other transport modes, nor has it been tied to the evolving settlement pattern of that region.[23] Nevertheless, port development has been an admittedly vital facet of southern economic growth during the post-war period.

Integrated Transport Development

Most studies of transportation in the Mezzogiorno appear to agree with such regional development theorists as Rosenstein-Rodin, Myrdal, Hirschmann, Friedmann, and Boudeville, that the creation of a transport infrastructure is a necessary stimulus for economic growth.[24] There is, however, considerable debate about its timing and scale, which need not concern us here. In addition, there appears to be general agreement on the need for a coordinated economic planning within which transport development would be a key dimension. In fact, no such plan was ever *implemented* by the central planning agencies. There were separate programs for road and rail development and possibly for the ports, but the reality appears to be a lack of coordination among the various agencies. Cagliozzi, in a stimulating essay prepared for SVIMEZ, has presented some exploratory thoughts on such a plan for integrated transport development, but his ideas are far from implementation.[25] Clearly, when one reads about traffic congestion in certain metropolitan areas, notably Naples, as well as complaints about the lack of adequate services in others, transport planning should have high priority. I should add that there have also been numerous stories in the Italian press about excessive expenditures on road construction in southern Italy, in which the road construction program is criticized using the "cathedrals in the desert" argument.[26] These are typically coupled with heated comments that such moneys might better have been spent on "more crucial" facilities such as hospitals, schools, universities, housing and mass urban transit. Evidently, these bolster the arguments for more effective and coordinated planning and a more judicious allocation of admittedly limited resources. I will return to this question in my concluding remarks.

NOTES

[1]ISTAT, *Censimento Generale della Popolazione, IX (1951) and XI (1971)* various volumes and pages.

[2]Guido Cella, "Industrializzazione e Emigrazione, Il Caso dell Mezzogiorno nel Decennio 1961-1971," *Rassegna Economica*, Banco di Napoli, Vol. 37, No. 4 (1974), p. 1079.

[3]Carmelo Formica, "Esodo Agricolo e Trasformazione dell Agricoltura nel Mezzogiorno," *Nord e Sud*, Vol. 22 (July-September 1975), p. 68.

[4]ISTAT, *1951 and 1971 Censuses of Population*, various volumes and pages.

[5]*11th Censimento Generale dell Agricoltura*, 1970, Vol. VI: Dati Generali Riassuntivi (Rome, 1976).

[6]"La Situazione della Bonifica," *Informazione Svimez*, Vol. 28, No. 19 (Oct. 15, 1975), p. 902.

[7]ISTAT, *Annuario di Statistica Agraria*, Vol. 2, 1955 (Rome, 1956), and Vol. 22, 1975, (Rome, 1976), various pages.

[8]King, op. cit., pp. 27-39.

[9]Ibid, p. 30.

[10]Gian Giacomo dell Angelo, "Problemi I Linee di Sviluppo dell Agricoltura dell Mezzogiorno, *Informazione Svimez*, Vol. 28, No. 18 (September 30, 1975), p. 814.

[11]These include Roberto Cagliozzi, *Prospettive del Traffico Marittimo e Problemi Portuali del Mezzogiorno*, (Rome: Guiffrè, 1970), Roberto Cagliozzi, *L'Ammodernamento delle Ferrovie ed il Ruolo del Trasporto Ferroviario nel Mezzogiorno*, (Rome: Guiffrè, 1975), and Roberto Cagliozzi, *Infrastture di Trasporto e Sviluppo' del Mezzogiorno*, (Rome: Guiffrè, 1975).

[12]D. Cecchini, *Trasporto Stradale e Struttura Insediativa nel Mezzogiorno*, SVIMEZ (Milan, Rome: Giuffrè, 1975), p. 112.

[13]Cecchini, 1975, op. cit., pp. 116-121.

[14]ISTAT, *Annuario di Statistico Italiano*, 1956, Rome and *Annuario di Statistico Italiano 1976* (Rome, 1976), various pages.

[15]Cecchini, op. cit., pp. 36-41.

[16]Ibid, p. 53.

[17]ISTAT, 1976, op. cit., pp. 248-249.

[18]Cagliozzi, Ferroviario, 1975, op. cit., and Attilio Celant, "Trasporti e Porti del Mezzogiorno nel Quadro della Politica Meridionalistica," *Rivista di Politica Economica*, Vol. 66 (Rome, 1976), pp. 1055-1095. Data provided by the Ferrovie Dello Stato and modified by materials in Celant.

[19]Cagliozzi, Ferroviario, 1975, op. cit., p. 3.

[20]ISTAT, *Annuario di Statistico Italiano*, 1953 (Rome, 1952), various pages.

[21]ISTAT, 1976, op. cit., pp. 263-264.

[22]Celant, op. cit., p. 1079.

[23]See Cagliozzi, Infrastrutture, op. cit. and Cagliozzi, Marittimo, op. cit.

[24] See the summary of their ideas in Celant, op. cit.

[25] Cagliozzi, Infrastructure, op. cit.

[26] "Cathedrals in the desert" is a term used by Italian writers to describe some of the huge new capital-intensive undertakings.

Chapter 7

DEMOGRAPHIC TRENDS

NUMBERS AND GROWTH

In 1951 the "resident" population of southern Italy (including the provinces of Frosinone and Latina) was roughly 18,500,000 or about 39% of the nation's inhabitants.[1] That share decreased to 36% by 1971 despite an absolute increase of more than 1,000,000 people. The changes by region recorded during this two-decade interval are demonstrated in Table 28 and province data are mapped on Figure 19.

There is clear evidence from these ISTAT "counts" of a much higher natural growth in the Mezzogiorno than in northern Italy. However, that increase was offset by an out-migration of over 4,000,000 people to northern Italy and abroad, as discussed in depth in the final section of this chapter. The map illustrates significant variations in regional growth patterns. Here the results of that out-migration are readily evident; note that with the exception of Pescara, on the northern Adriatic coast, there was an actual *decline* of population in 17 of the 34 provinces in the South. That decrease occurred in a belt extending from L'Aquila and Teramo on the northern border of the region down the Appenine spine through Avellino and Potenza to southern Calabria. Losses were also recorded in central and southwestern Sicily. In contrast, the areas that gained population were mainly confined to those provinces which received the heaviest industrial subsidies, i.e., Latina through Caserta and Naples to Salerno on the Tyrrhenian coast, Puglia, Palermo and southeastern Sicily, and all three provinces of Sardinia.

Table 28. Population Growth, by Region from 1951 to 1971[a]

Region	Population 1951 (thou.)	Population 1971 (thou.)	Relative increase (%)
Frosinone and Latina	752.3	798.8	6.2
Abruzzi and Molise	1,684.0	1,486.6	-11.7
Campania	4,346.3	5,059.4	16.4
Puglia	3,220.5	3,582.8	11.2
Basilicata	627.6	603.0	-3.9
Calabria	2,044.3	1,988.1	-2.7
Sicily	4,486.7	4,678.9	4.3
Sardinia	1,276.0	1,473.8	15.5
Southern Italy	18,437.7	19,671.4	6.7
Northern Italy	29,077.8	34,465.1	18.5
Italy	47,515.5	54,136.5	13.9

Region	Absolute increase (thou.)	Natural growth (thou.)	Net migration (thou.)
Frosinone and Latina	46.5	186.6	-140.1
Abruzzi and Molise	-197.4	256.4	-453.8
Campania	713.1	1,465.0	-751.9
Puglia	362.3	1,046.4	-684.1
Basilicata	-24.6	183.9	-208.5
Calabria	-56.2	635.4	-691.6
Sicily	192.2	1,243.3	-1,051.1
Sardinia	197.8	425.5	-227.7
Southern Italy	1,233.7	5,442.5	-4,208.8
Northern Italy	5,387.3	3,239.2	2,148.1
Italy	6,621.0	8,681.7	-2,060.7

[a]*Estimated from:* M. Natali, "Stima Retrospettiva della Popolazione Residente Provinciale nel Periodo 1951–1961," *Sviluppo della Popolazione dal 1861 al 1961, Annali di Statistica*, Series VIII, Vol. 17 (Rome, 1965), pp. 146–153, and *Popolazione e Movimento Anagrafico dei Comuni, 1973*, Vol. 17 (Rome, 1974), pp. 349–397.

Perhaps as important for this study has been the move to the cities of the South during the development era. There is, of course, no agreement among nations nor among students of urbanization on a single definition of the term "urban." In the Italian case, and particularly in the Mezzogiorno, a very large share of the rural population lives in overgrown agglomerated agricultural villages, some with populations exceeding 20,000. Yet by other than a size

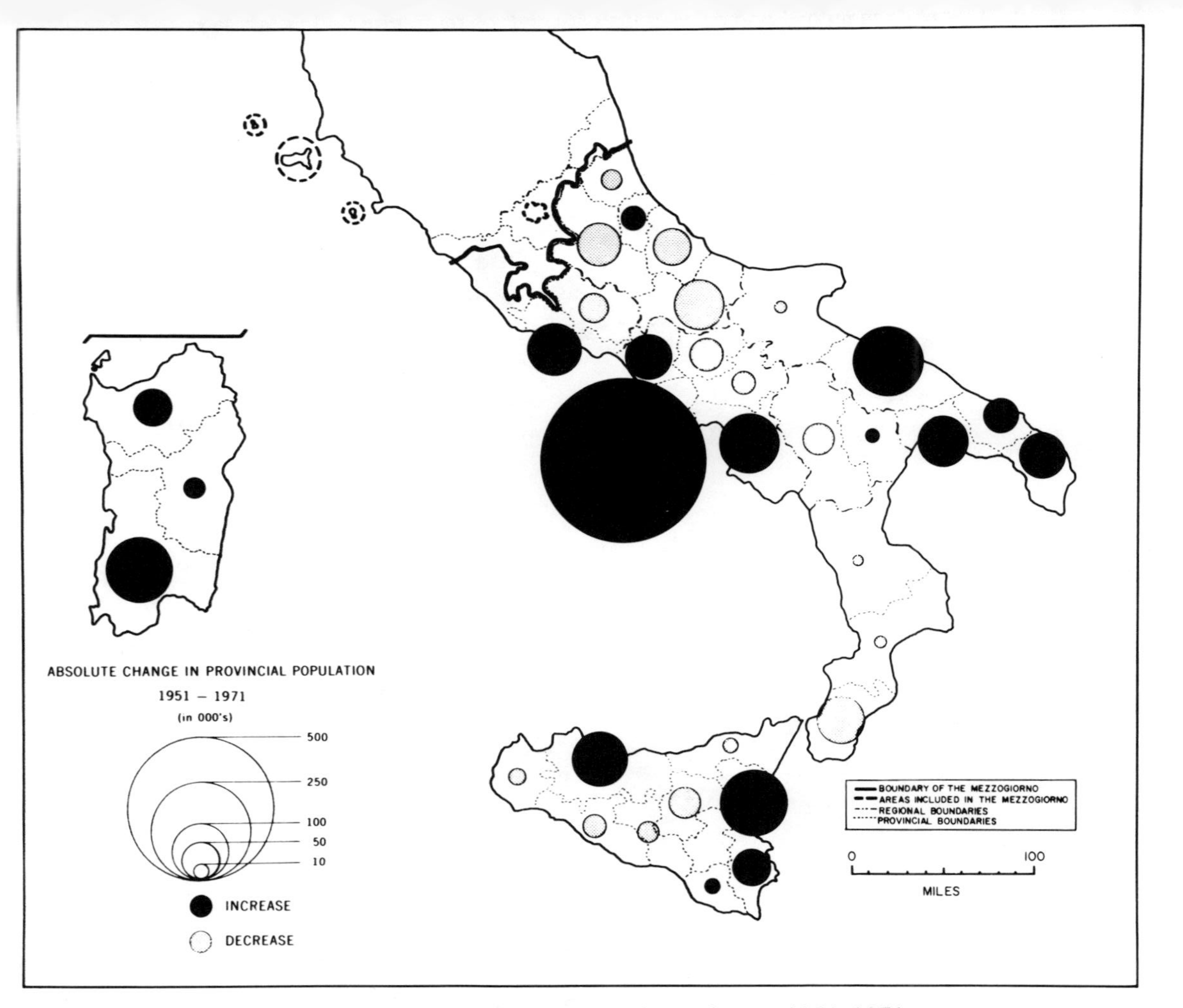

Fig. 19. Absolute changes in provincial population, 1951–1971.

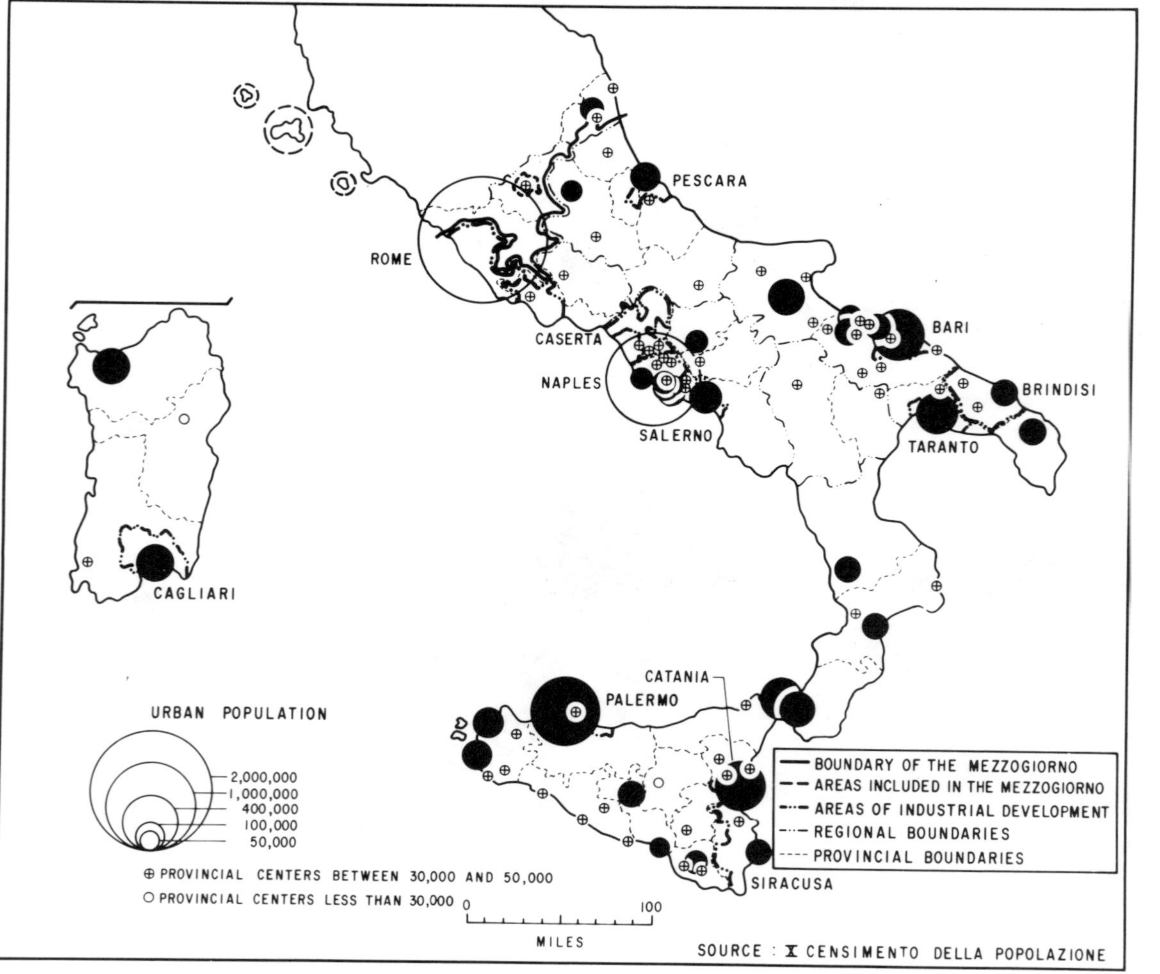

Fig. 20. Population of all communes with over 20,000 people in 1961.

criterion, these communes would probably not be classified as urban. They certainly lack key economic functions that are normally associated with cities. To test this hypothesis, I computed the proportion of the population living in communes with over 20,000 people for both northern and southern Italy in 1971 and found that the share was roughly half in *both* areas. These results clearly disagreed with my view of the reality of the urban landscape as seen in travels through all segments of the country over the past two decades. The North, all would agree, is far more urbanized than its southern counterpart. I, therefore, decided to map the urban population in 1961 (the *intermediate* census year) including all communes with over 20,000 people but using a different symbolization for those below 50,000 (Figure 20). The resulting distribution is far more illustrative of urban patterns in southern Italy than would be true of a map that used *solely* a 20,000 base level, but it is still clearly unsatisfactory!

ISTAT in Rome, has made a far more fruitful attempt to classify communes as to their levels of urbanization.[2] Their method is based on national averages for five indicators: the proportion of the active population engaged in agriculture, the proportion in tertiary activities, educational levels, the proportion of the population living in the principal center of the

Table 29. Urban Population of Italy in 1931 and 1961 as a Percentage of the Total Italian Population

Region	1931[a]			1961[b]		
	Urban population (thou.)	Total population (thou.)	Percent urban	Urban population (thou.)	Total population (thou.)	Percent urban
Abruzzi and Molise	253	1,545	16.4	378	1,564	24.2
Campania	1,698	3,509	48.4	2,486	4,761	52.2
Puglia	649	2,508	25.9	1,006	3,421	29.4
Basilicata	45	514	8.8	88	644	13.7
Calabria	244	1,723	14.2	403	2,045	19.7
Sicily	1,244	3,906	31.3	1,778	4,721	37.7
Sardinia	211	984	21.4	450	1,419	31.7
Southern Italy	4,344	14,689	29.4	6,589	18,575	35.5
Northern Italy	10,564	26,371	40.1	17,579	32,049	54.9
Italy	14,908	41,060	36.3	24,168	50,624	47.7

[a] F. Spagnoli, "Popolazione Urbana e Rurale" in *Sviluppo della Popolazione Italiana dal 1861 al 1961, Annali di Statistica*, Series VIII, Vol. 17 (Rome, 1965), p. 197. The term "urban" is based on a multiple criteria method developed by ISTAT in 1959–60.

[b] *11th Censimento Generale della Popolazione*, 1961, Vol. IX, Dati Generali Riassuntivi (Rome, 1969), pp. 20–23. Frosinone and Latina are not included in either the 1951 or 1961 values.

Table 30. Changes in the Population of Italian Metropolitan Areas from 1951 to 1971

Metropolitan area	1951 population[a] (thou.)	1971 population[b] (thou.)	Absolute change (thou.)	Relative change (%)
Naples	2,713.0	3,652.3	939.3	34.6
Pescara	219.7	303.9	84.2	70.3
Bari	283.9	373.9	90.0	31.7
Taranto	168.9	228.8	59.9	35.5
Reggio di Calabria	153.9	176.9	23.0	14.9
Messina	228.7	266.1	37.4	16.4
Catania	389.0	508.3	119.3	30.7
Palermo	535.1	701.9	166.8	31.1
Cagliari	138.5	224.4	85.9	62.0
Southern metropolitan areas	4,830.7	6,436.5	1,705.8	36.6
Southern non-metropolitan areas	13,707.7	13,324.9	-472.8	-3.4
Northern metropolitan areas	10,774.5	16,649.1	5,874.6	54.5
Northern non-metropolitan areas	18,302.7	17,704.7	-598.0	-3.3

[a] S. Cafiero and A. Busca, *Lo Sviluppo Metropolitano in Italia*, SVIMEZ (Rome, 1970), p. 25. Data for areal boundaries as of 1961.

[b] Calculated on basis of 1961 areal limits from XI Censimento Popolazione 1971, Vol. I, op. cit., various pages.

communes, and the quality of housing. In addition, all communes were classified as urban if they had a population of over 70,000 inhabitants. These criteria are obviously subjective and liable to serious question when viewed in dynamic perspective; nevertheless, in my opinion, their method is far better than one using a number or density criterion. Their results for 1931 and 1961 are shown in Table 29; unfortunately, data for 1971 are not yet available. I also computed the share of the urban population by their criteria for each of the southern provinces. Given an average for that region of roughly 36%, only Naples (78), Palermo (57), Catania (43), Taranto (42), Bari (40) and Pescara (40) exceeded that mean.

Southern Italy like the North is currently enmeshed in an urban crisis arising primarily from unusually heavy natural increase in its cities. Although there has also been a significant flow of rural migrants to the urban centers of the South, that movement has been offset by heavy out-migration. This mushrooming metropolitan growth has either been completely unplanned or at least dreadfully conceived. The press argues increasingly for the need to restrict urban development in Italy because of the social and economic diseconomies of metropolitan concentration. While air pollution, traffic congestion and crowded sub-standard housing are clearly most evident in such areas as Milan, Turin, Florence and Rome, they are increasingly true of the Mezzogiorno.[3] Naples is the classic case in southern Italy, but the same problems are becoming evident on the coast of Puglia, in eastern Sicily, and in the environs of Palermo and Cagliari. Table 30 demonstrates the growth of the key metropolitan areas in the South since 1951. While the increases are impressive, they are still far below those of northern Italy.

DISTRIBUTION

Figure 21 demonstrates the population distribution pattern as of 1951. No such detailed density map was available for 1971, and even this one suffers from its failure to separate urban from rural areas. It is beyond the scope of this study to discuss these extremely complicated patterns. What is important for our purpose, however, is the unusually high share of the population found in the hilly and mountainous interior. In 1951 69% of the inhabitants of the South lived in such difficult environmental areas and the change by 1971 had not been spectacular (see Table 31). It is true, as is evidenced by Table 32, that the population in what can be termed the "coastal communes" grew by 22%, while the population of the interior was essentially stable. However, the exodus to the coastal cities was hardly a flood, and the movement was mainly out of the region.

DEMOGRAPHIC TRAITS

No discussion of the demography of the South would be complete without an analysis of its vital rates, sex ratios, and age structure, for these have obviously played important roles in the growth of population in the past and will affect future increases, barring further outmigration.

In Table 1 of the introductory chapter, I documented the contrasts over time between the vital rates of northern and southern Italy. The data in Table 28 also showed that the rate of natural growth of the southern population was well above that in the North between 1951 and 1971. This trend is a relatively recent phenomenon dating from the 1920s when, concomitant with urbanization and industrialization, the birth rate fell rather rapidly in the North but much less so in the Mezzogiorno. At the same time the death rate, viewed on a macro-level, fell more modestly in both areas. By the 1950s southern mortality levels began to fall at an accelerated pace, presumably as a result of improved medical measures, sanitation and diet. By 1971 they were slightly lower than those of the North, perhaps reflecting differential age structures, but the contrasts in birth rates was even more impressive. Thus the natural increase in the South was quite high (11 per thousand vs. 5 in

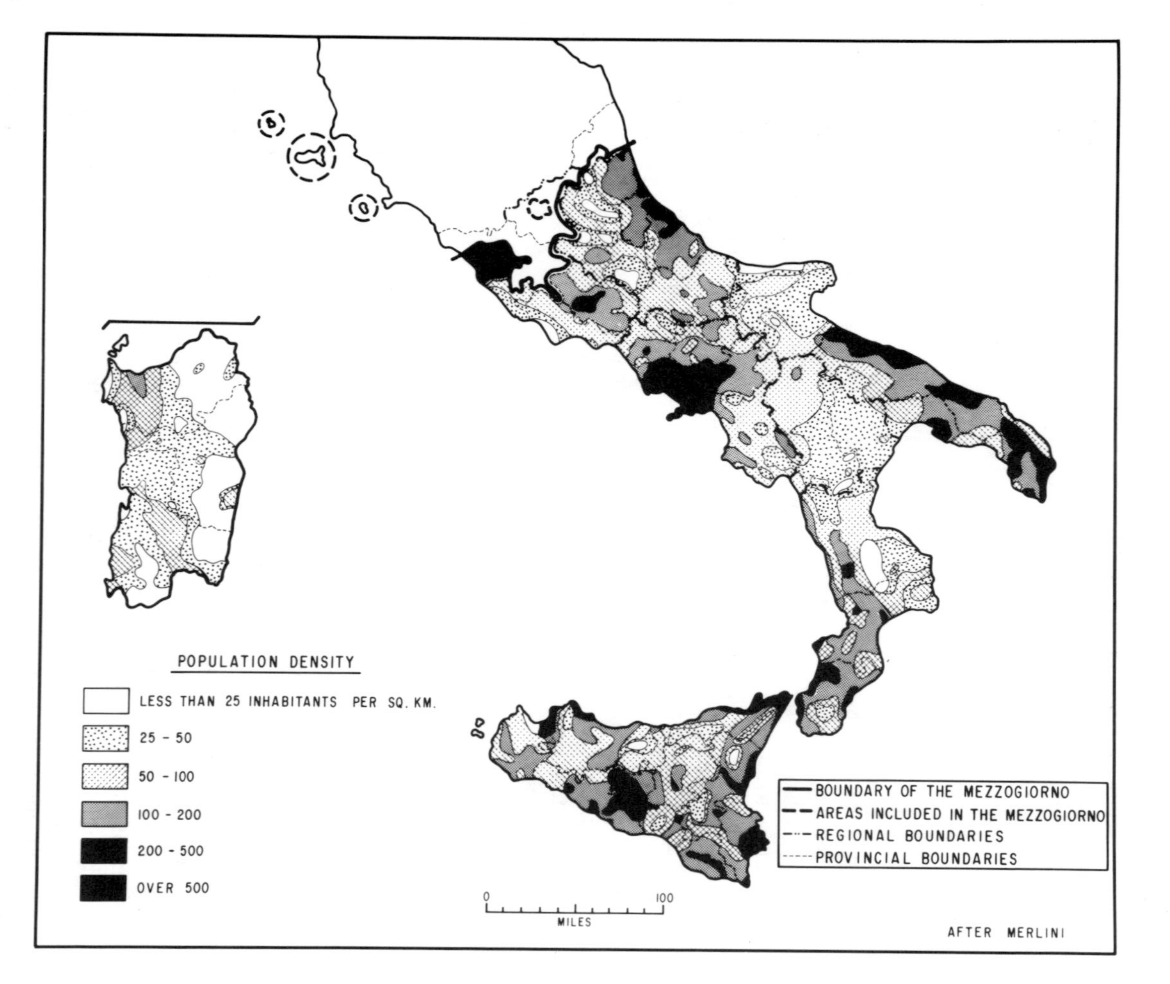

Fig. 21. Population distribution in southern Italy in 1951.

Table 31. The Share of the Southern Italian Population and Area by Landform Categories in 1951 and 1971[a]

Region	Population 1951 (%)				Population 1971 (%)			
	Mountain	Hill	Plain	Total	Mountain	Hill	Plain	Total
Abruzzi and Molise	44.8	55.2	0.0	100.0	38.3	61.7	0.0	100.0
Campania	11.0	64.5	24.5	100.0	7.8	64.9	27.3	100.0
Puglia	1.1	29.7	69.2	100.0	0.5	26.4	73.1	100.0
Basilicata	47.6	45.4	7.0	100.0	48.1	42.3	9.6	100.0
Calabria	30.4	60.0	9.6	100.0	25.5	61.8	12.7	100.0
Sicily	16.5	51.0	32.5	100.0	15.0	45.7	39.3	100.0
Sardinia	6.8	55.9	37.3	100.0	5.5	48.7	45.8	100.0
S. Italy (share of popul.)	17.1	52.0	30.9	100.0	13.2	50.2	36.6	100.0
S. Italy (share of area)	28.5	53.2	18.3	100.0	28.5	53.2	18.3	100.0

[a]Data from the 1951 and 1971 Censuses of Population, various volumes.

northern Italy). Were it not for out-migration, population pressure in the Mezzogiorno would have been overwhelming.

Regional differences in vital rates within the Mezzogiorno and comparisons with the North for 1971 are demonstrated in Table 33. With an average birth rate of 20, the regional range was from 14 to 22. Not surprisingly, when viewed on a provincial level (not demonstrated here), the poorest provinces had the lowest birth rates; it will be seen, however, that these were typically areas with the greatest out-migration of men and women in the age cohort with greatest potential fertility.

The question of age structure is obviously related to both the development stage of a region and, in this instance, to differential out-migration. The contrasts between North and South are illustrated in Table 34. It should be stressed that the 1971 data are for resident population while those for 1931 and 1951 are for "population present." This does affect the values presented. Nevertheless, the relative youth of the southern population is evident, reflecting its stage in the development transition. On the other hand, when the sub-regional data are examined for the South, the areas of great out-migration like Abruzzi and Molise and Calabria do have a larger share than the average in the over sixty age group. I suspect that data for the "population present" in 1971, by age group, when available, will be more reflective of age specific migration patterns.

Finally, a word about sex ratios is in order. There were only minimal

Table 32. Population Growth for Coastal and Interior Communes of Southern Italy between 1951 and 1971

Region	Population 1951[a] (thou.)	Population 1971[b] (thou.)	Absolute growth (thou.)	Relative growth (%)
Coastal communes[c]	5,632.0	6,862.0	1,230.0	21.8
Interior communes	12,805.7	12,809.4	3.7	0.3
Southern Italy	18,437.7	19,671.4	1,233.7	6.7

[a]*Popolazione Legale dei Comuni, (IX Censimento Generale della Popolazione), 1951* (Rome; 1955), various pages.
[b]*Popolazione Legale dei Comuni, (XI Censimento Generale della Popolazione), 1971* (Rome, 1973), various pages.
[c]Communes fronting on the sea.

Table 33. Vital Rates in Italy, by Region for 1971[a]

Region	Birth rates (per thou.)	Death rates (per thou.)	Natural increases (per thou.)	Infant mortality (per thou. live births)
Latina and Frosinone	17.3	8.2	9.1	27.0
Abruzzi and Molise	14.4	9.6	4.8	23.1
Campania	21.9	8.5	13.4	43.8
Puglia	20.7	8.1	12.6	33.5
Basilicata	17.2	8.0	9.2	37.5
Calabria	18.4	8.0	10.4	32.5
Sicily	19.3	9.3	10.0	34.1
Sardinia	20.1	8.4	11.7	26.7
Southern Italy	19.9	8.7	11.1	35.4
Northern Italy	15.1	10.1	5.0	23.0
Italy	16.8	9.6	7.2	28.3

[a]*Annuario Statistico Italiano, 1972* (Rome, 1972), pp. 25–27.

differences between the average values for both northern and southern Italy. However, within the South, given a mean of 94 males per hundred females (present population), the range ran from 91 in Calabria and 92 in Abruzzi and Molise (regions of heavy out-migration) to 97 in Sardinia (minimal net out-migration) and 99 in southern Lazio (with modest levels of net *in-migration*).

Table 34. Share of the Italian Population, by Region, from 1931 to 1971

Region	Year	Age groups (%)				
		0–14	15–39	40–59	60 and over	Total
North Italy	1931[a]	27.7	41.5	20.2	10.7	100.0
	1951[a]	23.4	38.5	25.0	13.1	100.0
	1971[b]	22.0	35.0	25.3	17.7	100.0
South Italy	1931[a]	33.5	37.5	18.0	11.0	100.0
	1951[a]	31.4	38.4	19.5	10.7	100.0
	1971[b]	28.8	35.1	21.4	14.7	100.0
Region of South Italy[b]						
Latina and Frosinone	1971	26.4	35.8	22.9	14.9	100.0
Abruzzi and Molise	1971	23.7	33.5	24.7	18.1	100.0
Campania	1971	28.5	35.6	22.9	13.0	100.0
Puglia	1971	29.7	35.6	20.8	13.9	100.0
Basilicata	1971	28.4	34.6	21.7	15.3	100.0
Calabria	1971	27.3	29.4	26.4	16.9	100.0
Sicily	1971	27.5	34.6	22.1	15.8	100.0
Sardinia	1971	29.5	35.9	19.9	14.7	100.0

[a] *Un Secolo di Statistiche Italiane: Nord e Sud: 1861-1961*, SVIMEZ (Rome, 1961), p. 27. This was the population present.

[b] *11th Censimento Generale della Popolazione*, Vol. II, Dati Riassuntivi-Parte Prima (Rome, 1974), pp. 10–13. This was the resident population.

MIGRATION

As noted earlier, there was only a minimal level of emigration during the inter-war years, and internal migration was clearly constrained. It is the movements between 1951 and 1971 that concern us here. My basic source materials were a report from ISTAT based on the 1971 Census of Population, which details place of residence and place of birth, by province and region, as of that date,[4] coupled with yearly anagraphical reports on transfers of residence during that two-decade interval.[5]

It should be recalled that roughly 3% of the resident population of the North were born in southern Italy as of 1951. A comparable ISTAT sample taken as of the 1971 Census indicates that this value had risen to 9%, or over 3,000,000 persons. If data were available on the "population present" as of that census, I suspect that this percentage would have been much higher. There are undoubtedly large numbers of southerners living and working in the

Table 35. Estimated Out-Migration from the Mezzogiorno through 1971, Based on Residence and Place of Birth Data[a]

Southern regions	Proportion of out-migration to the North (%)	Proportion of the resident population of the South in 1971 (%)
Abruzzi and Molise	10.9	7.9
Campania	19.2	26.8
Puglia	20.3	19.0
Basilicata	4.8	3.2
Calabria	15.1	10.5
Sicily	22.4	24.8
Sardinia	7.3	7.8
Southern Italy	100.0	100.0

[a]*Bolletino Mensile di Statistica*, No. 12 (December, 1976), Appendix I, pp. 276–280.

North who have not officially changed their residence despite the general relaxation of legal bars to migration that were implemented in 1961.

Table 35 is a compilation based upon the ISTAT place of birth and residence data. It demonstrates the relationship between the shares of the individual southern regions in the overall migration to the North and their proportions of the resident population of the Mezzogiorno as of 1971. Note the disparity between the two percentage values for both Abruzzi-Molise and Calabria. These areas clearly contributed more than their expected share of the movements to northern Italy. The results here are, of course, highly correlated with the data on population change that were previously demonstrated in Table 28.

The same source also demonstrated the minimal internal mobility of the population within the South. Thus there were no major movements between the various southern regions. The bulk of the population of the Mezzogiorno came from the same province or to a lesser degree from adjacent provinces of the same region. Although there had undoubtedly been notable flows from the countryside to the cities of the South, such movements, unfortunately, cannot be documented from this source which unlike the 1951 and 1961 tabulations does not include detailed data at the commune level.

These materials can be supplemented by the anagraphical statistics for the 1951 and 1971 period which record transfers of residence on a yearly basis. Despite the limitations of these records noted earlier, they are the most detailed materials available on Italian migration patterns. These will first be examined in summary form, that is, net migration; this will be followed by an analysis of the broad flow patterns.

First, I should explain that the anagraphical records are compiled initially on a monthly and then on a yearly basis for provinces, for provincial capitals, and for communes with over 50,000 inhabitants. They register both inbound and outbound movements (internal and foreign), but initially there is no indication of the actual geographical flows. At a later date, ISTAT then publishes a yearly geographical matrix by province of internal and foreign inbound and outbound movements.[6]

The net out-migration for the South, using the anagraphical records from 1952 through 1971, was approximately 3,500,000 people during that 20-year interval. If the resident population of the South for the mid-year 1962 (19,300,000) is taken as a base level, then the division results in a negative value of 18% for the entire region.

However, the net migration rates, when viewed on a provincial scale, varied greatly from region to region. The provincial data are plotted on Figure 22. Note that with one exception–Latina–all of the values are negative (or out-migration). This map is an obvious reflection of spatial variations in population pressure and economic differentials within southern Italy. There was clearly a belt of heavy out-migration which followed the highland spine of the Appenines from the northern tier of provinces (Teramo, L'Aquila, and Frosinone) to Reggio di Calabria in the South. Note, too, the very high negative values in central and southwestern Sicily and in the province of Nuoro in central Sardinia. In contrast, the net out-migration from eastern Sicily had been quite low. This was also the case on the western coast of the mainland in the zone stretching from Caserta to Salerno. The percentage for Naples was particularly small. Pescara, in the northeast, also had minimal outbound flows. Latina was the only province with a net *inbound* value. Intermediate positions on this scale were occupied by the provinces of Cagliari and Sassari in Sardinia and southern Puglia.

Turning to the flow patterns themselves, I have summarized the yearly matrices for the 1955 to 1971 period, and the *net movements* are presented in Table 36. This compilation only shows migration to adjacent provinces of the same region, to the rest of southern Italy, to the North and abroad. The intra-provincial flows, which dominate the inbound and outbound movements, cancelled themselves when net values were computed.

Had I included the inbound and outbound movements in this table, the strengths of the intra-provincial flows would have been apparent. Thus, on the average, 37% of the outbound movements or "cancellazioni" were within the same province. In the case of the inward flows or "iscrizioni" the share was even higher, an average of 49%. Returning to the net migration data, these were clearly dominated by movements to nothern Italy and abroad with the North as the most important destination. That region, of course, includes both central and northern Italy. Within that area, which contains 58 of Italy's 92 provinces, Milan, Turin, and Rome were the three major destinations absorbing the largest share of the southern migrants. As for migration abroad, two countries received the bulk of the southern expatriates–West Germany and Switzerland. Extra-provincial movements within the South were typically to adjacent provinces, particularly those with large urban centers such as Naples and Palermo. In an article published some years ago on migration and

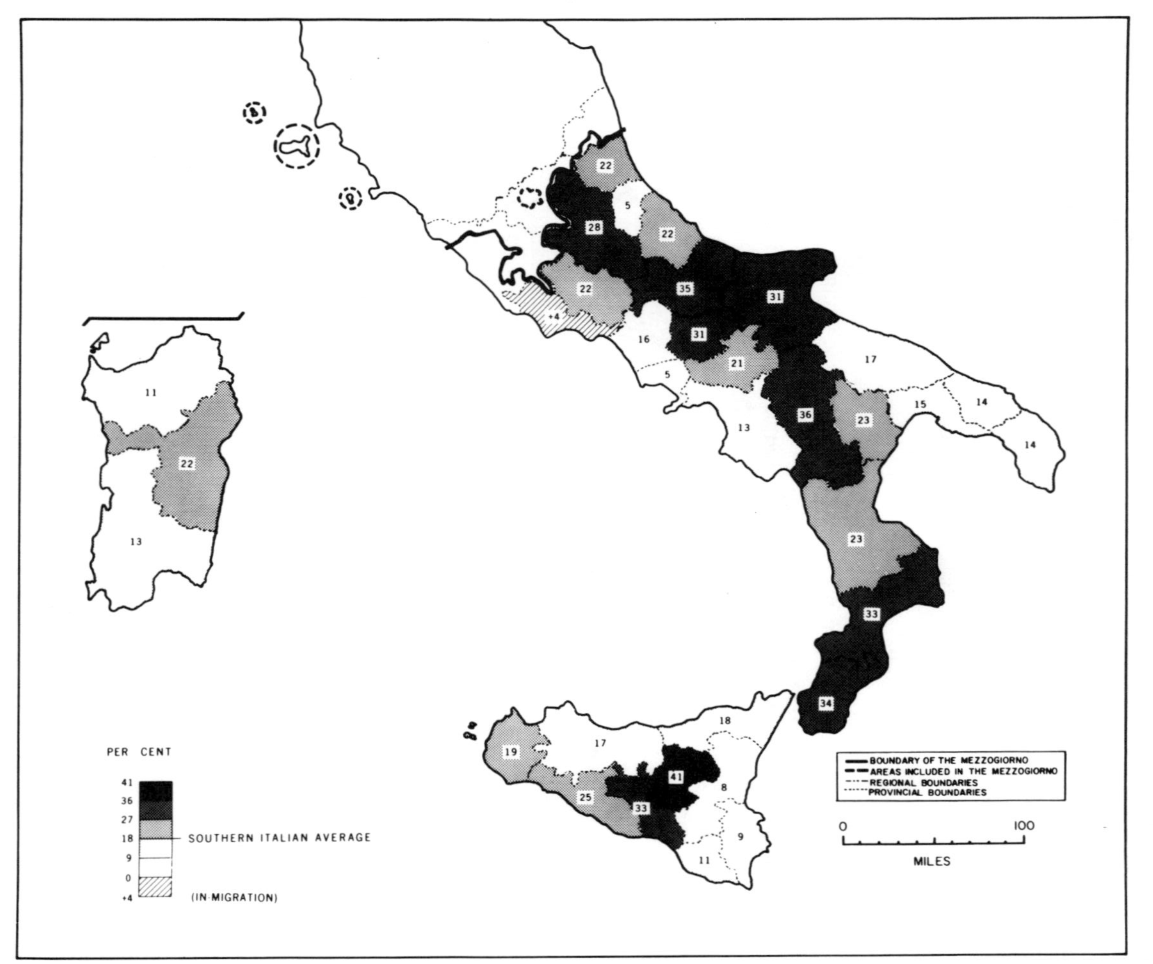

Fig. 22. Net out-migration, by province in southern Italy, from 1951 to 1972, over 1962 population.

Table 36. Net Migration Statistics for the Mezzogiorno, Based on Transfer of Residence Reports for 1955-71[a]

Province	Same region	Other Southern Italy	Northern Italy	Foreign	Total
Latina	6.5	10.5	-5.2	-5.1	6.7
Frosinone	-3.9	3.7	-60.1	-27.7	-88.0
Caserta	-4.6	-7.5	-44.9	-19.0	-76.1
Benevento	-6.7	-6.4	-33.3	-23.3	-69.7
Napoli	18.5	-14.0	-13.4	-9.0	-17.9
Avellino	-19.4	-10.3	-43.5	-28.6	-101.8
Salerno	6.0	4.2	-79.1	-18.1	-87.0
L'Aquila	-4.3	-0.3	-43.5	-20.3	-68.4
Teramo	-5.4	-1.8	-22.4	-11.4	-41.0
Pescara	14.1	2.3	-10.0	-15.9	-9.5
Chieti	-4.3	-0.9	-29.2	-31.1	-65.5
Campobasso	0.9	0.7	-112.4	-105.3	-216.1
Foggia	-6.3	2.0	-113.5	-25.5	-143.3
Bari	-2.7	-1.6	-103.9	-47.1	-155.3
Taranto	2.1	0.8	-46.4	-7.8	-51.3
Brindisi	-1.2	2.8	-29.3	-6.5	-34.2
Lecce	-6.0	2.5	-52.6	-25.8	-81.9
Potenza	-2.3	-9.5	-78.7	-23.3	-113.8
Matera	1.1	-5.0	-28.5	-5.7	-38.1
Cosenza	-1.6	-3.5	-76.0	-40.1	-121.2
Catanzaro	-4.7	0.6	-133.1	-53.8	-191.0
Reggio di Calabria	-0.6	-2.9	-15.4	-45.0	-143.9
Trapani	-0.8	0.4	-45.4	-15.6	-61.4
Palermo	5.8	0.1	-110.9	-34.1	-139.1
Messina	1.9	0.1	-59.0	-29.9	-86.9
Agrigento	-6.4	-2.4	-44.7	-31.4	-84.9
Caltanisetta	-2.3	-0.8	-50.7	-22.7	-76.5
Enna	-19.2	-1.3	-43.4	-4.0	-67.9
Catania	26.9	2.2	-59.9	-22.6	-53.4
Ragusa	1.2	-1.4	-13.5	-5.4	-19.1
Siracusa	11.4	-0.2	-15.1	-7.6	-11.5
Sassari	3.0	-0.5	-13.5	-5.4	-16.4
Nuoro	-13.5	0.8	-28.4	-9.6	-50.7
Cagliari	8.2	0.6	-86.1	-7.2	-84.5
Southern Italy	-42.8	-35.6	-1,817.1	-785.3	-2,680.8
Percent	1.6	1.3	67.8	29.3	100.0

[a]Derived from data published in the *Annuario di Statistiche Demografiche and Popolazione e Movimento Anagrafico dei Comuni*, Istituto Centrale di Statistica, 1955-72.

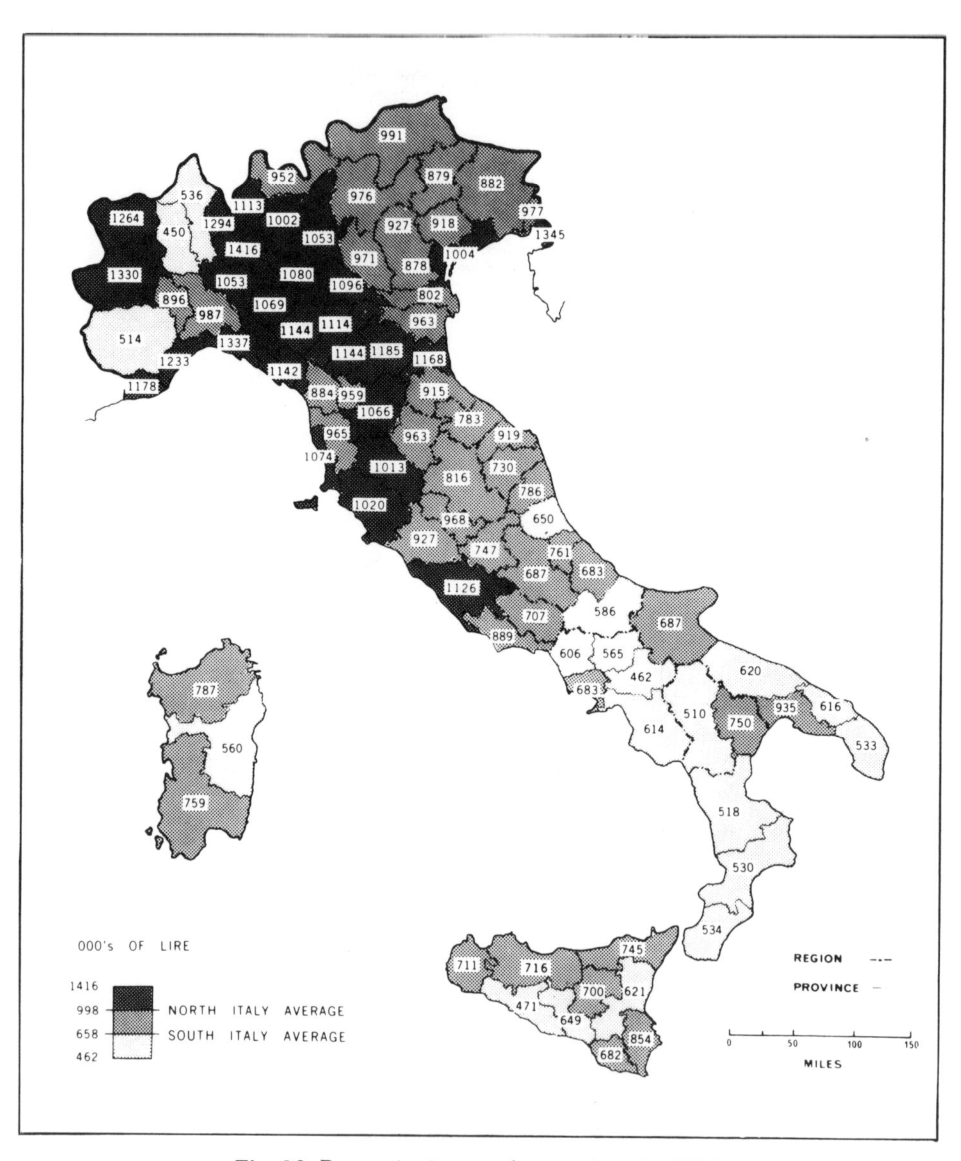

Fig. 23. Per capita income by province in 1971.

industrial development within the South,[7] I argued that there were apparently two types of migration patterns in operation in the Mezzogiorno. The first was obviously a direct and immediate flow from the South to northern Italy and abroad, while the second involved a two stage migration process. These involved movements from the countryside to nearby towns and villages, often paralleled by greater numbers of young adults who migrated from these centers to northern and foreign destinations.

It was my initial hypothesis that there would be a negative correlation between net out-migration by province over the 1952–71 period and measures of change in income during the same era. Figure 23 demonstrates the per capita income values in 1971. This map should be compared with that for 1951 (Figure 2). Incomes had clearly increased in the South but the gap between the two Italies had widened. The only positive note was the narrowing of income differentials within the Mezzogiorno. When I tested the relationship between migration and relative changes in total provincial income, the results were disappointing for the coefficient of determination for the 1951 to 1971 period was roughly 40%. I computed the provincial residuals from this simple regression and found that 15 provinces had values that were more than one standard error of the estimate (over and under prediction from the regression equation). Thus over half of the provinces of the Mezzogiorno fitted into these categories. Since migration is not the *major* focus of this book, I did not attempt (as I did in an earlier article),[8] to search, in any statistical sense, for additional explanatory variables. Detailed migration data for those provinces that were over or under predicted are shown in Table 37. The analysis that follows is based on more detailed data, from the same sources, than could be presented in this foremat, coupled with a somewhat subjective analysis of departures from predicted values.

RESIDUAL ANALYSIS

It should be stressed, at the outset, that my interpretations of these deviations from "expected" migration patterns, when regressed *solely* with one related variable, are not purely speculative. They are based on field interviews (but not with the migrants, themselves), detailed migration statistics provided by ISTAT and published socioeconomic data. I would best describe these explanations as "informed" judgments.

As shown in Table 37, there were eight provinces with greater than predicted out-migration (low positive residuals). In most of these instances, the answers appear to lie in the sheer poverty of these provinces and their extreme dependence on a depressed agricultural structure. The excessive out-migration from Teramo may be attributable to its position on the northern border of the Mezzogiorno (with heavy movements to Rome, accounting for 13% of its out-migrants) and the relative proximity of Pescara, a thriving economic center in its own region which may have offered job opportunities for some of Teramo's migrants (7% of its migrants moved to

Table 37. Migration Statistics for the Mezzogiorno, Based on Transfer of Residence Reports for 1955–71[a] (in thousands)

Province	Out-migration					In-migration					Net migration			
	Same province	Same region	Other S. Italy	N. and foreign	Total	Same province	Same region	Other S. Italy	N. and foreign	Total	Same region	Other S. Italy	N. and foreign	Total
Provinces with low positive residuals or greater than predicted out-migration[b]														
Teramo	86.6	16.8	7.1	66.4	176.9	86.6	11.4	5.3	32.6	135.9	-5.4	-1.8	-33.8	-41.0
L'Aquila	46.9	15.5	11.2	124.4	188.0	46.9	11.2	10.9	50.6	119.6	-4.3	-0.3	-63.8	-68.4
Campobasso	251.4	63.8	44.1	339.9	699.2	251.4	64.7	44.8	140.0	500.9	0.9	0.7	-217.7	-216.1
Potenza	66.3	8.2	47.5	138.3	260.3	66.3	5.9	38.0	36.5	146.7	-2.3	-9.5	-102.0	-113.8
Taranto	74.2	38.6	29.6	108.0	250.4	74.2	40.7	30.8	53.8	199.5	2.1	0.8	-54.2	-51.3
Catanzaro	139.9	29.6	30.3	256.6	456.4	139.9	24.9	30.9	69.7	265.3	-4.7	-0.6	-186.9	-191.0
Reggio di Calabria	88.9	18.7	31.5	209.8	348.9	88.9	18.1	28.6	69.4	205.0	-0.6	-2.9	-140.4	-143.9
Caltanissetta	32.0	45.2	6.4	111.6	195.2	32.0	4.9	5.6	38.2	118.7	-2.3	-0.8	-73.4	-76.5
Provinces with high negative residuals or lower than predicted out-migration[b]														
Latina	59.7	4.1	26.0	89.2	179.0	59.7	10.6	36.5	78.9	185.7	6.5	10.5	-10.3	6.7
Frosinone	69.8	11.4	21.0	160.6	252.8	69.8	7.5	24.7	62.8	164.8	-3.9	3.7	-87.8	-88.0
Pescara	64.5	25.1	8.1	63.8	161.5	64.5	39.2	10.4	38.9	152.0	14.1	2.3	-25.9	-9.5
Naples	567.4	121.1	96.4	185.9	970.8	567.4	139.6	82.4	163.5	953.0	18.5	-14.0	-22.4	-17.9
Lecce	118.9	35.3	16.7	120.1	291.0	118.9	29.3	19.2	41.7	209.1	-6.0	2.5	-78.4	-81.9
Catania	193.5	86.6	16.7	158.5	455.3	193.5	113.5	18.9	76.0	401.9	26.9	2.2	-82.5	-53.4
Enna	23.4	55.8	3.0	71.0	153.2	23.4	36.6	1.7	23.6	85.3	-19.2	-1.3	-47.4	-67.9
Southern Italy	3,993.4	1,394.4	813.1	4,624.3	10,825.7	3,993.4	1,351.6	777.5	2,021.9	8,144.9	-42.8	-35.6	2,602.4	2,680.8
Percent	36.9	12.9	7.5	44.7	100.0	49.0	16.6	9.5	24.9	100.0	1.6	1.3	97.1	100.0

[a] Derived from data published in the *Annuario di Statistiche Demografiche* and *Popolazione e Movimento Anagrafico dei Comuni*, 1955–1972, Istituto Centrale di Statistica.
[b] Provinces with residuals greater than one standard error of the estimate.

Pescara). Taranto is clearly a problem here, for it is one of the fastest growing industrial centers in the South. However, interviews there evidenced a two-stage migration process. In the first phrase, former displaced agricultural workers were attracted to this center to help build its massive steel mill. Lacking the requisite skills for steel production, these workers became unemployed after the plant went into operation. Since relatively few other industrial activities developed in the province, these workers then moved north, particularly to Turin and Milan. With respect to Caltanisetta, it should be recalled that it possesses the famous ENI petrochemical plant at Gela. The great bulk of the potential applicants for jobs at this facility from within the province did not possess the requisite skills for a petrochemical facility, so they too moved North with some pausing initially at Palermo and Catania prior to the jump to places like Turin, Milan, and Genoa.

As for provinces with high negative residuals (lower than expected out-migration), I found seven provinces had residuals that were greater than one standard error of the estimate. The answers for Latina and Frosinone appear to lie in the growth of employment opportunities within these provinces, their proximity to each other and the ready commuting accessibility to Rome. In the case of Pescara we have one of the most interesting and certainly one of the best documented cases because of the works of SVIMEZ. This province has been an important focus for *in-migration* from its own region—the Abruzzi—as is evident from Table 37. Yet Pescara has not benefited significantly from government industrial subsidies. Cafiero's excellent study of internal migration in the Mezzogiorno,[9] coupled with unpublished SVIMEZ materials, demonstrate Pescara's long history of balanced economic growth, subsidized in large measure by local capital. Other factors include the excellence of its entrepeneurial spirit (still a rarity in most areas of the South), the skills of its labor force, and above all its role as the political, social and economic capital of the Abruzzi. These have in essence been the stimuli for in-migration from the other provinces of this generally impoverished region.

Catania is another puzzle. It is clearly the destination for large numbers of migrants from adjacent provinces, double the average percentage for southern Italy. The province has had, in relation to the rest of Sicily, an unusually long tradition of commercial and industrial development. This economic growth has not only attracted in-migration, but it has also retarded outflows. Given the high level of industrial subsidization of the province of Naples, I am not surprised at the lower than predicted out-migration levels. In fact, I would have expected a net in-migration value. Note the large in-migration from neighboring provinces of the same region (Campania). Then, too, with improved communications in the region, as a whole, there is apparently significant daily commuting from these provinces to Naples, itself rather than true in-migration. Two provinces remain in this group, and I am frankly at a loss to explain their results. In the case of Lecce, there has been absolutely minimal industrial development and no significant growth in non-agricultural job opportunities; yet, Lecce has witnessed some agricultural investments and improvements. This may have dissuaded out-migration of farmers from that province; in addition there may be daily commuting to the more industrialized provinces of Puglia. As for Enna, I am truly at a loss for reasonable explanations. The lack of industry coupled with an impoverished agriculture should have produced greater out-migration. Yet the province may have already reached equilibrium levels by 1951. For even before that date, Enna was an important Sicilian exodus area. Then, too, there may be some daily commuting to take advantage of agricultural and non-agricultural opportunities in the Catanian plain.

My survey of demographic trends from 1951 to 1971 evidences a region in transformation. There was a movement through the so-called demographic transition, drastic changes in occupational structure and a vast out-migration to northern Italy and abroad. The years since 1971 have witnessed a weakening of these processes that is only in part a response to socioeconomic changes. These shifts will be treated in my review of recent trends in the socioeconomic development of the Mezzogiorno.

NOTES

[1] *IX Census of Population*, 1951.

[2] F. Spagnoli, "Popolazione Urbana e Rurale," *Sviluppo della Popolazione Italiana dal 1861 al 1961, Annali di Statistica*, Series VIII, Vol. 17, (Rome, 1965), pp. 184–187.

[3] See S. Cafiero and A. Busca, *Lo Sviluppo Metropolitano in Italia*, SVIMEZ (Rome; Giuffrè, 1970).

[4] ISTAT, Popolazione Residente per Provincia di Residenza e Luogo di Nascita, *Bollettino Mensile di Statistica*, Vol. 51, No. 12 (December 1976), Appendix I, (20% sample) pp. 275–280.

[5] Derived from data published by ISTAT in the *Annuario di Statistiche Demografiche and Popolazione e Movimento Anagrafico dei Comuni*, 1955–1972.

[6] These statistics were available for the entire study period, while the flow data were first released for 1955. Both sets of statistics will be used in my analysis. I should stress, from the start that there are some discrepancies between the data in Table 28 and these anagraphical statistics, but they were calculated on completely different bases, and, unfortunately the differences cannot readily be resolved. However, these variances do not appear to affect, in any significant sense, my analysis of migration patterns.

[7] Rodgers, 1970, p. 133.

[8] Ibid., pp. 123–128.

[9] Salvatore Cafiero, *Le Migrazione Meridionali*, SVIMEZ (Rome: Giuffre, 1964), pp. 55–57.

Chapter 8

THE EVOLVING STRUCTURE OF MANUFACTURING

Although numerous references were made in Chapter 5 to changes resulting from the implementation of the industrial development program in the Mezzogiorno, that analysis was primarily focused at the macrospatial level (with the South viewed mainly as a integral territorial unit). Nor was there particular attention paid to sectoral detail. We now move, where possible, to the regional and, if feasible, to provincial levels. Focus will be on the changing role of industry in the southern economy, industrial characteristics, as well as growth and lag sectors, all viewed in spatial perspective. It should be stressed, however, that the materials that follow do not attempt to discuss the detailed industrial geography of the South, for that is a subject which is well covered in the Italian and French literature. Nor will there be a focus in this chapter on locational patterns, because these have already been treated in Chapter 5 and will be reevaluated later in light of regional growth theory.

For many reasons, data on income, value added by manufacturing and investment at a high level of sectoral and spatial detail are either unavailable or of recent vintage in Italy. Therefore, in this chapter, at least, employment statistics (structural, temporal, and spatial) will be the main bases for my analysis.

CHANGING ROLE OF INDUSTRY IN THE ECONOMIC STRUCTURE OF THE SOUTH

If we examine the data in Table 38, there was obviously an increase in every southern region of the relative share of those employed in industry, but that growth was minimal compared to reductions in the agricultural labor force and expansion in the "all other" category. The latter group mainly comprises employment in commerce, services, and construction or what are commonly termed tertiary functions. In the case of the South, that increase was not necessarily a reflection of sharply increased demand. To a significant degree, the growth portrayed was concentrated in marginal activities which were clearly linked to the drastic decline of the agricultural population. Their relative and absolute growth then was not necessarily related to improvements

Table 38. Changes in the "Active Population" of the Mezzogiorno, by Sector and Region from 1951 to 1971

Region	Industry[a]	Agriculture	Other	Total	Number (thou.)
		Sectors, 1951 (%)			
Abruzzi and Molise	10.9	64.7	24.4	100.0	706.1
Campania	22.1	53.3	24.6	100.0	1,359.1
Puglia	16.1	58.2	25.7	100.0	1,270.7
Basilicata	8.3	73.1	18.6	100.0	287.2
Calabria	10.0	63.4	26.6	100.0	783.3
Sicily	13.7	51.3	35.0	100.0	1,482.9
Sardinia	15.2	50.4	34.4	100.0	438.8
Southern Italy	14.8	57.3	27.9	100.0	6,640.9
Northern Italy	28.9	34.5	36.6	100.0	12,936.2
		Sectors, 1971 (%)			
Abruzzi and Molise	18.7	32.0	49.3	100.0	504.3
Campania	24.1	23.9	52.0	100.0	1,481.9
Puglia	20.7	36.9	42.4	100.0	1,165.1
Basilicata	14.3	39.6	46.1	100.0	203.5
Calabria	14.5	32.6	52.9	100.0	609.2
Sicily	18.0	28.7	53.3	100.0	1,324.4
Sardinia	19.0	21.4	59.6	100.0	422.9
Southern Italy	20.0	29.5	50.5	100.0	5,966.1
Northern Italy	38.6	11.5	49.9	100.0	12,865.0

[a]Industry here includes manufacturing *plus* extractive activities.

Table 39. Changes in the Gross Income (in Current Prices) of the Mezzogiorno by Sector and Region from 1963 to 1971[a]

Region	1963 Sectors (%)[b]			1971 Sectors (%)[b]		
	Manu-facturing	Agricul-ture	Other	Manu-facturing	Agricul-ture	Other
Abruzzi and Molise	12.1	28.4	59.5	14.4	19.1	66.5
Campania	19.6	17.1	63.3	19.3	14.3	66.4
Puglia	12.8	26.8	60.4	16.0	21.4	62.6
Basilicata	8.4	32.2	59.4	14.3	22.9	62.8
Calabria	8.7	31.0	60.3	8.7	20.0	71.3
Sicily	11.5	22.0	66.5	13.4	16.0	70.6
Sardinia	9.9	22.2	67.9	12.9	15.8	71.3
Southern Italy	13.1	22.6	64.3	15.2	17.5	67.3
Northern Italy	32.4	10.7	56.9	31.8	7.0	61.2

[a]Unione Italiana delle Camere di Commercio ed Industria, *I Conti Economici Regionali, 1963–1971*, (Milan: Franco Angeli, 1973), pp. 164–1999.
[b]Percentages for each region total 100.0 (mfg. + agr. + other = 100.0).

in the socioeconomic health of southern Italy. Finally, an examination of this table reveals a general regional uniformity of the industrial changes (at least in proportional terms). This seems to belie our earlier emphasis on the spatial differentiation of industrial investment and employment shifts. However, the disparities reflect, in part, aggregation at regional rather than provincial or communal scales. Despite this limitation, these data do demonstrate an absolute growth in the share of industry in the Mezzogiorno, yet that ratio was still far lower than was the case in the North. Ideally, these employment figures should be supplemented by income data, by sector and region, for 1951 and 1971, but that was not possible, for the published 1951 industrial income statistics included sectors other than manufacturing. However, comparable data do exist for the period between 1963 and 1971, and the shares of income by sector and region for both years are demonstrated in Table 39. The notable decline of agriculture is apparent as is the modest growth in the share of income derived from manufacturing. Note the spatial variations in the regional values. Thus Campania was clearly the leading region in terms of manufactural income in both years; yet, despite heavy subsidization of industry in the form of loans and direct grants, the percentage of income derived from this sector remained remarkably stable. Note, in contrast, the relative growth of manufacturing in such regions as Puglia, Basilicata, and Sardinia.[1]

INDUSTRIAL CHARACTERISTICS

It has been noted earlier that despite the limited growth of manufacturing in the South during the study era, changes in the region's industrial structure might be more significant than increases in the numbers employed. Macro-level shifts were demonstrated by comparing Tables 7 and 13.

That comparison clearly demonstrated an improvement in the structure of manufacturing in southern Italy whether it be size of firm, proportion of employees in firms using electric motive power, or the share of those employed in first order industries.

We now turn to the regional values for the two-decade interval. Table 40 demonstrates the size of firm data. Note that the employment in plants with over 100 employees increased from 19 to 35% for the South as a whole, while

Table 40. Percentage of Employees in Manufacturing by Region and Size Group in 1951 and 1971[a]

1951					
Region	0–10	11–100	101–500	501 plus	Total
Abruzzi and Molise	72.6	15.2	6.2	6.0	100.0
Campania	45.4	19.2	14.9	20.5	100.0
Puglia	65.1	15.6	11.3	8.0	100.0
Basilicata	89.9	10.1	0.0	0.0	100.0
Calabria	79.8	13.2	4.2	2.8	100.0
Sicily	74.6	16.8	4.8	3.8	100.0
Sardinia	74.0	16.9	5.3	3.8	100.0
Southern Italy[b]	64.3	16.6	9.2	9.9	100.0
Northern Italy	25.9	23.2	22.5	28.4	100.0

1971					
Region	0–9	10–99	100–499	500 plus	Total
Abruzzi and Molise	39.5	29.1	11.8	19.6	100.0
Campania	29.1	25.6	18.9	26.4	100.0
Puglia	42.8	22.1	13.6	21.5	100.0
Basilicata	50.3	17.1	15.4	17.2	100.0
Calabria	62.9	23.0	8.7	5.4	100.0
Sicily	51.4	21.4	12.6	14.7	100.0
Sardinia	46.2	26.1	18.7	9.0	100.0
Southern Italy	41.2	24.0	15.1	19.7	100.0
Northern Italy	20.5	32.3	23.4	23.8	100.0

[a]Note the slightly different size groupings in 1951 and 1971. Regretfully, no exactly comparable data were published but the differences are clearly minimal.
[b]Data exclude southern Lazio (Frosinone and Latina).

the share of those employed in the smallest factories (less than 9 to 10 workers) declined sharply from 64 to 41%. The raw data from which these percentages were derived demonstrated an absolute decrease in employment in the small workshops (on the order of 50,000 workers), and a significant increase in those employed in the larger and presumably more modern factories (over 160,000 workers). To keep these values in perspective, it should be recalled that the overall increase in manufacturing employment in the Mezzogiorno during this period was only 200,000. Thus these changes can be viewed as a positive shift in the structure of southern industry, for many of these small plants were marginal activities that were commonly undercapitalized with low productivity per worker. Most were, in essence, handicraft workshops. On a regional basis, the largest reductions of employment in these small firms were registered in Calabria and the less industrialized segments of Sicily. In contrast, as might have been expected from the analysis in Chapter 5, the major gains in the larger plants occurred in Campania and Puglia. It would have been useful to examine regional data on artisan employment in 1951 and 1971. Unfortunately, the published values for these two census years are not comparable because of definitional changes. Were such materials available, it is my educated guess that they would reveal a significant decline of artisan employment in the South paralleling the reduction in the numbers employed in small workshops.

There are several other measures which provide us with additional evidence of industrial transformation and structural change. Here, again, I have used the Census of Industry data for 1951 and 1971.

The first criterion to be tested is the changing share of the manufacturing employment in plants that lacked electric motive power. Such establishments would presumably still be in the handicraft stage of production. The results of

Table 41. Characteristics of Manufacturing by Region in 1951 and 1971

Region	Percent of employment in firms without motive power		Percent of employment in first stage industries[a]	
	1951	1971	1951	1971
Abruzzi and Molise	52.1	24.6	98.2	80.7
Campania	35.8	26.3	79.7	61.7
Puglia	52.7	29.1	88.3	70.1
Basilicata	75.3	29.1	98.5	66.7
Calabria	62.3	44.1	94.7	84.0
Sicily	53.1	35.5	89.9	69.3
Sardinia	51.6	24.7	93.0	77.2
Southern Italy	48.9	29.5	87.3	68.3
Northern Italy	15.5	10.7	61.9	52.3

[a]First stage industries were defined in Chapter 2.

my compilation are portrayed, by region, in Table 41. For the Mezzogiorno that share declined from 50 to 30% in this two decade interval. In absolute terms, this meant a reduction of roughly 60,000 employees in these workshops with the bulk of the contraction confined to Abruzzi and Molise, Calabria and Sicily. Apparently, judging from changes in the numbers of these establishments, many small workshops failed during this era. Their closing may have resulted from intense competition with firms in northern Italy and, conceivably, with newly established plants within the Mezzogiorno itself.

Another criterion worth testing is the changing proportion of employment in "first stage" industries. Here I am considering the composition of manufacturing by region. As industrialization progresses in an underdeveloped region, the process typically begins with activities which require limited capital, need no unusual technical skills, and depend on either proximate raw materials or inputs whose transport costs are a relatively small share of total costs. Such industries might include food processing, wood products, textiles, clothing, leather and shoes, machine repair shops, and stone, clay, and glass fabrication. Obviously, there may be assembly-line, mass production establishments in all of these sectors, so that this grouping is of necessity relatively arbitrary. Nevertheless, such modern plants in these industrial branches are still uncommon in the South.

Despite the evident decline in the relative importance of these traditional industries in every region of the South, the absolute data on which these percentages were based do not provide conclusive evidence of significant shifts to the more modern sectors. In fact, in three regions–Abruzzi and Molise, Calabria, and Sicily–there was an absolute increase in the numbers employed in the more traditional sectors. There are two possible conclusions that can be drawn from these data; the first obvious deduction is that my grouping and classification procedure was inappropriate, for it clearly conceals modernization and the creation of new assembly-line facilities within the first order industries themselves. Secondly, it should be recalled that a large share of the loan and direct contribution funds were allocated to capital intensive industries; therefore, the modern sectors did not have commensurate increases in employment. It is my admittedly subjective judgment that both of these explanations are reflected in these data. A sample survey of firm size and investment per employee in the traditional branches does appear to support these conclusions, but more detailed analysis would be required to verify them.

GROWTH AND LAG SECTORS

Our attention now shifts to the statistics by industrial branch and region which are demonstrated in Table 42. The sectors that experienced notable growth in employment were textiles, clothing, stone, clay and glass, machine repair shops, metallurgy, fabricated metals, machinery, and chemicals. In contrast, there were major declines in the food, shoe, lumber and furniture

Table 42. Absolute Changes in Manufacturing (Employment) in Southern Italy by Sector and Region from 1951 to 1971

Sector \ Region	South Lazio	Abruzzi and Molise	Campania	Puglia	Basilicata	Calabria	Sicily	Sardinia	Total
Foods and tobacco	3,098	-4,786	-8,758	-16,131	-1,212	-8,857	-15,445	-352	-52,416
Textiles	3,031	1,013	-6,561	9,002	434	1,284	2,802	4,443	15,448
Clothing	1,722	4,449	7,044	10,359	199	-3,536	1,547	-1,295	20,489
Leather	183	1,202	1,604	-639	-74	-80	-603	-118	1,475
Shoes	-1,034	-3,212	-3,345	-3,187	-2,038	-3,298	-10,334	-2,528	-28,976
Lumber	1,240	-27	-760	2,410	-882	-3,900	-2,053	433	-3,539
Furniture	527	49	-2,049	-848	-809	-2,359	-2,007	-517	-8,013
Stone, clay and glass	3,872	6,518	6,248	7,315	542	2,084	5,803	3,511	35,893
Mach. repair	2,864	2,656	12,305	11,095	355	1,742	9,166	3,304	43,487
Metallurgy	849	655	6,316	14,672	484	243	549	-705	23,063
Nonelectr. machinery	4,887	325	5,862	1,489	291	33	860	-219	13,528
Electr. machinery	5,082	4,180	16,024	2,283	589	66	4,723	257	33,204
Transport equipment	837	417	9,079	4,816	407	-298	4,063	152	19,473
Fabricated metals	2,093	2,089	6,039	5,703	355	1,535	5,114	1,986	24,914
Chemicals	4,339	238	5,153	3,281	3,193	673	10,133	4,445	31,445
Rubber	2,542	235	1,135	2,391	471	216	2,036	138	9,168
Paper	177	766	952	-270	107	137	1,223	523	2,661
Printing	860	334	1,347	1,018	79	219	517	444	4,818
Varied	722	679	4,805	1,619	346	215	2,362	581	11,329
Total	38,791	17,780	62,444	56,378	2,837	-14,311	20,456	1,483	197,958

industries. These data reinforce the notion that essentially all of the modern sectors demonstrated significant growth. The converse was true of four of the "traditional" industries in the South. Of course, these growth and lag patterns varied markedly by sub-sector, by region and by province. Although data at the provincial scale will be used in my written analysis, they are not demonstrated here in a systematic tabular format.

The greatest absolute reductions in employment came in the food and tobacco industries. With the exception of southern Lazio, every region of the South shared in these declines. It might be hypothesized that these losses were highly correlated with the declines in agricultural employment, by province and commune, in the Mezzogiorno. However, in fact, the chief reductions occurred in the plains rather than the hills. Not unexpectedly, major contractions were found in the numbers of small food processing workshops which had traditionally dotted the southern lowlands.

Second, in order of magnitude, was the sharp decline in the shoe industry. Again, the losses were spread rather evenly through the South. Here, too, it was the artisan workshops that suffered so dramatically. Losses in the lumber and furniture branches were far less striking. It should be noted, however, that there was a rise in employment in the lumber industry in Latina (southern Lazio) coupled with increases in Bari and Lecce (Puglia) which ran counter to the general trend for that industrial branch.

Turning now to the growth sectors, for some the increase was true of all regions, while for others the regional and provincial values are quite mixed. For example, there was an overall growth of more than 15,000 employees in the textile industry; yet that increase came mainly in synthetic fibres, wool and knit goods rather than in the traditional silk or cotton industries, which, in fact, declined. In addition, the overall positive increment masks significant *losses* in Naples and Salerno (Campania) contrasted with major additions in Bari and Lecce (Puglia). The stone, clay and glass sector grew dramatically in all regions but that increase was confined mainly to cement and glass production. Employment growth in the metallurgy branch came almost exclusively in the integrated steel plants of Bagnoli, a suburb of Naples (Campania) and Taranto (Puglia). The increase of employment in machinery production came primarily in Caserta and Naples (Campania) and southern Lazio. While the transport equipment industry did grow notably in Naples, later data would have shown a far greater increase with the entry into production of the Alfa-Sud automobile plant. As noted in Chapter 5, the growth of the petrochemical sector was one of the highlights of industrial change in the Mezzogiorno. Note that in this highly capital intensive branch of manufacturing practically all regions shared in the expansion of employment. Yet, in fact, a commune map would show that the growth was confined to Mellili and Gela in Sicily, Brindisi (Puglia), Porto Torres (Sardinia), Matera (Basilicata), and Naples. More recent data would have shown a far greater growth in Sardinia than evidenced in this table. It is abundantly clear from this brief analysis that the growth and lag patterns were extremely complex, whether by branch or area.

In sum, it has been argued in this chapter that, despite minimal absolute changes in industrial employment in the South, there had been a notable relative growth of the role of industry in the economy of the Mezzogiorno. Then, too, there had been significant alterations in the spatial, sectoral, and structural facets of southern industry. However, I hasten to add that a reduction in employment in traditional industries and a decline in the number of handicraft workshops and their employees are not necessarily positive features of economic development in an underdeveloped region. This is particularly true when those employment opportunities are not replaced by suitable options in factory-type, assembly-line enterprises. Thus, I do not accept the view that artisan activities are necessarily unproductive or marginal in nature. Any assessment clearly depends on local circumstances.

My analysis to this point has confined itself primarily to the first two decades of the development program and the impact of that effort on socioeconomic change within the South. I have also attempted to appraise the degree to which the differences between the two Italies had narrowed. The results, clearly, were mixed. Unquestionably the gap had not been bridged to the dismay of southern advocates. Concomitantly, however, socioeconomic conditions within the South had distinctly ameliorated but only in a relative sense. Such improvements were certainly not in accordance with the hopes and perhaps the dreams of the southerners. Unquestionably, the results had fallen far short of their expectations.

NOTES

[1] I am convinced that data on a provincial scale would be far more revealing. Such statistics have been calculated for "industry," a far too inclusive category, but, unfortunately, not for manufacturing alone.

Chapter 9

DEVELOPMENTS IN THE 1970s

Seven years have passed since the census enumerations of 1971. These years have witnessed marked economic, social, and political change in Italy. While these developments have affected the nation as a whole, their influence has been felt most notably in the Mezzogiorno. This has been an era of major increases in investment in southern Italy, but contemporaneously there have been significant alterations in the development program, itself; economic stagnation; the virtual cessation of out-migration; social and political change. It is the purpose of this chapter to describe and assess the impact of these events insofar as they have affected the economic development of the South.

RESTRUCTURING THE DEVELOPMENT PROGRAM

There have been so many shifts in the development legislation since 1970 that it is difficult to keep track of such changes. A major modification of the development policy was approved in October of 1971 (N. 853), and this new law was subsequently amended in May of 1976 (N. 183).[1] This legislation assured financing through 1980 (16,000 billion lire) and extended the program until 1990. Thus, its implications will presumably affect the South for many years to come.

Given the primary focus of the study on the 1951 to 1971 period and the fact that this new legislation has been thoroughly discussed in the Italian literature, I will only summarize the key innovations. Since the two laws are so highly interrelated, they will be referred to here as the new legislation or the new program.

In broad terms, the new legislation supports and strengthens the role of "planning" in southern economic development. Investment in the Mezzogiorno *must* now take place within the overall framework of national social and economic goals. Thus, the role of CIPE (Interministerial Committee for Economic Planning) in the determination of the direction of the development program has been sharply increased. However, at the same time much more power has been delegated to the "regioni" or regional governments. Thus these two elements of the new program could conceivably run counter to each other. I should add that the role of the Cassa has been markedly reduced.

More specifically the following are the chief elements of the new program as of 1978:

(1) Private firms whose assets are over 5 billion lire (roughly $8,000,000 at current exchange rates) and all State enterprises in Italy must submit their planned investment programs to the Ministry of Budget and Economic Planning. It is now obligatory for firms proposing to construct new industrial plants or expand existing ones, where the investment exceeds 7 billion lire, to submit their proposals to the same authority. CIPE, working closely with this Ministry, can issue a negative decision if such a project would contribute to further congestion or if the new plant is not located in an area of significant unemployment. Firms which ignore such negative assessments could theoretically be subject to severe fiscal penalties. I should add, however, that I have no personal knowledge of the enforcement of such fines. In contrast, all State organizations *must* adhere to CIPE decisions. Such entities must also submit 5-year plans for investment in the South to the central planning authorities.

(2) State-owned or controlled industries must now devote at least 80% of their new investments and 60% of overall investments to the Mezzogiorno. In addition, 40% of the investments programmed by public administrations and 30% of the value of all contracts involving purchase of industrial products must be allocated to the South. Both provisions are to be far more strictly enforced by CIPE than was true in the past.

(3) The support of industrial growth has been sharply increased at the expense of agriculture. Infrastructure, however, does remain a relatively high priority, particularly in what are termed the "Special Projects" (such as the decontamination, now underway, of the Bay of Naples).

(4) The loan and grant program has been restructured so as to give greater priority to small and medium sized firms. Grants are increasingly becoming a major feature of the investment program. Their size can range from as high as 40% for investments up to $2,400,000 (2 billion lire) down to 15% for investments as high as $18,000,000 (15 billion lire). In addition, firms planning to invest in the South can continue to get low interest (4 to 6%)–long term (10 to 15 years) loans as in the past. The normal upper limit for both forms of aid is 70%, but this value can be raised to as much as 86%

if the projects are located in particularly depressed areas that have suffered severe depopulation and if the proposed industry is in a "priority sector." In the latter instance, the move is away from the support of basic industries like primary petrochemicals and steel (which received the lion's share of the financing in the past) so often with minimal multiplier effects, to industries that because of market conditions and technological efficiency were needed in the South. These are the so-called growth sectors like machinery, metal fabrication, electronics and a host of secondary chemical products. As in the past, the "traditional" industries will not be encouraged except where a major technological innovation is envisaged. First order industries in those sectors considered to be saturated (with respect to market potential) are to be actively discouraged. Similarly, as noted above, proposed new plants in the congested areas of southern Italy are to be curbed. However, despite continued discussion of the possible adoption of an "Industrial Development Certificate" program, similar to that in Britain where it was designed to curtail further industrial growth in London, this method has not been adopted for the heavily industrialized areas of the Po Valley. In fact, modest sums are still available, under the new legislation, for industrial loans (not grants) for the so-called depressed areas of the Centro-Nord. Proposals for aid for large plants (with investments over 15 billion lire) must now be approved by CIPE. Another change is the greater emphasis on labor-intensive industries. Establishments with low investments per worker are to receive priority. There is also continuing discussion of providing a cash subsidy for each additional worker to be gained through the constructions of a new plant or by modernization or expansion of existing facilities.

(5) "Leasing" is to be encouraged, but this feature, common to programs in western Europe and the U.S., has still not attracted the imagination of prospective investors in southern Italy.

(6) The rebates on social insurance, designed to reduce labor costs in the Mezzogiorno, have now been strengthened; now *all* such charges are to be waived.

(7) The most perturbing part of the new legislation is the essential elimination of the locational advantages of the "agglomerati" of the areas and nuclei. The current program, at least on paper, appears to be even more diffusive in character than the previous legislation. The focus on the most depressed regions which have suffered "severe out-migration" appears to be another response to political pressure. It is clearly not an economically rational approach, given the scarcity of investment funds and the impact of the economic recession which remains to be discussed. The other disturbing element of the new program is the decentralization of the development efforts by reducing the power of the Cassa and transferring many of its responsibilities to the "Regioni."

ECONOMIC PERFORMANCE IN THE 1970s

The years since 1970 have been an era of erratic economic growth in Italy, punctuated by intervals of minimal increase and one period of absolute decline. Studies by Moore and Klein[2] have demonstrated that Italy's economic performance during this epoch has lagged behind that of all of its partners in the European Economic Community. In fact, it proved necessary for Italy to borrow huge sums from West Germany and the International Monetary Fund. Although these debts are being repaid, the nation has yet to surmount these recurring economic crises. The 1970s have also been an era of severe unemployment, particularly in the South, coupled with inordinately high inflation which, though still excessive, has now been reduced to more acceptable levels. Explanations for this series of events are beyond the scope of this study, but undoubtedly they were triggered by the burden of vastly increased payments for Middle Eastern and Libyan oil, escalating labor costs due in part to labor contract provisions, low productivity, lagging entrepeneurial initiative and, I suspect, fiscal mismanagement.

The history of sluggish and erratic growth could be illustrated by a variety of temporal indicators such as: gross domestic product, value added, investment, unemployment, etc. I will use only one of these measures here to document these trends. Table 43 illustrates the growth of gross domestic product from 1970 to 1977, at 1970 constant prices, in the two Italies.

Economic fluctuations in the South were clearly not as drastic as those in northern Italy and in some instances ran counter to the experience of its far more prosperous counterpart. This differential economic growth would appear to indicate that in the short run, at least, southern Italy's performance was far more stable. Podbielski[3] has argued, most effectively, that this variance could reflect the structural backwardness of the Mezzogiorno with its greater dependence on the production of consumer goods and agricultural products for local consumption. This, she argues, would make it far less sensitive to national and international economic fluctuations than northern Italy with its far greater emphasis on producer's goods. Her reasoning with respect to investment is that the more favorable position of southern Italy in 1975 may reflect the maturation of earlier investment decisions by private and public agencies.

On the positive side, industrial employment in southern Italy has grown at a faster rate than in the North (see Table 44). While these data are not comparable with those from the 1971 Census, the growth was roughly 10% in 5 years. However, that increase has hardly been commensurate with the decline in agricultural employment.

It is also noteworthy that southern industrial growth surpassed that of the North in both relative and absolute terms. As might have been expected, the absolute increases on a "regioni" basis were greatest in Campania and Puglia. What is perhaps surprising is the growth in the northeastern segment of the region (Abruzzi and Molise) and the minimal change in Sicily. This contrast is reinforced by the net shift data (in which "regioni" growth is compared to that of the South as a whole). Note the large negative shift for Sicily contrasted with the positive values for Puglia and Abruzzi and Molise. Notice too the slight negative shift for Campania which indicates that its growth barely kept up with that for the entire region.

Table 43. Changes in Gross Domestic Products in Italy from 1970 to 1977[a]

Year	Italy	Northern Italy (billions of lire at 1970 constant prices)	Southern Italy
1970	57,937	44,445	13,492
1971	58,836	44,759	14,077
1972	60,689	46,494	14,195
1973	64,905	49,718	15,186
1974	67,459	51,693	15,766
1975	65,086	49,392	15,695
1976	68,752	52,641	16,111
1977[b]	70,226	?	?
Years	**Italy**	**Northern Italy (% change)**	**Southern Italy**
1970–71	+1.6	+0.7	+4.3
1971–72	+3.1	+3.9	+0.8
1972–73	+6.9	+6.9	+7.0
1973–74	+3.9	+4.0	+3.8
1974–75	-3.5	-4.5	-0.5
1975–76	+5.6	+7.1	+3.7
1976–77	+2.1	?	?

[a]"Principali Aggregati Economici a Livello Territoriali, Anni 1970–76," Istituto Centrale di Statistica, *Bolletino Mensile di Statistica*, January 1978, Appendix, p. 245. Northern Italy includes southern Lazio.

[b]*Relazione Generale sulla Situazione Economica del Paese*, 1977 (Rome, 1978) as reproduced in *Mondo Economico*, Vol. 33, No. 14, April 8, 1978, pp. 33–68.

THE CHANGING ROLE OF THE STATE-CONTROLLED FIRMS

I commented in Chapters 3 and 5 on the role of State agencies in the industrial growth of the South during the initial two decades of the development program.[4] There I stressed the fact that these agencies had been potent forces in the development effort, for a large share of the new investments between 1951 and 1971 and a significant proportion of the growth in industrial employment in the Mezzogiorno could be attributed to investments by IRI, ENI, EFIM-BREDA etc. However, since these investments were almost exclusively in highly capital-intensive industries (steel, petrochemicals, cement, etc.), their contribution to the increase of employment was not

Table 44. Estimated Changes in Industrial Employment by Region in the Mezzogiorno, 1971 to 1976[a]

Region	1971	1972	1973	1974	1975	1976	Absolute change	Net shift[b]
			(in thou.)				(1971–1976)	
Abruzzi and Molise	89.4	90.0	96.4	105.6	107.6	108.0	+18.6	+10.0
Campania	299.4	299.3	311.3	323.6	322.6	326.2	+26.8	−0.6
Puglia	194.5	199.3	205.5	214.5	218.6	222.6	+28.1	+9.6
Basilicata	29.0	28.6	28.9	30.1	31.1	31.3	+2.3	−0.5
Calabria	55.3	53.4	53.6	55.0	55.7	57.8	+2.5	−2.8
Sicily	202.1	198.3	200.6	202.8	204.2	204.3	+2.2	−17.0
Sardinia	54.5	55.2	56.5	60.4	61.2	62.2	+7.7	+2.5
Southern Italy	924.2	924.1	952.8	992.1	1,001.0	1,012.4	+88.2	
Index	100.0	100.0	103.1	107.3	108.3	109.5		
Northern Italy	4,560.6	4,497.8	4,557.6	4,646.8	4,625.9	4,633.9	+73.3	
Index	100.0	98.6	99.9	101.9	101.4	101.6		

[a] *Occupati per Attivita Economica e Regione, 1970–1976*, Istituto Centrale di Statistica, Collana d'Informazioni, Vol. I, No. 3 (Rome, 1977),various pages. These values are estimates rather than a true census, but they are considered reasonably reliable. The classification code for industry is based on standard EEC categories. Note the major difference between these values and those for the 1971 Census (924,000 vs. 715,000).

[b] Net shift values based upon the average employment growth rate for Southern Italy during this period.

nearly as great as the proponents of southern development had envisaged. There was virtually no spin-off or multiplier effect such as the growth of small- and medium-sized establishments that were linked to the basic industries. Nevertheless, these expenditures, unquestionably, did lead to a growth of industrial output and gross income in the region.

At this point, I plan to discuss the performance of the state sector in the 1970s. It should be recalled that the 1971 legislation had decreed an increased role for State-owned or controlled facilities. It was now "mandatory" that all of these industrial establishments allocate at least 80% of their investments to new facilities (as opposed to 60% under previous legislation) and 60% of their overall investments (compared to 40%) in the South.

Table 45 shows the investments by these firms in southern industry in relation to total industrial investments during this era. There was clearly an absolute and relative decline in investments by state firms in the Mezzogiorno during the 1972 to 1976 period. Part of this decrease was undoubtedly attributable to the completion of the major expansion program at the Taranto

Table 45. Industrial Investments by State-Controlled Firms from 1954 to 1976 in Southern Italy in Relation to Total Industrial Investments in that Region
(billions of lire at current prices)

Year	Total investments by state firms in Italy[a]	Investments by state firms in the South[a]	(%)	Total investments in the South[b]	Percent of total investment by state firms in the South
1957–70[c]	401.5	182.2	45.4	492.6	37.0
1971	1,224.3	738.0	60.3	1,585.5	46.6
1972	1,338.9	869.0	64.9	1,852.4	46.9
1973	1,361.2	789.2	58.0	1,856.8	42.5
1974	1,720.8	876.4	50.9	2,454.1	35.7
1975	1,620.9	733.6	45.3	2,318.2	31.6
1976	1,792.1	699.8	39.0	2,395.7	29.2
Total 1957–76	14,679.4	7,256.2	49.4	19,358.4	37.5

[a] *Annual Reports* of the Ministero delle Participazioni Statali. The term "industrial" in this instance includes production of energy and the distribution of petroleum and petroleum products.

[b] "Conti Economici Territoriale" in *Bolletino Mensile di Statistica, Supplemento Straordinario*, Istituto Centrale di Statistica, No. 9 (Rome: Sept. 1974), and "Principali Aggregati Economici a Livello Territoriali, Anni 1970–76" in *Bolletino Mensile di Statistica*, No. 1 (Rome: Jan. 1978), p. 249. South does not include southern Lazio.

[c] Average annual values for the 1957 to 1970 period.

steel mill. However, as Saraceno and Podbielski[5] have noted, the economic crises of the 1970s forced IRI and other State-holding companies to engage in major rescue efforts to aid faltering public and private firms in northern Italy. These, in turn diverted resources from the Mezzogiorno. Thus investments by State agencies in southern Italy during the 1970s fell far below the requirements of the 1971 legislation. Nevertheless, expenditures by State agencies in the region did continue and there was a modest increase in the industrial employment of these firms. The most important of these investments financed the building of the Alfa-Sud automobile plant.[6] That facility, alone, increased employment in the Province of Naples by roughly 16,000 workers. In this instance there has been a modest multiplier effect in the construction of plants to provide components to Alfa-Sud; many of these are also State-owned. However, this facility has suffered severe financial losses in each year of its operation. These losses are mainly attributable to excessive absenteeism by its work force and low labor productivity. As of June 1978, management believed that these problems were under control, but similar optimism in the past proved erroneous. The State-owned integrated steel mill at Bagnoli, located near Naples, has also been a financial burden for IRI, but here the problem has been the lack of space, for the plant is surrounded by residential quarters.

PERFORMANCE OF THE AGRICULTURAL AND TRANSPORT SECTORS

I indicated in Chapter 6 that agricultural performance in the late 1960s had exhibited some evidence of improvement, but the resulting pattern was mixed. There had been the beginnings of a long overdue structural transformation of agriculture, yet that change was accompanied by a continued out-migration from the land. There were also marked regional variations in these outbound flows which now meant labor surpluses in some areas and labor scarcities in others. These imbalances retarded fuller use of the best farmland. Nevertheless, the land reclamation expenditures of the Cassa, and, in particular, the expansion of irrigated acreage had finally begun to bear results.

By the early 1970s, southern agriculture had become far more intensive and productivity had clearly improved. At first, the greater focus on the production of fruits and vegetables brought sizeable financial returns, but, in time, southern farmers began to face intense foreign competition in both export and domestic markets. Their main rivals appear to be Israel, Spain, Greece, and Morocco, and that competition is based both on lower costs and higher quality. The immediate result has been a virtual stabilization of fruit and vegetable production in the South. In the reverse sense, with changing income and dietary patterns in northern Italy and to a lesser degree in the South, there have been increased demands for meat and dairy products from northwestern Europe and even from New Zealand. On a national scale this may mean an actual food deficit.

Conceivably, southern Italy may have the potential to counter both of these problems. First, there is an evident need to improve the quality of its fruit and vegetable products and market them at competitive prices. This means that greater technical assistance must be made available to the farmers; production and marketing cooperatives should be encouraged and subsidized; and unquestionably there must be a better utilization of the irrigated acreage, which is in some instances still devoted to cereal crops. Then, too, it appears conceivable that some of the newer irrigated lands could be used for the production of fodder crops. This might make it possible to increase the numbers and quality of beef and dairy cattle in the Mezzogiorno and thus help to satisfy both regional and national markets.

Turning to achievements in the transport sector, I have already covered the limited available statistical evidence for the 1970s in Chapter 6. While I found no detailed regional railroad data beyond 1973, I had hoped that the 1975 Censimento Stradale (road census) would be available before the completion of this study. Regretfully, the data have not been released. Therefore, of necessity, I will treat recent transport developments in broad brush fashion.

First, it is clear that minimal change had occurred in the rail network in this era. Although there were modest improvements in trackage, electrification and equipment, railroad traffic whether measured in tons or ton-kilometers actually declined from 1970 to 1975.

Figure 15 is still a reasonable portrayal of the main highway network. With respect to the evolution of the road network, developments in the 1970s in the Mezzogiorno were truly monumental. It was in this era that a unified system of superhighways was completed in southern Italy, providing the region with a transport infrastructure that was vital for its future economic development. Some southerners argue that the massive resources used for the building of this system could have been better utilized in other sectors. I find myself in total disagreement with that viewpoint, for it is extraordinarily short-sighted. I would admit that so far the network is underutilized, but that is a short term perspective. What is still badly needed is the building of additional high capacity access roads and peripheral superhighways which would permit the bypassing of congested urban metropoli. Such a series of roads has now been completed in the Naples area. Similar roads are needed elsewhere in the South.

IMPACT OF THE ECONOMIC CRISES OF THE 1970s

Among the best measures of economic performance are the absolute level of unemployment and the ratio of the unemployed to the total labor force. These data for southern and northern Italy are presented in Table 46. I should caution the reader that these materials (based on an ISTAT sample) must be used with care, for Italian unemployment statistics are notoriously underestimated. They are, however, of considerable value in a relative sense (i.e., for regional comparisons).

Podbielski's[7] explanations of this underreporting effectively summarize the main arguments. One is the "discouraged worker" thesis. It argues that workers withdraw from the unemployment rolls because the demands of labor, particularly in the South, are so limited. She also reasons that with the massive outmigration from agriculture and in many instances the retention of farm holdings, women have abandoned the job market because of continuing labor requirements on the farm. In any event, it is amply clear that unemployment ratios were crudely twice as high in southern Italy as in the North.[8] Note the fact that there is a fair visual correlation between these data and those on gross domestic product presented earlier.

I should add here that part of the current political unrest in Italy results from the distress caused by the adoption of a policy of virtual open admissions to institutions of higher education without an appreciable expansion of staff or facilities. For example, the University of Rome is said to have enrolled 160,000 students in its various faculties with staff and quarters

Table 46. Unemployment in the Two Italies from 1970 to 1976[a]

Unemployment (thou.)	1970	1971	1972	1973	1974	1975	1976
Southern Italy	292	275	324	319	274	313	334
Northern Italy	317	340	373	349	286	341	398
Unemployment ratios (% of labor force)	**1970**	**1971**	**1972**	**1973**	**1974**	**1975**	**1976**
Southern Italy	5.2	4.5	5.8	5.6	4.8	5.5	5.8
Northern Italy	2.4	2.5	2.9	2.6	2.2	2.6	3.0

[a]These values were computed from *Annuario di Statistiche del Lavoro*, Vol. XVLLL, Istituto Centrale di Statistica (Rome, 1977), pp. 15–16. The data in this table do not include southern Lazio (Frosinone and Latina).

for only a mere fraction of that number. In time, through the relaxation of standards, most will eventually graduate with very little prospect of job opportunities. The same is true for most universities, including the new centers in southern Italy. Clearly, the economic, social and political ramifications of this educational policy are explosive. So far, their impact in the Mezzogiorno has been minimal. More critical is the broader problem of unemployed youth (with or without higher education). I was told in Naples in the fall of 1977 that 40% of the young people of that Commune (between the ages of 18 and 25) were unemployed. It should be recalled that Naples was an area that received a disproportionate share of the development funds. Such data do not typically appear in the published unemployment statistics, but they are, nevertheless, closer to reality. Thus in this major metropolitan center, and I suspect that it would also be true for Palermo, unofficial unemployment rates for that age group resemble those common for black youth in the inner ghettos of a number of the largest American cities such as New York, Newark, and Detroit.

POLITICAL REPERCUSSIONS

The past two decades have been an era of major political change in Italy. They have witnessed an erosion of the political power of the Christian Democratic Party, and a precipitous decline in the strength of the minority groups which had helped to form innumerable centrist coalitions in earlier years and more recently center-left alliances. Finally and most important, there has been a notable rise in vigor and popularity of the Communist Party. Most recently, the Christian Democrats have ruled with the tacit support of their arch rivals, but that arrangement is still tenuous. Although the Communists still do not have a plurality on a national scale, they already control the local governments of most of the major Italian cities, but the purse strings and police powers still rest in Rome. So far the South has continued its long-term conservative tradition, but both Naples and Palermo are run by the Communists or by leftist coalitions. The increased industrialization and urbanization of the South coupled with unemployment and unfulfilled dreams have boosted Communist support in the entire region. It is noteworthy that southerners who have migrated to northern cities like Milan, Turin, and Genoa have shifted their votes to the left. Despite American and possibly West European opposition, it is not inconceivable that the Communists may end up as the nation's ruling party. This would be particularly possible if economic conditions do not improve, and no economic miracle is on the immediate horizon.

This background leads me to the logical question—what would be the possible impact on the Mezzogiorno of such a shift in political power? I have failed to find a current statement of the Communist Party perspective with respect to the development program. Nevertheless, using the Informazioni SVIMEZ Bulletins for 1957, 1965, 1971, and 1976, I was able to trace their evolving viewpoints. These were expressed by party leaders in discussions in the Chamber of Deputies and the Senate on the proposed development legislation in each period. In a Bulletin published in March of 1971, I found the text of a legislative proposal by the Party on "Public Intervention in the Mezzogiorno."[9] My research indicated that in almost every respect the current laws follow their original recommendations. To illustrate, they fought for a marked reduction of the role of the Cassa because of its purported "clientilismo" (favoritism to special interests or investors); this possibility was reduced in the 1971 legislation and reinforced in 1976, for the role of the Cassa has been greatly diminished. The Party also supported the creation of CIPE and argued that it should be given a key role in national and regional planning. This too was accomplished. In addition, I should stress that it was the Communists and Socialists who fought for the shift of power to the "regioni." Then too the Party favored the support of labor-intensive small- and medium-sized enterprises. A review of the current legislation indicates that most of their proposals are now integral parts of the development laws. However, there are several areas where their demands have yet to be fulfilled. They argued for fiscal disincentives to be applied in the congested industrial areas of the North, and for the elimination of the local consorzia (which planned the development of each area and nucleus) with the relegation of their roles to the "regioni." Neither proposal has been accepted. Finally, as good politicians they proposed a major expansion of funding for the South. In the current fiscal crisis, this proposal has not been adopted. In summary, to answer the original question, if the Communists were to come to power, I envisage no drastic changes in the development program. I do suspect, however, that a larger share of the available resources, over time, would be allocated to the Mezzogiorno.

THE CLOTTING OF THE HEMORRHAGE

In the 1970s, the natural population growth of southern Italy, as measured by vital rates, continued to outstrip that in the North (see Table 47). Note the continued contrast in natural increase (per thousand) between the two Italies.

More meaningful perhaps to the reader are the absolute and relative increases, for the same period, by "regioni." These are demonstrated in Table 48. It is noteworthy that unlike the 1951 to 1971 period no region exhibited a decline during this era and Southern Italy's rate of increase was nearly double that of the North. The contrast between the two periods is attributable to "the clotting of the hemorrhage."

In recent years the southern press has increasingly evidenced its concern about the problem of "return migration."[10] There has, in effect, been a virtual cessation of the population drain. Net out-migration has been reduced from an annual average of 122,000 from 1952 to 1961 and 192,000 from 1962 to 1971 to a yearly mean of roughly 7,000 from 1972 to 1977. In fact, in 2 years, 1973 and 1975, there was a small net in-migration (see Table 49). The question can be posed in this manner, given these shifts in migration: what have been their patterns, can these changes be explained, and what trends can be discerned? Finally, there is the more crucial problem which plagues southern politicians and planners–"what if the emigrants return?"

Table 49 not only illustrates yearly migration trends but also differentiates internal from foreign movements, diaggregated by province. Note the contrasts between internal out-migration and inward movements from abroad. The geographic pattern of internal flows during this period replicated those of the 1950s and 1960s. I base this judgment on ISTAT, transfer of residence, flow data for 1972 to 1975 (not reproduced here). Those movements, as in the past, were predominantly to the industrialized areas of the Northwest and to

Table 47. Regional Variations in Natality and Mortality Levels (per thou.) in Italy, 1971, 1974 and 1977[a]

Year	North			Mezzogiorno		
	Birth rates	Death rates	Natural increase	Birth rates	Death rates	Natural increase
1971	15.2	10.1	5.1	19.6	8.4	11.2
1974	13.7	10.0	3.7	19.3	8.5	10.8
1977[b]	11.3	10.3	1.0	17.2	8.6	8.6

[a]Computed from various yearly volumes of the *Annuario di Statistico Italiano*.

[b]Special tabulation compiled for me through the courtesy of Dr. Pinto, the Direttore Generale of ISTAT. North includes southern Lazio.

Table 48. Absolute and Relative Increases in Resident Population, by Region, in Italy between 1971[a] and 1977[b]

Region	Population 1971 (thou.)	Population 1977 (thou.)	Absolute increase (thou.)	Relative increase (%)
Abruzzi and Molise	1,436.5	1,559.7	123.2	8.6
Campania	5,059.3	5,378.8	319.5	6.3
Puglia	3,582.8	3,856.4	273.6	7.6
Basilicata	603.1	619.1	16.0	2.7
Calabria	1,988.1	2,057.9	69.8	3.5
Sicily	4,680.7	4,936.2	255.5	5.5
Sardinia	1,473.8	1,582.1	108.3	7.3
Southern Italy	18,824.3	19,990.2	1,165.9	6.2
Northern Italy	35,312.2	36,610.2	1,298.0	3.7

[a]Computed from data in the *Annuario di Statistico Italiano* for 1976.
[b]Special tabulation provided by ISTAT.

Rome. Thus, despite reduced or possible stable employment opportunities there (resulting from stagnent or deteriorating economic condition), the drawing power of the "North" persisted. The net inbound foreign flows reflected sharply reduced labor demands in northwestern Europe, particularly in Germany. This reduction was first felt by such non-EEC countries as Turkey and Yugoslavia. Eventually, it affected the Italian "guest workers." In the case of Switzerland, explanations are necesarily more complex. There the outward movements of Italian workers reflected both a saturation in the tertiary sector (construction and services) coupled with the general rise of xenophobia which resulted in the tightening of immigration controls and stricter regulation in the granting of work permits. This has particularly affected the Italian segment of the labor force, which still comprises a major share of the overall population of Switzerland.

As to the provincial patterns, in numerical terms the net outward flows were dominated by Naples, Foggia, and the three provinces of Calabria. In contrast, Latina, Frosinone, Caserta, Pescara, and to a lesser degree eastern Sicily were areas of significant net inbound movements. Seventeen of the 34 provinces of the South had positive values. Despite significant industrial growth, Puglia exhibits a surprisingly mixed pattern. What is particularly disturbing is the fact that many of the returnees from abroad had held highly skilled jobs, yet there were now only limited employment opportunities available for them in the Mezzogiorno. A decade or two earlier, their skills and work attitudes had been desperately needed in southern Italy; now openings for these returnees were rare or at best infrequent. Of course, this labor reservoir could presumably be tapped in the future, if there were a substantial growth in the southern economy. An examination of data for the

Table 49. Net Migration Statistics for the Mezzogiorno Based on Transfer of Residence Reports for 1972 to 1977[a]
(in thou.)

Province	1972	1973	1974	1975	1976	1977	1972–1977			
							Internal	Foreign	Total	Net migration rate[b]
Latina	3.8	3.8	3.9	3.2	2.5	1.8	14.5	4.6	19.1	4.73
Frosinone	4.2	5.4	2.7	1.6	1.3	1.6	8.5	8.3	16.8	3.78
Caserta	0.4	1.6	1.5	2.3	2.0	0.9	-0.6	9.3	8.7	1.23
Benevento	-0.6	-1.2	-0.8	-0.3	-0.5	-0.8	-8.5	3.9	-4.6	-1.59
Napoli	-11.6	-5.4	-7.2	-6.6	-6.0	-5.9	-48.9	6.2	-42.7	-1.52
Avellino	-1.5	-1.3	-1.9	-1.1	-0.5	-1.3	-13.5	6.3	-7.2	-1.66
Salerno	-3.6	0.2	-0.8	-0.5	-1.0	-1.5	-14.9	7.7	-7.2	-0.73
L'Aquila	1.6	0.9	0.5	0.7	0.4	0.1	-0.9	5.1	4.2	1.41
Teramo	-0.2	0.9	0.3	1.0	0.9	0.8	-0.4	4.0	3.6	1.37
Pescara	2.4	3.0	2.3	1.8	1.9	1.4	7.2	5.6	12.8	4.59
Chieti	1.9	1.2	0.7	1.0	1.0	0.9	-0.4	7.1	6.7	1.85
Campobasso	0.5	3.0	0.3	0.2	-0.2	0.1	-2.7	6.6	3.9	1.19
Foggia	-5.7	-1.2	-4.2	-3.2	-2.7	-3.1	-24.1	4.0	-20.1	-2.99
Bari	-0.8	-1.7	0.1	1.4	1.6	-0.5	-6.1	6.3	0.2	–
Taranto	0.7	3.0	1.4	1.2	1.0	-0.4	1.9	5.0	6.9	1.28
Brindisi	-0.2	0.6	0.4	0.8	0.1	-0.2	-2.3	3.7	1.4	0.39
Lecce	1.6	1.8	2.4	5.4	4.3	1.7	-1.8	18.9	17.1	2.36
Potenza	-2.2	-2.2	-3.2	-1.6	-2.1	-2.4	-16.9	3.2	-13.7	-3.23
Matera	-0.2	-1.1	-0.3	-0.8	-0.7	-0.8	-5.6	1.7	-3.9	-1.95
Cosenza	-4.3	-1.1	-1.2	-0.5	-0.4	-2.2	-16.6	6.9	-9.7	-1.37
Catanzaro	-7.3	-4.4	-3.6	-4.4	-1.9	-3.8	-27.0	1.5	-25.5	-3.49
Reggio di Calabria	-3.2	-0.8	-1.9	-2.2	-3.4	-3.6	-16.1	1.0	-15.1	-2.57
Trapani	-0.6	-0.2	-0.1	0.8	0.6	–	-3.7	3.1	-0.6	-0.12
Palermo	-1.9	1.4	-4.4	-1.2	-0.2	-1.4	-14.4	6.8	-7.6	-0.67
Messina	-0.5	0.4	0.5	1.5	0.3	-0.2	-4.1	6.0	1.9	0.30
Agrigento	0.5	0.8	-0.5	0.4	0.3	-0.1	-6.8	8.2	1.4	0.30
Caltanisetta	-0.5	-1.2	-0.2	-0.8	-0.6	-1.5	-8.7	4.0	-4.7	-1.66
Enna	-0.8	-1.2	-1.5	-1.0	-0.9	-1.2	-9.8	3.2	-6.6	-3.24
Catania	-3.3	-1.0	1.8	-0.4	0.2	0.2	-9.7	7.2	-2.5	-0.26
Ragusa	0.3	1.0	0.6	0.5	0.7	0.6	1.2	2.5	3.7	1.40
Siracusa	0.6	1.6	0.7	1.4	–	0.3	1.5	3.0	4.5	1.21
Sassari	0.5	0.9	1.4	1.2	1.0	1.5	3.8	2.7	6.5	1.57
Nuoro	-0.6	-0.2	-0.7	-0.6	-1.1	-1.0	-6.6	3.0	-3.6	-1.32
Cagliari	2.3	2.0	0.9	0.9	0.6	–	3.0	3.7	6.7	0.79
Southern Italy	-28.3	9.3	-10.1	2.1	-1.5	-20.0	-229.5	180.3	-49.2	-0.29

[a]Derived from data published in the *Annuario di Statistiche Demografiche, Popolazione e Movimento Anagrafico dei Comuni*, 1955 and *Bolletino Mensile di Statistica*, Istituto Centrale di Statistica, 1973–1978. Minus signs signify net out-migration, while the absence of a sign indicates net in-migration.

[b]Net migration for 1972–1977 over 12/31/74 provincial population values.

same 7-year period by provincial capitals and communes with over 50,000 inhabitants and aggregate provincial values indicates that most of these repatriates returned to their original small towns and villages. The remainder apparently terminated in the largest towns of their provinces presumably hoping for industrial jobs, but "temporarily" employed in marginal tertiary activities. Some presumably joined the growing list of the unemployed. So far this return movement has not reached alarming proportions, but southern politicians clearly fear the possible social, economic and political consequences of what might conceivably become a vastly increased flow.

NOTES

[1] IASM (Istituto per L'Assistenza allo Sviluppo del Mezzogiorno), "La Nuova Legge per il Mezzogiorno," *Documenti sul Mezzogiorno*, No. 3, 1971, 44 pp., and ISAM, "Legge, May 2, 1976, No. 183, Disciplina dell Intervento Straordinario nel Mezzogiorno nel Quinquennio 1976-1980", *Documenti sul Mezzogiorno*, No. 5 (1976), 48 pp.

[2] Jeffrey Moore and Philip Klein, "International Economic Indicators," National Bureau of Economic Research (New York, 1978).

[3] Podblielski, op. cit., p. 130.

[4] See also Rodgers, op. cit., 1976.

[5] P. Saraceno, *Il Sistema delle Imprese a Participazioni Statale nell' Esperienza Italiana*, SVIMEZ (Milan, 1975), 132 pp.

[6] For a more detailed account, see A. Rodgers, "The Alfa-Sud Industrial Complex: A Regional Sub-System of the Italian Automobile Industry," paper presented at the *International Geographical Union's Commission on Industrial Systems*, Krakow, 1977.

[7] Podbielski, op. cit., p. 113.

[8] These include the unemployed (those who lost their jobs) and young men and women under 21 years of age seeking their first jobs. The Minister of Labor also publishes data for three other groups: housewives without previous work experience seeking jobs, retired persons on pensions seeking employment, those who are employed but are searching for better jobs.

[9] "Proposta di Legge Comunista sull' Intervento Pubblico nel Mezzogiorno," *Informazione SVIMEZ*, Vol. 24, No. 6 (1971), pp. 219-224.

[10] One example is E. Corsi, "Campania Se Ritornano gli Emigrati," *Nord e Sud*, Vol. 22 (1974-75), pp. 221-233.

Chapter 10

A CRITICAL ASSESSMENT

In this chapter we return to the central theme of my study: the assessment of the growth of manufacturing in southern Italy since 1951 and the relationship of the development program to that change. The restriction of this evaluation primarily to industrialization reflects my own central research concerns. However, it is also related to the fact that other foreign scholars such as Schachter, Podbielski, Allen, Lutz, and a host of Italian economists and planners have already covered the broader ground of socioeconomic development,[1] and I should add that they have done so with considerable insight and success. Then, too, many other facets of economic and demographic change have already been treated in previous chapters and do not warrant detailed repetition. Finally, unlike most students of development in the Mezzogiorno, my interests are mainly oriented to the spatial aspects of industrialization. It is this viewpoint that has received far too limited attention in the economic literature on southern Italy.[2]

After an initial examination of changes in major socioeconomic indicators followed by an analysis of Cassa expenditure patterns during the development era, I propose to assess southern industrial progress from three perspectives.

(1) *Southern vs. northern Italy:* Do quantifiable measures of industrial performance show a narrowing of the gap between these two areas?

(2) *Southern Italy as an integral economic unit:* Are there signs of industrial achievement and positive structural change, and does this region show indications of the creation of a basis for self sustained industrial growth?

(3) *Regional development theory, locational practice and industrial development in the Mezzogiorno:* To what extent have differences in levels of industrial development emerged and what are the implications of such geographical diversity?

CHANGES IN LEADING SOCIOECONOMIC INDICATORS FROM 1951 TO 1971

I have deliberately selected only a limited number of indicators for my calculations of socioeconomic change over time (Table 50). However, I believe that these measures are representative of a broader range of phenomena which would demonstrate the same trends. Of course, as I have stressed repeatedly, these are macro-regional data which conceal considerable intraregional variation.

Table 50. Selected Socioeconomic Indicators for Southern and Northern Italy from 1951 to 1971

Indicator	1951		1971	
	South	North	South	North
Instruction[a]				
Illiterate or lacking elementary certificate (%)	48.5	20.9	35.9	29.6
Health[a]				
Infant mortality (deaths in first year per thou. live births)	78.9	50.8	35.6	23.2
Housing[a]				
Dwellings with inside drinking water (%)	28.7	43.3	82.6	87.2
Dwellings with inside toilet (%)	19.6	48.6	83.4	82.6
Electric energy consumption for illumination (kwh per inhabitant)[b]	23.9	56.3	116.3	196.9
Economic[b]				
Net annual per capita income (thou. of lire at 1963 prices)	131.5	254.1	876.4	1551.4
Other				
Index of motorization (per 100 inhabitants)[b]	2.9	10.3	14.8	25.5

[a]Censuses of Population for 1951 and 1971.

[b]Taglica.ne op. cit., yearly indicators by province and region prepared for the National Union of Chambers of Commerce and Industry.

Note that in the case of illiteracy and retrograde illiteracy (the latter defined here as those *not* holding an elementary school certificate), there had been a rather marked improvement in the South from nearly a half down to one-third. Though a gap still existed between the two regions, the contrast had narrowed to minimal differences. I frankly do not understand the rise in the value for the North, unless it possibly reflects in-migration from the South (or possibly the use of differing criteria between 1951 and 1971). For the measurement of improvement in health provision, I chose infant mortality because of the lack of data on doctors and hospital beds for 1951. Note the marked change in both regions in the two-decade interval. In the case of housing, three indicators were used; here again the measures showed decided improvement in both northern and southern Italy. Note, in particular, the values for "internal" drinking water; in this instance, by 1971, no appreciable differences was discernible between the two regions. The data on the proportion of homes or apartments with inside toilets in the South are now, surprisingly, comparable with those in the North. One of the most impressive changes readily apparent to those familiar with southern Italy in the 1950s is that area's entry into the motor age. The statistics on motorization indicate that one-quarter of the inhabitants of the North owned a private vehicle compared to one-seventh in the Mezzogiorno. However, if these data are expressed in terms of the cost and quality of cars, northern Italy would still be clearly in the lead. The most important measure presented in this table is, obviously, that on per capita income (expressed in constant 1963 prices). The percentage gap had narrowed to a modest degree, but the absolute growth in the South, itself, was most impressive. Nevertheless, average per capita income in southern Italy are still less than 60% of those in the North. Here, Figure 23, which shows the provincial values (in current lire), demonstrates striking spatial variations within the Mezzogiorno. To conclude this brief temporal survey of selected socioeconomic indicators for the two Italies, I would simply note that if these measures are at all representative of other characteristics, then clearly there had been major improvements in southern Italy. While it is true that the gap between the two regions still persisted, the socioeconomic health of the Mezzogiorno had clearly ameliorated. However, I should stress, again, that these data are merely regional averages. Micro-regional data at the province or preferably at the commune level would demonstrate the heterogeneity of both regions, but in a relative sense that geographic variability would be far greater within the South. As I noted in the preface, inter-provincial income differentials are still very great as are urban-rural contrasts. Then too, intra-urban differentials remain immense. All that one has to do is to walk the streets of Naples or Palermo to be aware of the congested urban slums vis-a-vis pockets of extreme wealth in these two major metropolitan centers. A large share of the people of southern Italy have been minimally affected by the development program. They still live at marginal levels where incomes are low and precarious; unemployment and underemployment are persistent phenomena. It must be remembered, too, that part of the reason for the improvement, measured on a per capita basis, is attributable to the vast out-migration of the 1950s and 1960s. Remittances have also helped to ameliorate the economic problems of the South.

CASSA EXPENDITURES AND SOUTHERN ECONOMIC DEVELOPMENT

Before assessing industrial growth in the South, it is important to place the overall investment program by the Cassa in broader perspective. Podbielski[3] has attempted to do so by comparing Cassa expenditures, which totaled more than $20 billion (at constant 1975 prices) over the 2½ decades between 1950 and 1975, with gross national product and gross internal product for the South itself. Her calculations show annual values averaging less than 1% in the first instance and between 3.5 and 4.5% for the latter measure. Thus, in Podbielski's view, the funds available to the Cassa were very modest compared to the nation's total resources and those of the South alone. Although I agree with her conclusion, I do not find this type of comparison to be very meaningful, despite the fact that the same analytical technique is commonly used by economists. In my opinion, a comparison between the overall fixed investments for the nation and those allocated to the South is more instructive. That ratio was 29% for the 1951 to 1975 period.[4] The comparable value for the share of population in the South was 36%. Thus it would appear, based on demographic values alone, that the Mezzogiorno did not receive an equitable share of national investment funds (both public and private). Many southerners would argue on the basis of their proportion of the nation's population that these values in no way demonstrate a major national commitment to economic development in the Mezzogiorno. This is, of course, an "equity" type of argument. There is another perspective, however, if we view investments in terms of industry *alone* during this era, the share was not much higher—31%. But given the fact that the South had, as an average, only 15% of the nation's manufacturing employment and only 13% of its gross manufacturing output, these industrial investments did represent a significant national commitment, at least in this sector. We could argue the merits of both arguments endlessly, but there is no question that the development program has clearly not been a "herculean" effort.

Table 51, which shows the relationship between Cassa investments and total investments in southern Italy, provides another view of the development program. These data make it evident that, over time, the Cassa played an increasing role in the development of the Mezzogiorno. Its share of the total investments in the South had nearly doubled since the early 1950s, reaching a level of 32% in the 1971 to 1975 period. Of course, these values cannot be equated with the overall role of State agencies in that region, because many State organizations had also allocated an increasing share of their expenditures to the South. Unfortunately, such data were unavailable. However, it is commonly argued in the southern press that, over time, the role of the Cassa has too often been substitutive rather than additive to the normal

Table 51. Investments Funded or "Induced" by the Cassa in Relation to Total Investments in the South from 1951 to 1975[a]

Period	Cassa investments (billions of current lire)	Total investments in the South[b] (billions of current lire)	Share of the Cassa (%)
1951–55	567	3,232	17.5
1956–60	1,079	5,097	21.2
1961–65	2,592	9,414	27.5
1966–70	3,573	14,225	25.1
1971–75	9,094	28,408	32.0
Total (1951–75)	16,905	60,376	28.0

[a]Cassa per il Mezzogiorno and ISTAT, *Bollettino Mensile di Statistica*, Vol. 52, No. 11 (November 1977), pp. 272–274.

[b]The "South," in this instance, is the ISTAT statistical region which does not include southern Lazio. Thus these comparisons must be used with some caution for the area encompassed by the Cassa includes southern Lazio, parts of Rome, Rieti-Cittaducale, Ascoli Piceno, and several offshore islands.

expenditures by other government agencies in southern Italy. If so, the role of the Cassa has been less than would be indicated by their investments.

We now turn to the sectoral structure of Cassa investments during the same period. These are demonstrated in percentage terms, by 5-year periods, in Table 52. The absolute data at the bottom of the table are expressed in current lire. In terms of constant 1970 prices the investments of the Cassa increased roughly sixfold during that 25-year period. There are, however, two problems with the statistics in this table: first, the data are aggregated so that year to year fluctuations are "erased"; second, I am not happy with the term "induced," for such investment data may be purely speculative, yet there was no way of partitioning them into funded as opposed to "induced" investments.

Nevertheless, the data on sectoral changes over time are still very useful. There was a striking reduction of expenditures on infrastructure during this era. The only major exception to this generalization was the growth of investments for infrastructure facilities in the "areas" and "nuclei" of industrial development which began in the 1961 to 1965 period. As for the expenditures on incentives whose increase has been the inverse of that on infrastructure, here, again, there have been structural shifts. Note the disappearance of land reform investments and the decline of expenditures on land tenure. Yet these mutations are dwarfed by the extraordinary growth in industrial incentives (subsidized interest on loans and direct capital contributions). By 1975, these incentives accounted for about two-thirds of the Cassa's investments.

Table 52. Investments Funded or "Induced" by the Cassa from 1951 to 1975[a]

Sector	1951–55 (%)	1956–60 (%)	1961–65 (%)	1966–70 (%)	1971–75 (%)
Infrastructure	55.5	35.6	23.3	22.7	23.6
Land reclamation	28.5	17.6	11.3	6.8	5.1
Acqueducts and drainage	8.3	7.9	6.5	8.3	6.3
Tourism	1.4	1.3	1.0	0.7	0.6
Roads	13.4	4.0	2.7	3.8	3.9
Railroads	3.9	4.8	1.0	0.2	–
Hospitals	–	–	–	0.9	0.4
Industrial areas	–	–	0.8	1.8	4.6
Other	–	–	–	0.2	2.7
Incentives	44.5	64.4	76.7	77.3	76.4
Land reform	27.0	10.6	–	–	–
Land tenure improvements	7.6	14.2	8.3	3.5	2.8
Industrial incentives[b]	9.0	35.3	58.8	63.3	68.0
Tourism	0.9	0.7	1.8	3.7	2.6
Fishing and handicraft	–	1.5	4.9	4.9	1.5
Other	–	2.1	2.9	1.9	1.5
Total (%)	100.0	100.0	100.0	100.0	100.0
Amount (billions of current lire)	566.7	1,079.3	2,592.2	3,572.8	9,094.0

[a]Annual Reports (Bilancio) of the Cassa per il Mezzogiorno.
[b]Industry includes mining and utilities.

The results of a special survey by the Cassa of its direct contributions for industrial development for the period from 1969 through 1975 are presented in Table 53. It provides some tangible information on the average size of investment per grant and the average investment per worker during this interval. I only used monetary data earlier, because the normal employment data released by the Cassa are *overestimates* for they report on jobs anticipated. In contrast, the statistics in this survey are apparently far more securely based. Unfortunately, only sectoral (non-spatial data) have been published to date. Although investments per employee were higher for enlargements than for new plants, the former values reflect the heavy monetary contributions for the expansion of the Taranto steel mill. Note the contrasts by industrial branch; thus the investment coefficients were far lower for the first stage (traditional) industries. This could have been expected, but the low values for the machinery industry are less readily explainable. In

Table 53. The Dimensions of Plants Subsidized via Direct Industrial Contributions by the Cassa per il Mezzogiorno, from 1969 to 1975[a]

Initiative	Subsidized initiatives (*N*) [A]	Employed (*N*) [B]	Investment (millions of lire) [C]	Capital coefficient per employed [C/B]	Average *N* of employed per initiative [B/A]	Average investment per initiative [C/A]
New plants	4,757	213,592	2,992,259	14.0	44.9	629.0
Enlargements	4,562	135,910	3,467,343	25.5	29.8	760.0
Total	9,319	349,502	6,459,602	18.5	37.5	693.2
Subsidized						
1969–1972	4,485	148,792	4,488,870	13.2	33.2	439.4
1973–1975	4,834	200,710	1,970,732	22.4	41.5	928.6
Total	9,319	349,502	6,459,602	18.5	37.5	693.2
of which:						
Food	2,608	39,836	498,158	12.5	15.3	191.0
Clothing	232	16,724	48,902	2.9	72.1	210.8
Furniture and wood	644	15,389	111,574	7.3	23.9	173.3
Metallurgy	175	29,355	1,783,938	60.8	167.7	10,193.9
Machinery	1,448	116,242	1,101,811	9.5	80.3	760.9
Building materials	1,698	36,162	446,577	12.3	21.3	263.0
Chemicals	653	26,805	1,417,921	52.9	41.0	2,171.4
Plastics	283	9,641	100,292	10.4	34.1	354.4

[a] *Bilancio 1975*, Cassa per il Mezzogiorno, Vol. II, Appendice Statistica (Rome, 1976), pp. 72–73.

contrast, the very high coefficients for chemicals and metallurgy do follow expectations. I will return to the results of this survey in my assessment of the geographical impact of the industrial development program.

THE NORTH-SOUTH DIFFERENTIAL

To repeat the question posed at the beginning of this chapter: "Had the development program succeeded in narrowing the industrial gap between the two Italies, and, if not, could the slower growth of the South be explained?" Here I propose to use a variety of quantifiable measures of industrial performance to answer that pivotal query. I have already probed the problem of the North-South gap in industrial development in Chapters 5 and 8; now those findings will be supplemented by additional measures of industrial performance. In addition, I will treat more directly the question of explanations and implications of industrial development trends in the two Italies since 1951.

Table 54. Changes in Employment and Output of Manufacturing in the two Italies from 1951 to 1971

Employment and output	1951		1971	
	South	North	South	North
Employment in manufacturing[a] (*N* in thou.)	574	2,925	774	4,513
Share (%)	16.4	83.6	14.6	85.4
Gross output of manufacturing[b] (billions of lire at constant 1963 prices)	441	3,112	1,514	9,835
Share (%)	12.4	87.6	13.3	86.7

[a]For sources see Table 12.
[b]*Annuario di Contabilità Nazionale*, Vol. III, Tome II, ISTAT (Rome, 1973), pp. 29–30. North includes southern Lazio.

It should be recalled that in Chapter 5 (Table 12) I analyzed changes in manufacturing employment for southern and northern Italy from 1951 to 1971. Those data demonstrated that, despite an absolute growth of roughly 200,000 employees in the Mezzogiorno over the two-decade interval, the region's share of the national total had actually declined. It was, in fact, far below the South's proportion of Italy's population in both years. However, other statistics by sector, size of firm and employment in firms with motive power (Chapters 5 and 8) all appeared to show positive evidence of structural improvement over time. Table 54 compares the changing share of manufacturing employment in the two regions with their relative proportions of the nation's gross industrial output. It should be noted that while the latter values do show a slight percentage increase, clearly the share of the South in industrial output was still below that in employment. Thus these two indicators of industrial development do not reflect any major reduction in regional disparities.

A logical extension of this analysis would take this form: Were there significant differences in industrial investments in the two regions that might account for the comparatively poor performance of the South? To answer this question, I have assembled a variety of investment statistics originally published by ISTAT. These are presented in Tables 55 and 56.

Table 55 shows the gross fixed investment, by economic sector, from 1951 to 1971. Note that the share of industrial investments in the overall capital outlays in the South had grown to 38% (compared to 30 in the North). These industrial expenditures had increased over four times in the South compared to a very modest growth in northern Italy. However, it should be kept in mind that these values were quite low in southern Italy in 1951, so that *relative* increases are greatly magnified. Despite this reservation, there is no question that there had been major increases in industrial investment in the Mezzogiorno during that era. Over time, *industrial investments per employee*

Table 55. Gross Fixed Investment by Economic Sector from 1951 to 1971[a]

Sector	1951		1971		Growth 1951–71 (1951 = 100)	
	South (%)	North (%)	South (%)	North (%)	South	North
Industry[b]	26.0	42.0	38.4	30.3	418	121
Agriculture	16.3	9.0	9.6	5.6	167	104
Services	40.9	42.5	41.3	56.5	286	223
Public administration	16.8	6.5	10.7	7.6	180	196
Total	100.0	100.0	100.0	100.0		
Absolute amount (billions of lire at 1963 prices)	1,981	7,064	5,601	11,838	283	163

[a]*Annuario di Contabilità Nazionale*, 1973, Vol. III, Tome II, ISTAT, (Rome, 1973), p. 127. North includes southern Lazio.
[b]Industry includes manufacturing, mining, utilities, and construction.

increased significantly, and by 1971 they were twice those in the North.[5] As for Table 56, these 1974 data by industrial group show striking variations in investment by region and by branch. Thus the ratios for the metallurgical and chemical industries reinforce my generalization, voiced earlier, that many of the new southern plants and industries were highly capital-intensive. In contrast, the values for the more traditional industries were far lower, and in some instances their levels were below those in the North. In sum, the statistics in these two tables would appear to dispel any notion that the low industrial productivity of the South (compared to northern Italy) reflects a lag in capital investments in that region.

Given these positive indications of the level of industrial investment in southern Italy, why hadn't there been a significant growth of employment and output? What variables can help to account for this retardation? Cafiero[6] feels that the explanation lies mainly in the very high investment-output ratios in the South; his findings are reproduced in Table 57. Notice how high those ratios were for southern Italy during both the 1951–63 and the 1964–74 periods, compared to the values for the North. Why did such variations exist between the two Italies? My speculations, that follow, are based in part on the literature coupled with the results of interviews with knowledgeable government economists. It also reflects the views of the management of major corporations that have created branches in the Mezzogiorno.[7] First, there is the obvious fact that the building of new plants is almost invariably more costly than the expansion and or modernization of preexisting facilities, which was so common in the 1950s and 1960s in northern Italy. Second, for a host of reasons, the building of new plants in the South is a much slower process

Table 56. Investment per Employee in Manufacturing in 1974[a]

Sector	South	North[b]
	(thou. of lire at current prices)	
Food and tobacco	1,424	1,673
Textiles	2,516	914
Clothing, furs, leather and shoes	371	338
Wood and furniture	935	1,149
Metallurgy	7,157	2,008
Machinery	1,085	907
Transport equipment	1,294	1,402
Stone, clay and glass	2,509	1,749
Chemical and artificial fibers	10,603	2,080
Rubber	1,106	1,034
Paper	1,535	1,674
Printing	1,104	703
Other	1,554	994
Average for all manufacturing	3,135	1,199

[a]"Il Prodotto Lordo e Gli Investment delle Imprese Industrali nel 1974," *Bollettino Mensile di Statistica, Supplemento Straordinario*, No. 11, 1976. Based on a sample of 35,000 firms only including those with 20 or more employees, which affects the southern values.

[b]Southern Lazio data included.

than in the North. Thus an initial investment often takes several years to mature in the form of an operating facility. Third, as we have seen, the growth of new factories in southern Italy was paralleled by a major decline of traditional small workshops (40,000 establishments with less than 6 employees closed between 1951 and 1971). Fourth, there is increasing evidence that the building of new industrial establishments like Alfa-Sud has contributed far more to industrial development *outside* of the South than it has to the growth of component plants within Campania or in other parts of the Mezzogiorno. Fifth, Amendola and Baratta[8] have argued persuasively that many of the newer and larger plants in southern Italy, though modern, were built in "isolation." External economies for such facilities were either absent or weak. In contrast, in the North, where there are so many multi-plant enterprises, external economies are accepted economic facts of day to day operations. Complex inter-plant linkages are often the rule. Thus, in that advanced industrialized region there were far greater opportunities for innovation, for increases in productivity, and for the lowering of production costs. Such opportunities were rare, at best, in the South.

As a logical next step, we can now ask: Are there additional variables which may help to explain the low industrial productivity of the South? Given the shift-share analysis in Chapter 5 and the analysis of growth and lag

Table 57. Relationships between Industrial Investment and the Increment of Gross Industrial Product (at constant 1963 prices), from 1951 to 1974[a]

Region	Investment-output ratio	
	1951–1963	1964–1974
South	3.47	4.67
North	2.51	2.45

[a]Salvatore Cafiero, *Sviluppo Industriale e Questione Urbana nel Mezzogiorno*, SVIMEZ (Milan, Rome: Giuffrè, 1976), p. 59. Calculated from ISTAT data. North includes southern Lazio.

Table 58. Manufacturing Employment and Product in the two Italies from 1951 to 1971[a]

Sector	1951 (%)				1971 (%)			
	Employment[b]		Product[c]		Employment[b]		Product[c]	
	South	North	South	North	South	North	South	North
Food and tobacco	29.7	7.6	32.7	10.4	15.2	6.4	20.2	9.2
Textiles	4.6	19.1	4.4	17.0	5.3	11.0	2.8	8.2
Clothing and shoes	20.6	12.1	17.1	8.1	14.6	10.6	12.5	6.9
Leather and skins	0.9	1.0	1.2	1.2	0.9	1.1	0.7	1.4
Wood and furniture	14.2	8.4	9.4	4.5	9.1	7.2	7.9	5.4
Consumer goods	70.0	48.2	64.8	41.2	45.1	36.3	44.1	31.1
Metallurgy	1.8	4.1	2.9	7.4	4.6	4.6	6.6	5.6
Chemicals and related	3.2	5.6	7.9	9.9	6.3	5.8	13.6	12.1
Stone, clay and glass	6.6	5.2	6.6	4.9	9.6	5.6	8.1	5.2
Machinery	12.5	24.7	10.7	21.1	23.7	30.6	16.2	28.2
Transport equipment	2.8	5.3	3.6	5.9	4.8	6.6	5.8	8.0
Basic industries	26.9	44.9	31.7	49.2	49.0	53.2	50.3	59.1
Rubber	0.1	1.2	0.1	2.4	1.0	1.7	0.9	2.1
Paper	0.8	1.8	1.5	3.1	1.1	1.9	1.6	2.0
Printing and various	2.2	3.9	1.9	4.1	3.8	6.9	3.1	5.7
Other industries	3.1	6.9	3.5	9.6	5.8	10.5	5.6	9.8
Total	100.0	100.0	100.0	100.0	100.0	100.0	100.0	100.0

[a]I have used the ISTAT categories in its national accounts volumes. Although the breakdown is not ideal, it does permit sector comparability over time. North includes southern Lazio.

[b]Employment data were derived from the 1951 and 1971 Censuses of Industry.

[c]Data were derived from the *Annuario di Contabilità Nazionale, 1973*, Vol. III, Tome II, ISTAT, (Rome, 1973), pp. 31–2. These data are termed "gross product of manufacturing (at factor cost)."

Table 59. Value Added per Employee in Manufacturing in the Two Italies in 1976[a]
(millions of lire at current prices)

Sector	South	North
Ferrous and nonferrous minerals and metals	10.0	10.1
Nonmetallic minerals and products	6.4	7.4
Chemical and pharmaceutical products	13.2	12.9
Metal products, machinery, and electric equipment	6.2	8.0
Vehicles	7.5	9.7
Food, beverages, and tobacco	6.6	9.0
Textiles, clothing, furs, and leather	2.7	5.4
Paper and printing	6.5	8.4
Other[b]	3.7	6.4
Average for all manufacturing	5.5	9.7

[a]"Principali Aggregati Economici a Livello Territoriali, 1970–76," *Bollettino Mensile di Statistica*, Vol. 52, No. 1 (January 1978), pp. 246–7; southern Lazio data included in the North.
[b]Other includes wood and furniture, rubber, plastics and varied.

sectors in Chapter 9, can these approaches strengthen our explanations? Table 58 demonstrates changes by sector in employment and product for the two Italies during our study period.[9] I should emphasize again that within many of these sectors there are modern as well as traditional branches and firms. For example, the machinery sector includes machine repair workshops and modern assembly-line plants; similarly the textile industry includes both artisan facilities and innovative large enterprises.

However, even given the rather arbitrary nature of this sectoral division, there is no question that there had been significant positive shifts in the structure of southern industry during this two decade interval. Note in particular the growth, in relative terms, of the basic industries both in employment and production. However, it must be conceded that, despite this relative improvement, a large share of southern industry was still in the traditional sectors! Then, too, given our analysis of size of firm in Chapter 8, despite marked improvement there as well, the average firm was still far smaller than its northern counterpart, and I suspect far less productive. This is demonstrated by the data in Table 59 which show value added by manufacturing branch in the two Italies for 1976. Note that in the modern sectors like metallurgy and chemicals southern values were comparable to those for the North, but on the average the value added per employee in the South was 56% of that in northern Italy.

The question of industrial change can also be approached through shift-share analysis. This type of analysis begins with the premise that growth in an economy is attributable to two components of change. The first is termed "mix" and the second is designated "competitive." As I noted earlier, mix effects reflect the fact that some sectors in a nation are expanding more rapidly than others (secular trends). Therefore, those regions that have specialized in slow-growth industries have negative shifts, and those that have concentrated their manufacturing activities in rapid-growth sectors experience positive shifts. In contrast, the competitive component involves a different perspective. Here negative shifts are traceable to the failure of specific industries in an area to keep up with employment gains at national or sub-national levels. The reverse is true of positive competitive shifts. The problem for the Mezzogiorno is that it is still dominated by slow-growth industries despite sectoral changes that were evidenced in Table 58. In terms of the competitive component, Table 59 demonstrated the failure of many of the industries of the South to keep up with the productivity of their northern counterparts. Again, chemicals and metallurgy were the key exceptions to this pattern.

Allen and Stevenson[10] working with 1969 value added per labor unit data for 1969 and a varient of shift-share analysis calculated that only 25% of the North-South overall manufacturing productivity differential could be explained by the relatively poor structure (mix component) of southern industry and some 75% by nonstructural factors (competitive component). My own calculations based on employment changes, by province, between 1951 and 1971 were drastically different from those by Allen and Stevenson. They gave far greater weight to the mix component. Nevertheless, despite the differences between our results, it is still evident that *both* components are involved in North-South industrial productivity contrasts. So far, we have extracted those elements that appear to play a role in this disparity.

My next step is to speculate further about the reasons for the variance. I said earlier that the *size of firm* and *the nature of the industrial facilities* differ greatly between the two Italies. Without question, these are two of the causes for the productivity differential. It used to be argued that one of the key assets of the South was its far lower labor cost per unit of output. This advantage is no longer as great because regional disparities in pay scales (except for artisan workshops) have been generally eliminated. Although this change has been partially offset (since 1968) by social insurance subsidies accruing to enterprises in the Mezzogiorno, that region's labor cost advantage has still been reduced. Another element which is far more difficult to assess is the question of work attitudes and labor productivity. It is clear that this question is fundamental to an understanding of the North-South gap. There is still a problem of shortages of skilled labor in the Mezzogiorno. This is not a universal concern and can be overcome by training programs both in school and out (witness Taranto and its record of success). However, in other areas it is still a major handicap. Far more important and less amenable to ready solutions is the question of worker attitudes. These are, of course, more difficult to quantify. Here the key problems are absenteeism and wild-cat strikes. As an example, Alfa-Sud has been a prime illustration of financial losses and low productivity resulting from absenteeism. I am not arguing that such problems do not exist in the North, but they are, in a relative sense, more serious for a developing industrial area like southern Italy. To reverse my focus, there is also the question of managerial initiative and ability that are termed entrepreneurial skills. This problem is less acute for managers of branch plants whose headquarters are in the North or abroad. Typically, members of this group are Northerners or they have lived and been educated outside of the Mezzogiorno. I should note, in passing, that this pattern is slowly changing, particularly at the middle management level. For the manager of a branch plant in Italy, key decisions on purchases, operations and marketing are commonly made "in the home office." It is for the "true" southern entrepreneur who typically oversees an operation whose scale is small and whose equipment is often not the most advanced that the problem of entrepreneurial skills is most applicable. It is among this group of entrepreneurs that innovation and adaptability to changing technology and changing market conditions are frequently sub-optimal. Often this has led to bankruptcy and failure. Part of the problem is their position outside the industrial "contact system"[11] of the more advanced regions of northern Italy and, in turn, western Europe. Finally, just a word about location, for I am reserving that discussion for the concluding section of this chapter. It is conceivable that the lower productivity of the South continues to be attributable, in part, to its "locational disadvantages."

SOUTHERN ITALY AS AN INTEGRAL ECONOMIC UNIT

It should be recalled that in the introduction to this chapter I asked: "Are there signs of industrial achievement and positive structural change, and does this region show indications of the creation of a basis for self-sustained industrial growth?"

Professor Saraceno (the undisputed leader in development planning for the Mezzogiorno) has commented that at this stage in the South's development it is its absolute growth rate rather than the difference between North and South that is critical. Allen and Stevenson have phrased it as follows:

> It is far better that the South grows rapidly even though more slowly than the North" [even though the differential widens] "than that it grows slowly though more rapidly than the North" [with a narrowing of the gap].[12]

My discussion, here, will be restricted because these questions have already been covered, even though tangentially, in Chapters 5 and 8 and in the preceding analysis of the North-South differential. I will treat each of these issues separately, even though they are clearly interrelated.

First, as to the query about signs of industrial achievement, the growth of industrial output in southern Italy measured in percentage terms has been quite respectable, particularly in the light of the post-war experience of many developed nations. It has also been somewhat better than the performance of northern Italy. Thus, according to ISTAT data, the average annual increase in the output of manufacturing per worker (in constant 1963 dollars) was slightly lower than that of the North between 1951 and 1962 (5.6 to 6.1%) and considerably higher than that of its northern counterpart in the 1962 to 1971 period (7.9 to 5.4%). However, in terms of employment alone, its performance was far less notable. Here the lag reflects the capital-intensive nature of so many of the new industries of the Mezzogiorno discussed earlier.

The question of structure has been amply covered before, but again, in summary, my assessment was positive, although there is still the need for greater structural change. Here the South, particularly in some of its new capital-intensive industries such as oil refining, petrochemicals (both basic and secondary) and possibly synthetic textiles and plastics will face increasing competition from similar developments in many of the OPEC nations. Having lost most of its initial advantage in labor costs, it is already encountering severe competition on foreign markets from industrial producers in the Far East as is the North.

Has the South, in Rostow's[13] terms, reached a point where self-sustained industrial growth can take place without further net transfers of funds from the North? In my view, the answer is still firmly negative and may continue to be so for decades to come. Some southern scholars now readily concede that their original time horizon for the economic development of their region was far too short. Clearly, despite admittedly significant change, much remains to be accomplished before the South can be said to have reached "critical mass."

REGIONAL DEVELOPMENT THEORY, LOCATIONAL PRACTICE AND INDUSTRIAL DEVELOPMENT

It should be recalled that in the preface to this book, I contended that as an economic geographer I am primarily interested in the locational or spatial facets of industrial growth within the broader framework of regional development theory. In Chapters 5 and 8 I explored, in detail, spatial variations in levels of industrial growth that had emerged in the South during the first two decades of the development era. Now my focus shifts to the question of regional development theory and the degree to which southern Italian industrial development fits that body of theory.

Most chapters of this book were intended to comprise evaluations of regional industrial development policy and practice in a particular setting. Therefore, I do not plan to probe in depth the vast theoretical literature that dates back in some instances to the 1950s.[14] Nevertheless, a brief review of the main thrusts of those contributions is necessary to provide a broader framework for my analysis of the Italian experience.

Regional Development Theory

Although it could be argued that there are many theories of regional economic development, for my purpose they can be classified into two clusters. The first has been termed a convergence or equilibrating group of theories. Essentially, those who have followed this approach believe that over time, through the market mechanism, regional income inequalities should disappear. The central view of the second cluster is that economic development is spatially unbalanced, and over time there should be *increasing* regional economic differentiation, particularly with respect to income levels.

Others, like Williamson, take an intermediate historical perspective. He argues[15] that regional inequalities or spatial imbalances are characteristic, in Rostow's terms, of the industrial "take-off" stage of economic development, while in the "mature" phase such inequalities would either be reduced or conceivably eliminated. Richardson,[16] a noted specialist in this field, has written that the differences between the two opposing schools of thought could possibly be resolved by distinguishing between inter-regional and intra-regional growth. Thus there could be polarization *within* a region, but some degree of convergence *between* regions.

The most often cited example of the spatial divergence school is the work of Myrdal[17] who is noted for his theory of "circular and cumulative causation." Myrdal's model is one that focuses on spatial or regional inequalities. His basic starting point is with a country that though populous has a low level of economic development, and he assumes a free market economy. For Myrdal, market forces tend to *increase* rather than *decrease* regional income inequalities. Once a particular region or sub-region has by virtue of some initial advantage moved ahead of others, new increments of growth will focus on the already expanding region because of their *derived advantages* rather than in the rest of the country. Here, then, is the notion of cumulative causation or cumulative concentration. To this theorist, the main explanation of differential regional growth lies in the resulting spatial interchanges between the *growing* and the *lagging* regions of a nation. In the former regions once development has been sparked, this will eventually result in flows of labor, capital and commodities from the stagnant areas which support the further growth of that region. Myrdal called these *backwash effects* because of their impact on the backward hinterland of a country. These lagging regions tend to lose their most enterprising workers (migration) and local capital to the growth areas. At the same time, or perhaps with somewhat of a temporal lag, markets in the backward areas are swamped by cheap manufacured goods which throttle local industry such as traditional artisan manufacturing. The backward areas also suffer noneconomic effects such as an inferior social infrastructure (poorer educational services and inadequate health provision). So, in the overview, backwash effects stifle economic growth in the stagnating areas and stimulate and sustain further development in the growth regions.

The other facet of the Myrdal model maintains that there may ultimately be *spread effects* which diffuse economic momentum from the growth areas to the lagging regions. This, he believed, would most likely take place (down the urban hierarchy) in neighboring regions, thereby stimulating a new cumulative causation process. This should then lead to self-sustained economic growth in these newly-developing economic centers.

One of the key assumptions of the Myrdal model is the absence of governmental intervention. He concedes, however, that if the state intervenes and it makes a major commitment to a development effort, then the spread effects would be more rapid and more marked. As we have seen, from our historical survey of Italian regional economic development, this model is obviously relevant. Most scholars would agree that Myrdal's ideas are cogent and useful and that they do fit past development trends in many advanced western societies. There has, however, been very limited empirical verification of the notion of spread effects, particularly without government action.

Closely linked with Myrdal's formulation is the notion of *growth poles* or growth centers. This approach can be traced back to the writings of Francois Perroux and his colleagues, particularly J. R. Boudeville, in the 1950s and 1960s.[18] Perroux, a noted French economist, coined the term *growth pole–(pôle de croissance)*. However, he used it in an aspatial, functional framework. His focus was on growth sectors and propelling industries (l'industrie motrice). The propulsive industry or industries would be large and dynamic enterprises with high levels of technology and considerable functional interrelations with other firms (multiplier effects). These would result from backward (supply) linkages and forward (output) linkages. Ultimately firms producing similar products are attracted because of localization economies and urbanization economies.

Boudeville's contribution, among others, was the translation of the notion of a "functional pole" into that of a "geographical pole." For him, a spatially-focused growth pole would be an urban center of economic activity that might have one or more propulsive enterprises. It should have achieved a high level of self-sustained growth which permits development to be diffused outward into its immediate hinterland and to the less developed region which surrounds it. Although for Boudeville inter-industry linkages are still important, just as critical is the relationship between the pole and its "region." Thus he and others talk of the need for a highly developed economic infrastructure (particularly the provision of transport facilities), a well-endowed social infrastructure, and an intellectual mind set in the business community which can be termed a "growth mentality."

There are many technical difficulties which have arisen with the adoption of growth pole strategies for developing countries and for the backward regions of developed nations. They key problem is still the question of *location*. Despite the development of multiple location algorithms,[19] the determination of an optimum number and location of a system of growth points *in the real world* is still extremely complex. Other problems include the determination of criteria for the selection of individual growth poles, and among those criteria is clearly the threshold size of the population of such proposed poles. Nor do we know enough about the means by which growth at a pole is then diffused to its hinterland. So far regional analysts have failed to provide adequate solutions for these and other problems involved in growth pole strategy.

My review of the theoretical literature has been necessarily brief, and it has focused on spatial models of development. In that critique, I treated only those elements of theory which appeared necessary for understanding regional development policy and practice. However, the approaches that I summarized evidently have a strong economic bias, yet there are other dimensions which are just as significant. Of these noneconomic behavioral elements, the political and social dimensions are by far the most noteworthy.

The choice of areas that would benefit from the varied incentives available under a given development program typically involves selectivity because of scarce resources. Thus it follows that there must be the choice of some locations over others. These decisions may conform to some of the tenets of regional growth theory and may be rational in cost-benefit terms, but they could be "irrational" in a given political context. Then, too, we find counterarguments framed in terms of "spatial balance" and "equity" which are socially based. Often, these political and social arguments, in the real world, outweigh reasoning that is grounded in terms of economic rationality. I am contending, therefore, that regional planning cannot operate in a social and political vacuum. The ultimate result is that development decisions are often "boundedly rational" in Simon's terminology[20] and may conceivably run counter to purported goals of maximization of regional and national economic growth.

Location Policy in the South and Regional Development Theory

I propose next to review, in summary fashion, the evolution of location policy in the development program and its relationships with the principal constructs of regional development theory.

It should be recalled that in the early 1950s the main efforts of the Cassa were focused on the "pre-industrialization" of the Mezzogiorno. The North-South dichotomy was clearly recognized, and, in fact, development theorists have traditionally viewed it as perhaps the best illustration that we have of *spatial imbalance* in a country, to use Myrdal's term. Adopting an equilibrium approach, development officials argued that expenditures on social and physical infrastructure were necessary initial steps that would lay the groundwork for the eventual industrialization of southern Italy. This they maintained would take place via the market mechanism. However, industrialization was not completely ignored and a limited set of incentives was authorized. Given the minimum expenditures on industry and its relatively low priority, there could be no locational design, and the few new plants that were established during this era naturally gravitated to the preexisting industrial centers.[21] In effect, the disparity between the few established industrial nodes of the South and the nonindustrial areas widened. However, even the growth in the former areas was minimal.

With the passage of the new development legislation of 1957, followed by that of 1965, priorities shifted. It was now agreed that infrastructure improvements alone were not enough to attract new industry. Clearly, the market mechanism had not fulfilled exceptations. The solution adopted was the deliberate subsidization of industrial development in the South. This then was government intervention in both Myrdal's and Hirschman's framework.

A locational design slowly emerged. There were at least two divergent schools of thought on location policy. The first reflected the political and social rationales discussed earlier. Politicians who represented this viewpoint demanded that funds be allocated in terms of "equity" or "social justice"; some even argued that priority be given to the most impoverished and backward regions. A second perspective reflected the writings of Perroux and Boudeville. These were the proponents of a growth pole strategy, and they were ultimately more persuasive. We have seen the results in terms of the creation of areas and nuclei of industrial development which were to receive higher levels of support under the loan and grant program. I contended earlier that the proliferation of these development zones was counterproductive, impairing any growth center strategy. Of course, that expansion again reflected the political dimension in regional planning. However, if we return to the ILSES report of 1968, reviewed earlier, its authors argued that "the gradation of support on the basis of location was not a meaningful incentive for spatially differentiated investment within the Mezzogiorno."

Essentially, I am reiterating here my support for a location policy that calls for spatially concentrated investments in a *limited* number of growth centers such as the five "aree di sviluppo globale" designated in the 1966-70 Plan. This should be coupled with a *meaningful* gradation of support for plants that locate in these centers. However, despite my support for this type of spatially concentrated investment, I must admit that the evidence for spread effects from growth poles in Italy is far from conclusive.

The term "cathedrals in the desert" is now somewhat out of vogue in Italy. Although it was always an oversimplification of a complicated issue, I still find the idea provocative and worth pursuing. The argument stated simply is this: Far too high a proportion of the investment funds has gone for the support of capital-intensive industries, particularly those in the petrochemical and metallurgical sectors. I hesitate to include metallurgy in this group, for the Taranto steel plant is now one of the largest in western Europe and may employ as many as eighteen thousand workers. Nevertheless, it shares with the petrochemical industry a common problem. That is, considering the huge investments required, the repercussive or multiplier effects of these plants, to date, have been minimal. Podbielski has argued that the numbers of such plants is quite limited and that the investment funds allocated to them is still only a small fraction of the total loan and grant program.[22] She also questions how much real "desert" there is around these plants? I would argue "considerable." A visit to Gela in southern Sicily, a petrochemical facility, is highly instructive. Despite the creation of this major primary chemical plant, its influence has been extraordinarily small. The province of Caltanisetta, of which Gela is a commune, still has a very high level of outmigration. Hytten's study of Gela called: "Industrialization Without Development" is most illuminating.[23] In contrast, Alfa-Sud is typically cited as an example of a plant that is labor-intensive, for this automobile plant employs over 15,000 workers and it has had some repercussive effects. The sad fact, however, is that its suppliers in Campania and the rest of southern Italy employ only 2,000 workers (whose production is directly related to components for that automotive facility). Thus its multiplier effects have been felt more *outside* of the Mezzogiorno than *within*. There are, unfortunately, other "cathedrals," that are perhaps less imposing, which have also had minimal repercussive effects.

To blame the development planners or the regional growth theorists for this problem would serve no purpose. Intuitively, I too would have expected that growth industries like steel, automobiles, and even chemicals would attract other plants because of localization economies and probably because of urbanization economies. Similarly, on an intuitive basis, I would expect to find geographical spread effects outward from a growth pole, and there seems to be some evidence for these spread effects in the northwestern margins of the Mezzogiorno, but elsewhere the data are far less conclusive.

There is another facet of regional planning in southern Italy that deserves discussion here, not because of its success, but because of the lessons it has provided for other development programs. I am referring here to the much publicized attempt to create an interrelated industrial complex or industrial growth pole in Puglia.[24]

This experiment was probably unique for its time (the 1960s) in terms of industrial development projects outside of the socialist world. It arose from recommendations by the European Economic Community to the Italian government in 1965. Disappointed with the results of their development effort, the Italian planning authorities had asked for technical advice. The report was prepared by Italconsult. It argued for the need to use *new criteria* to promote industrial development in the South. It cited the fact that the existing program, which provided in essence a *2 to 5% subsidy* for new plants in the South, was insufficient to offset the *10% margin* that the North already possessed because of existing external economies in that region. We have noted previously that these economies are the benefits that accrue to modern multiple-cycle industrial establishments from linear and circular linkages with associated industries.

In accordance with this argument, regional theorists have written that the development of integrated industrial complexes, in appropriate economic environments, should reduce the time and cost of operations, lower inventory requirements, and improve the range and quality of industrial outputs. In essence, the Italconsult model was couched in an input-output framework which had as its major feature the notion that demand from the main plants would give sufficient economies of scale to their suppliers to ensure their economic viability. The same would also apply to fabrication plants that would purchase the output of the principal establishments.

The plan also argued that the creation of such a pole might overcome the reluctance of northern and foreign firms to invest in the South. Thus it was envisaged that aside from the linked plants that would be simultaneously inserted into the "pole," the improved "industrial climate" might produce the multiplier effects that had not materialized elsewhere.

Italconsult's recommendations called for an overall investment of $166,000,000 for the *simultaneous* insertion of a series of machinery and metal fabrication plants into a development pole in the Mezzogiorno. Those industrial branches were selected because at the time the market prospects for their products in Italy and abroad were good to excellent. As far as location is concerned, the plan called for the selection of a site that already had a reasonable infrastructure, an industrial tradition and a potential labor pool. These machinery and metal fabrication plants would be assisted by heavy state investments designed to improve their infrastructure, particularly transport facilities and to aid in the creation of vocational schools. In addition, these plants would be eligible for the maximum subsidies already available in the "areas of industrial development." The site ultimately selected lay in the provinces of Bari, Brindisi, and Taranto in Puglia and the adjoining province of Matera in Basilicata. This zone, which accounts for over one-tenth of the population of southern Italy, has been a region of relatively rapid economic growth during the development era. Its manufacturing employment has nearly doubled since 1951, reaching a possible level of 140,000 workers in 1976. The development era had witnessed the building and subsequent enlargement of an integrated steel mill at Taranto, which was central to the plans for the industrial complex, a cement plant, and a sizable oil refinery. Brindisi is the site of a massive petrochemical facility, while Bari had a long tradition in the machinery and metal fabrication sectors. Matera's industrial growth is linked to the availability of natural gas which is used for chemical production at Pisticci and Ferrandina. Per capita income in the region as a whole is well above the average for southern Italy. Major infrastructure investments have promoted the region's transport connectivity. Bari and Taranto are now linked by autostradas to Naples and to Pescara on the Adriatic coast. In addition, a link is beyond the planning stage to tie the region to the toll-free autostrada that runs from Salerno to Reggio di Calabria. These external linkages are now complemented by a greatly improved road network within the region itself. There have also been major improvements in the region's social and physical infrastructure. It should be emphasized that these infrastructure improvements had been planned long before the selection of the region as the location for a development pole, but they were clearly accelerated. The development plan called for 8 major facilities and 23 smaller supply plants with an estimated *initial* employment of 10,000 workers. However, it was also assumed that there would be multiplier effects, once the first industries were established, not only in manufacturing but also in the tertiary sector.

The plan was approved in 1965 and IASM (the promotion agency for industrial development in the South) was given the responsibility for the recruitment of suitable firms. Production was to have begun by 1970. It is generally agreed by IASM, CIPE, and outside observers that the program, in its initial form, never materialized. The reasons cited to me by IASM are these: (1) The recession in Italy and western Europe after 1966 caused a number of firms that were apparently initially interested to withdraw; (2) the State-controlled firms like IRI and EFIM-BREDA initially, at least, did not participate to the degree anticipated; (3) the planners failed to take sufficient account of changing technology and changing market conditions (the plan was too inflexible); (4) the planners overestimated the response of local industry and the local bureaucracy to the program; and (5) the planners placed so much emphasis on input-output analysis and economies of scale that they tended to ignore other factors that played roles in industrial location.

However, four groups are still currently involved in the program: Fiat and Ignis in the private sector and EFIM-BREDA and IRI which are State-controlled. The number and character of plants has been drastically altered because of reevaluations of original cost estimates and changed market conditions. Some facilities which do not fit the original concept are now planned such as plate glass and cold rolled steel, but key investments for plants which will produce compressors, pumps, automobile parts, and agricultural machinery are still in accordance with the original proposal. There is also continued discussion of linked facilities in the petrochemical field. So, the general plan has not been followed, but the growth of industrial employment continues. *However, the conceptual ideas have never been truly implemented.* The basic premise is still worth testing, but only as *one* tool in regional development planning. The same general ideas have been adopted elsewhere. Witness the Fos-sur-Mer development near Marseilles, but the notion of linked industries simultaneously inserted into a development area has yet to be truly tested and evaluated either in Italy or elsewhere.

The New North-South Dichotomy

Leaving the fields of regional theory and locational policy, for the moment, I should like to conclude this chapter with some brief reflections on what is perhaps the most important emerging locational phenomenon in the Mezzogiorno. I am speaking here of the evolving disparity between the northern portion of the Mezzogiorno and the rest of that region. This question was first broached by Mori in 1965 and Mazzetti in 1966; it surfaced again in the writings of Costa and Cafiero in 1975 and 1976.[25] However, Cafiero's comments are without question the most insightful.

All that one has to do is study several of the maps in this book, particularly Figures 14 and 23, to detect a *new* twofold regionalization of the South. If I were to draw a line from Salerno, via Matera, to Taranto that would now be a rough boundary between the two new regions. True, there are large impoverished areas in the "osso" or the Appenine spine running from L'Aquila through Campobasso, Benevento, Avellino and Potenza. Then too, as we have seen, there are still critical problems facing both Naples and Salerno. I should add that Cafiero places the boundary further north on a line from Salerno to Bari, based upon Costa's urban accessibility study. He also argues that the industrial installations in or near Matera (chemicals), Brindisi (chemicals) and Taranto (steel) are "industrie di base" (basic industries) similar to those in Sicily and Sardinia. Therefore, he contends that in terms of their industrial structure they are really part of the new "South." I might have agreed with his reasoning some years ago, but now it is highly debatable. The completion of a new set of autostradas and superstradas has resulted in a far higher level of transport integration in that region. Thus the centers I have cited are no longer isolated (as they are further south); for me, at least they are part of the new "North."

I should also add that there are several discontinuous enclaves south of my boundary line which are still prosperous. These would clearly include parts of the coastal region of eastern Sicily, in the Syracuse-Mellili-Augusta area plus the Cagliari and Sassari nodes of Sardinia. However, most of that island, western and central Sicily and Calabria belong in the "new Mezzogiorno." I can readily forsee the day when the northern boundary of the development region will follow that crude demarcation line.

Cafiero argues that the division of the South into these two zones is not based solely on the distribution of manufacturing, although its boundary is clearly evident on the map, by commune, of manufacturing employment in 1971. It is also grounded on the prospects for further industrialization in both regions. The separation of the Mezzogiorno into two regions rests also on the relatively close linkages that now exist *within* the northern area and *between* it and the larger metropolitan areas of central and northern Italy. As I argued earlier, Latina and Frosinone, in particular, though legally part of the Mezzogiorno, as are the southernmost communes of the province of Rome, are entitled to all of the subsidies and incentives of the development program. Their inclusion was undoubtedly based on political pressure, for there is no economic rationale. The same reasoning cannot be used for the Abruzzi. There, Pescara had a long tradition of economic leadership, but its growth has now been surpassed by that of Teramo and Chieti. On the west, Caserta is the new magnet of industrial development resulting in a new axis of industrial development leading from Latina directly to Salerno. In any event, the cities of the northern segment of the South can now provide the level of urban services that are essential for modern industry.

In contrast, the new South is remote and definitely on the periphery. It cannot overcome its locational handicap. True, there is now an autostrada leading to Reggio di Calabria and there are several others in Sicily. However, the cruel fact is that commodity traffic on these superb highways is largely local; long-haul movements are almost exclusively those by northern truckers who are taking advantage of reduced transport costs. Thus northern firms are using these new linkages to penetrate the formerly isolated markets of the South. The major cities of this region–Palermo, Catania, Messina, etc.–have a far lower level of industrial development (in relation to their working age population) than any of their counterparts in the North. The only industries that prosper are those in the basic chemical sector (or metallurgy if we were to include Taranto in the new South). These industries import raw materials (petroleum) by sea and reexport their products, because their production levels are far above the market requirements of the Mezzogiorno. Such basic industries are far less sensitive to the backward industrial climate of the new South. Nor have they contributed in any substantive sense to an industrial "take-off" of the region. There have been scarcely any significant multiplier effects in terms of the growth of satellite fabrication plants like drugs, plastics, paints, etc. The Cassa has recently subsidized several synthetic fiber and rubber facilities in the new Mezzogiorno and that is one encouraging development. However, students of the South would agree that given the competition of textile producers in the Far East and the creation of petrochemical plants in the OPEC countries, the possibilities of future growth are highly limited. The future economic integration of the southern part of the Mezzogiorno into the national economy seems very remote, but the Italian government can hardly write off a region that contains 8,000,000 people. Nevertheless, its future is uncertain.

NOTES

[1]Schachter, 1965, op. cit.; Podbielski, 1978, op. cit.; Allen, 1970 and 1974, op. cit.; Vera Lutz, *Italy, a Study in Economic Development* (London: Oxford, 1963); and Alan B. Mountjoy, *The Mezzogiorno* (London: Oxford, 1973).

[2]Excellent reviews of the Italian geographical literature are found in Calogero Muscarà, *La Geografia dello Sviluppo* (Milan: Comunità, 1967) and *La Società Sradicata* (Milan: Franco Angeli, 1976).

[3]Podbielski, op. cit., p. 75.

[4]Sources: ISTAT, Cassa per il Mezzogiorno, and SVIMEZ.

[5]Annuario di Contabilità, 1973, op. cit., various pages.

[6]Salvatore Cafiero, *Sviluppo Industriale e Questione Urbana nel Mezzogiorno*, SVIMEZ (Milan, Rome: Giuffrè, 1976), p. 59.

[7]See "Le Imprese Industriale del Mezzogiorno," *IASM Notizie Documenti sul Mezzogiorno*, Vol. 9, December 1976 (Rome, 1976).

[8]M. Amendola and P. Baratta, *Investimenti Industriali e Sviluppo Dualistico*, SVIMEZ (Milan, Rome: Giuffrè, 1978).

[9]A more detailed breakdown, by branch, would have been more useful, but published ISTAT data do not permit such comparisons.

[10]Allen and Stevenson, 1974, op. cit., p. 188.

[11]Gunnar Tornqvist, *Contact Systems and Regional Development, Lund Studies in Human Geography*, No. 35 (Lund, 1970).

[12]Allen and MacLennon, 1971, op. cit., p. 121.

[13]Walt Rostow, *Stages of Economic Growth*, (Cambridge, 1960).

[14]See note 1 in Preface for highly useful bibliographical sources.

[15]J. G. Williamson, "Regional Inequality and the Process of National Development," *Economic Development and Cultural Change*, Vol. 13, No. 4 (July 1965), pp. 1-84.

[16]Harry W. Richardson, *Regional Economics* (New York: Praeger, 1969); *Regional Growth Theory* (London: Macmillan, 1973); "Regional Development Policy in Spain," *Urban Studies*, Vol. 8 (1973), pp. 233-53.

[17]Gunnar Myrdal, *Economic Theory and Underdeveloped Regions* (London: Duckworth, 1957).

[18]Francois Perroux, *L'Économie du XXe Siecle*, 2nd edition (Paris: Presses

Universitaire de France, 1964). J. R. Boudeville, *Problems of Regional Economic Planning* (Edinburgh: University Press, 1966).

[19]Gunnar Tornqvist, Stig Nordbeck, Bengt Rystedt and Peter Gould, *Multiple Location Analysis, Lund Studies in Human Geography*, No. 38 (Lund: Gleerup, 1973).

[20]Herbert Simon, *Models of Man* (New York: Wiley, 1957), p. 196.

[21]Friedrich Vöchting, "Considerations on the Industrialization of the Mezzogiorno," *Banca Nazionale del Lavoro Quarterly Review* (September 1958), pp. 356-360, and Rodgers, 1960, op. cit., p. 171.

[22]Podbielski, 1978, op. cit., p. 179, and P. Saraceno, "Le Cattedrali nel Deserto," *Corriere della Sera* (September 15, 1974).

[23]Evind Hytten and Marco Marchioni, *Industrializzazione Senza Sviluppo: Gela: Una Storia Meridionale* (Milan: Franco Agnelli, 1970).

[24]Rodgers, 1969, op. cit., pp. 58-60, and IASM, *Presentation of the Study for the Establishment of an Industrial Development Pole in Southern Italy* (Rome: July 1965).

[25]Cafiero, 1976, op. cit., various pages, Ernesto Mazzetti, *Il Nord del Mezzogiorno* (Milan: Comunità, 1966), and P. Costa, "Urban Agglomeration Economies: Some Evidence from the Italian Case and Many Perplexities," *Ricerche Economiche*, Nos. 3-4 (Venice, 1975).

Chapter 11

IMPLICATIONS OF THE ITALIAN EXPERIENCE FOR REGIONAL PLANNING IN WESTERN SOCIETIES

It is my objective in this chapter to evaluate the implications of the Italian experience for western nations which contain both developed and underdeveloped regions within their borders. I have used the term "Western" because I believe that this study has greater relevance for market economies than for centrally-planned socialist systems. However, I would concede that all socialist states have problem areas within their confines, but their "professed" goals and certainly their options are rather different from those in market oriented economies.

My objective can be rephrased in this fashion: what can be learned from the Italian experiment? This query has validity because the past few decades have witnessed a growing interest on the part of advanced Western societies in the geographical distribution of economic activity with a particular concern for regional income inequalities. That interest has been reflected in the creation of development programs which were designed to aid problem regions.

Before I begin my analysis of the implications of the Italian experience for Western nations, I plan to summarize regional development policies and practices in Spain, France, the United Kingdom and the United States. These countries have been chosen because of the diversity of their regional problems, their varying levels of economic development, the length of their experience and the nature of their political and economic systems. The summaries will cover the following topics: (1) setting, (2) regional problems, (3) regional development policies and methods of implementation, (4) results to date, and (5) prospect.

REGIONAL DEVELOPMENT POLICIES AND PRACTICE IN SELECTED WESTERN NATIONS:

As noted above, the material presented in this section is designed to provide selected reviews of development practice in the West. Those who wish to expand their knowledge of the experience of Western countries, can partake of an abundant literature describing their goals, practices, accomplishments and errors.

Spain

Spain's[1] population in 1977 was estimated to be 36,000,000. Despite the heavy loss of life resulting from the Civil War of the 1930's, the Spanish population has been growing at a faster pace than those of its neighbors in southern Europe. Its comparatively youthful population coupled with relatively high birth rates all portend more rapid growth in the coming decades.

Worthy of emphasis is the fact that the average density of population in 1976 was only 70 persons per square kilometer, one of the lowest in Europe. That low population density is mainly a result of physical constraints such as aridity, infertile soils and rugged terrain, all of which, in combination, have retarded the development of agriculture, still the nation's most important economic activity. Equally significant, perhaps, has been the persistence of a semi-feudal socio-economic system reinforced by the general isolation of Spain from the main currents of development in Western Europe. This, of course, has clearly changed with the death of Franco.

Contrary to the complex Italian pattern, the broad outlines of population distribution in Spain are relatively simple. With the exception of the comparatively isolated node of higher density around Madrid, there is a striking contrast between the sparse, predominantly rural population of the dry interior uplands or the Meseta and the far higher densities of the periphery.

Several of the coastal regions, particularly those around Valencia, on the east, and the zone from Murcia to Algeciras, on the south, are areas of highly productive irrigated agriculture which support notable rural population densities. More important, in a quantitative sense, is the heavy clustering of urban population on the north and east coasts. These regions have witnessed an extremely rapid growth of population in recent decades, attributable particularly to economic opportunities associated with developing industry. The North is Spain's only significant center of heavy manufacturing. Here the chief cities are Santander, Bilbao and San Sebastian. In contrast, the Catalonian area of the east coast with its major metropolis of Barcelona is a region with a wide range of labor-intensive industries, particularly textiles. Over time, the disparity between the center and the periphery has widened. Depite the generally low level of urbanization in Spain, the postwar era had witnessed an explosive growth of the largest urban centers, especially those whose population exceeded 100,000 inhabitants. Within this group, Madrid with 4,000,000 inhabitants and Barcelona with nearly 2,000,000 are the outstanding growth centers.

While I would hardly classify Spain as a highly industrialized nation today, its share of the active population that is engaged in manufacturing had risen by 1976 to a respectable level of 26% (compared to 27% in France). However, it must be stressed that much of its industrial employment, as was true in the Mezzogiorno, is still found in small-scale, first-order establishments with limited capital investments and low productivity. Paralleling the growth of industry, there has been a decline in agricultural employment accompanied by a striking rural exodus.

It is evident from this review that there are in a sense *two* Spains as there are *two* Italies. In the Spanish case the contrast is between the Center (excluding Madrid) and the Maritime Periphery as opposed to the North-South dichotomy in Italy. However, the difference is not merely locational, for despite the growth of the periphery and Spain's rather striking urbanization in the postwar period, the country is still at an intermediate level of development.

Both Lasuen and Richardson[2] have expressed strong reservations about the quality of Spanish income data. Nevertheless, broad generalization are still valid. Although the average income per capita in Spain is still below that of Italy and the United Kingdom, the gap is closing. It is also clear that, just as in Italy, there are significant regional income inequalities. The regions with the highest per capita incomes (Madrid, Cantabria and Catalonia) have values that are three times those of the lowest regions (Andalusia and Galicia). However, the major high-income urban centers are widely separated from each other by areas of low population densities and low incomes. The weak linkages between these centers are still a major development problem, but new super-highways are rapidly increasing inter-urban connectivity. This explains, in part, the modest convergence of inter-urban per capita income levels and the growing divergence of urban-rural levels in the past two decades.

Spanish regional development programs really began in 1964 when the first development plan was adopted; however, modest investments designed to improve the nation's infrastructure (land reclamation, irrigation and transportation) were initiated as early as the 1930's. In fact, it could be argued that the 1950s and the early 1960s were periods of "pre-industrialization" in Spain paralleling the pre-1957 development policy in Italy. However, the assumption that productive investment would follow infrastructure improvements in backward areas was not borne out either in Spain or Italy. Regional development policy in Spain has consistently been subordinated to the main goal of sustained national economic growth. In fact, regional planning was conceived to be the geographic counterpart to sectoral planning. Decision-makers in Spain adopted a "Williamson" type model which, as noted earlier, argues that as national development progresses, regional income inequalities should disappear. Efficiency rather than equity was the main planning criterion. By the 1960's that national goal had resulted in a spatially concentrated pattern of development, for the government deliberately limited investment to the leading regional centers: Madrid, Bilbao-San Sebastian, Barcelona and Valencia.

The first development plan (1964 to 1967) adopted a growth pole strategy which called for the diffusion of development to several second order regional centers. Those selected were Zaragoza, Valladolid, Seville, La Coruña and Vigo. Several years later Burgos and Huelva were added to the priority list. In the second plan (1968 to 1971) four additional centers were included: Granada, Cordoba, Oviedo and Laguna. The planners ruled out major investments in the truly backward areas of the nation. Paralleling the Italian model, the main instruments designed to attract industry were grants (which ranged from 10 to 20%), reduced interest rates on long term loans, tax breaks and easy access to credit. In general, those fiscal incentives were far lower than those then available in Italy. I should add that no negative disincentives were adopted to discourage new industrial investment in the developed regions, although the third plan (1972–1975) called for the creation of decongestion zones to relieve pressures in centers like Madrid and Barcelona and for the support of industrialization along key transport axes. The decongestion policy appears to have been a failure primarily because the zones selected were located too far from the first order centers. Another element in the regional development policy which resembles the Italian experience was the "encouragement" of INI[3] (the state holding corporation) to build plants in the larger second order regional centers. The dispersion of private and publicly financed industrial development is now well underway. Studies of the establishments that were successful indicate that the great majority were branch plants of multi-product concerns located in the developed regions rather than completely new operations or an outgrowth of the expansion or modernization of pre-existing facilities. These branch plants were designed to take advantage of labor cost differentials in the underdeveloped regions and initially, at least, to serve regional rather than national markets. This diffusion of industrial development has apparently brought over 50,000 jobs to these second order cities. It has clearly helped to reduce interregional income differentials, while in contrast urban-rural disparities have been accentuated. The incentives discussed above were designed to attract capital-intensive industries, and in fact half of the funds were devoted to the petrochemical and metallurgical sectors duplicating the experience of southern Italy. So far multiplier and spread effects have been minimal.

France

As of the 1975 Census, France[4] had a population of nearly 53,000,000. Her demographic growth slowed during the past decade after a post-war spurt. This reflected the nation's low birth rates and her comparatively high death rates, typical of mature industrialized societies. France's per capita gross national product in 1976 was three-quarters of that for the United States and well below those of West Germany, Sweden and Switzerland. In contrast, the

French level was significantly above those of the United Kingdom, Italy and Spain.

The nation is dominated to an extraordinary degree by the Parisian agglomeration. Roughly 20% of the inhabitants of France live in this metropolitan area, and if the surrounding departments are included that value rises to 30%. In no other country on the continent does one agglomeration play such an overwhelming political, economic and social role. Most observers would rate the Parisian agglomeration as one of the two or possibly three leading economic regions in the EEC (European Economic Community). However, within France there are considerable income variations between the capital and the provinces; these reflect differential levels of economic development and economic health. Per capita income values in the metropolis are nearly double those of the poorest areas of the nation. A map of per capita income values by region would demonstrate levels well below the national average in the West, particularly in Brittany (the poorest region) and Poitou, in the "Central Massif" (Limousin and Auvergne), and in the South in Languedoc and the Pyrenees. Aside from Paris, incomes above the national mean are found mainly in northern and eastern France. These areas include upper Normandy, Picardy, the Nord (North), Lorraine, Alsace, the Rhone-Saone valleys (dominated by Lyons), and Provence-Cote d' Azur.

Although agriculture is by no means the leading national economic activity, it is far more important in France than the norm for northwestern Europe. The role of this sector has clearly been reduced during the post-war era. Productivity has not significantly improved and farm incomes are quite low even in the more prosperous areas. Despite the declining importance of agriculture, nearly one-eighth of the labor force is currently engaged in this primary activity which is so strongly supported and subsidized by the state.

As might be expected, given France's high level of development, about 75% of her population is classified as urban, somewhat lower than the percentages for West Germany, the United Kingdom and the United States–but high nevertheless. The Parisian agglomeration has a larger population than that of all the other French urban centers in combination. Until 1968, the capital was the key center of in-migration in France. Other major urbanized areas include Lyons, Marseilles and Toulouse in the South and the Lille conurbation in the northern French coal field.

I would have no hesitation in classifying France as a developed nation, for she fits all of the recognized criteria. Yet within her borders, as is true of most advanced nations, there are a number of problem regions. Unlikely Italy or for that matter Spain, France contains both types of problem regions commonly discussed in the development literature. She possesses underdeveloped areas located in the South and West, and stagnating industrial regions found in the North and Northeast. The latter include the Sambre-Meuse coal field with associated industries like textiles and chemicals as well as the iron ore mining region of Lorraine and its linked metallurgical centers. Although the per capita income of these two industrial regions is above the national average, Tagliacarne,[5] using an unweighted summation of economic indicators including income, unemployment and out-migration for all of the 111 officially designated regions of the EEC, ranked the Nord and Lorraine at the median position of that group of regions.

As regional planners and decision-makers in the developed world have attempted to cope with the difficulties arising from the urbanization diseconomies of metropolitan areas, there has been an increasing tendency to treat them as a third type of problem region. The Parisian agglomeration is clearly such an area. It would not be an exaggeration to claim that French planners have viewed the overwhelming dominance of Paris as their overriding regional problem. For them, the reduction of the economic and social roles of their metropolis is the critical issue. Gravier[6] was the first to use the terms "Paris et le Désert Francais." Clearly in the view of the French development authorities a diminution of the dominance of Paris would, on balance, stimulate the economic development of Gravier's "desert". Ultimately this would result in the economic growth of France's underdeveloped regions and the rejuvenation of her stagnating industrial areas. This change, they believed, could not take place as a result of the free operation of the market mechanism. It could only be an outcome of the direct intervention of the state.

I will confine my review of French regional planning to developments in the 1960s and 1970s, despite the fact that modest initial steps were taken earlier to induce manufacturing industries to move out of the capital. Planning in that era was mainly confined to the national level, and regional development had a far lower priority. It was in the fourth five year plan (1960 to 1965) that a strategy for reducing the dominance of Paris was first outlined. Not only was the policy enunciated, but a development agency under the direct supervision of the prime minister was created in 1963. It was called DATAR (Délégation à l Amenégement du Territoire et l' Action Régionale). Concurrently the necessary instruments were also authorized. Initially, DATAR was not given the authority to formulate regional development policy. Over time, it assumed that role mainly because of the quality of its staff and the active support of the prime minister. To a degree its authority increased by default, for the Planning Commission was primarily concerned with national goals. This left a void, so that regional planning was left to DATAR. However, it was not until the fifth plan (1966 to 1970) that this agency truly assumed that responsibility.

Gradually a regional strategy evolved. If the main goals, as noted earlier, were the stemming of the flow of migrants to Paris and the reduction of its economic, social and cultural role, there would have to be a decentralization of industries and tertiary activities from the capital, the economic development of the rural western portion of France, and the rejuvenation of the older industrial areas of the North and Northeast.

The spatial strategy adopted had two complementary thrusts. Initially, France was divided into five zones for the purpose of graduated incentives. Zone 1 comprised the western and southwestern periphery which, as indicated earlier, is a low income region in which agriculture is the chief economic activity. Zone 2 included a number of stagnating mining and industrial areas in the North and Northeast where state aid was needed for industrial adaptation, modernization and diversification. Both areas would be eligible for investment grants up to 10% as well as a variety of tax incentives coupled with subsidized long term-low interest loans. The size of the grant would be determined by the number of jobs to be created, and preference would be given to growth industries. The rest of France, outside of the Paris agglomeration, was divided between zones 3 and 4; these presumably would be regions where industrialization needed to be stimulated and diversified. There, only graded tax benefits would be available. Zone 5 was the Paris basin where industrial development was to be actively discouraged.

The second element in the regional development strategy called for the selection of a series of growth poles or growth centers. Eight urban agglomerations were selected based on a point system of 13 criteria which reflected size of population and central place functions. They were termed "metropoles d' équilibre" (equilibrating metropolises). It was judged that these nodes would have the potential to act as counter-magnets to Paris in attracting investments, employment opportunities and population. An additional argument for a growth policy was that the development of these centers would in turn, by spread effects, influence the economic growth of their urban and rural hinterlands. The centers chosen were Lille-Roubaix-Tourcoing in the North, Nancy-Metz and Strasbourg in the Northeast, Lyons-St. Etienne and Marseilles in the South, Toulouse and Bordeaux in the Southwest and Nantes-St. Nazaire in the West. These centers would be eligible for physical, economic and social infrastructure support and would receive twice the level of incentives available in their regions. The result would be that government investments would be concentrated in these eight regional centers which would then aid the economic development of the problem regions of France and also stimulate the spatial integration of her economy. As noted by MacLennon,[7] the French policy at this stage represented a major victory for those who believed in a growth pole strategy for regional economic development.

It should be added that unlike the Italian or the Spanish patterns, French development policy called for negative controls or disincentives with respect to further development in the Parisian region. Over time, the fiscal penalties for locating in that region were broadened to include both industry and tertiary activities. It was argued that some tertiary activities designed to serve the French market could be readily diverted to the provinces, while others serving an international market could best operate in Paris. I should also note that a major element in the decentralization process from Paris to the metropoles d' équilibre appears to have been the role of persuasion rather than penalization. Its success has been attributed to the close ties that exist between the government and the leaders of the French industrial establishment. This is a type of development strategy that has never been successful in Italy.

By the 1970s the fiscal incentives had been strengthened to a maximum grant of 25% and a decentralization bonus was implemented to reimburse firms moving out of Paris for up to 60% of the costs of moving industrial equipment. Allowances were also available for the transfer expenses of those employees who agreed to move with their employers to one of the development regions.

Had the constraints on the growth of the Parisian agglomeration in the 1960s been successful? Here the evidence is certainly positive for the growth of industrial employment in Paris had clearly been checked. As for the industrial growth of the problem regions of France, here too the response is positive, for 80,000 new jobs were added in the depressed areas of the North and Northeast from 1968 to 1975. The growth of industrial employment in the western region was also positive but on a more modest scale. However, far too large a share of the nation's industrial growth has been concentrated in the ring of provinces bordering on the Paris region. This occurred despite the fact that these provinces received no investment subsidies. Obviously proximity to Paris was still a major importance to many French manufacturers. On a more positive note, the remarkably modest cost of the development program should be stressed. On a cost-benefit basis, the French effort has been far more successful than comparable programs in the EEC.

Finally, it is imperative to stress that there were two major shifts in development policy in the 1970s. One change, inaugurated with the adoption of the seventh five year plan in 1976, was a shift in the character of the investment program. Regional development grants are now directly related to jobs created; they are graded from 12 to 25% of the total investment according to zone; small and medium sized plants now get higher priority. Grants to the tertiary sector are also keyed to the creation of new jobs in the development regions.

The second change has been in the area of locational policy. Some have interpreted this shift as a retreat from the growth center concept. It has long been argued in France that there were dangers in its locational policy because that policy could conceivably result in a series of regional imbalances that might substitute for the Paris and its desert notion, sub-national overconcentrations. As noted by Sundquist,[8] once economic growth has shifted to the regional metropoles, they in turn may have to be restrained to prevent overconcentration at that intermediate scale. Perhaps as a result of economic rationality or as a reflection of political pressures and equity considerations, the French development program has begun to focus on the medium sized cities in the urban hierarchy. This shift has meant that the regional centers have lost their advantages in terms of the gradation of incentives and infrastructure investments. By 1976, the country that so affirmatively adopted a growth pole strategy, had shifted to a diffuse decentralization policy. One could argue that shifting expenditures down the urban hierarchy is a logical extension of the growth center concept. However, the current program resembles in many ways its Italian counterpart in that there no longer appears to be any meaningful locational design.

The United Kingdom

The estimated population of the United Kingdom[9] in 1976 was 56,000,000.[10] Like France, Britain experienced a post-war demographic surge which prompted planners to envisage a population of 65,000,000 by the turn of the century. However, in the late 1960s and 1970s her birth rate slowed dramatically. This drop coupled with high death rates associated with an aging population has resulted in a slight decline in the number of the inhabitants of the United Kingdom since 1973. Demographers now feel that this trend will continue with such consequences as increasing expenditures on social insurance and medical care and a decline in her labor force for the conceivable future. This industrialized society is, of course, highly urbanized. Over 77% of the British population is now classified as urban. With only 3% of her labor force engaged in agriculture, the non-farm population is presumably within the 90th percentile like the United States.

The average per capita gross national product of the United Kingdom is now the third lowest in the EEC only surpassing those of Italy and Ireland; its current level is roughly one-fifth of our own. Britian is a nation whose political and economic strength has deteriorated. Its decline has resulted from many factors. Among these are the loss of her empire coupled with a decline in overseas investments and markets, the depletion of her resources (particularly coal), and the declining competitiveness of her traditional industries: textiles, steel, machinery and shipbuilding. Although there has been a significant increase of employment in the "growth industries," the British economy is still weaker than those of many of her partners in the Common Market. That weakness not only reflects the factors noted above, but it is also a result of a decline in entrepreneurial initiative and labor productivity compared to her economic rivals. Then too, much of Britain's industrial plant is obsolescent, and this has exacerbated her problems. Finally, the economy of the United Kingdom must be classified as "mixed," for some sectors have been nationalized like coal, steel and the railroads, while others operate as they would within a true market system.

Throughout the long era of British political and economic supremacy, the distribution of her industries reflected her overseas ties. Her export industries were clustered along the coasts, especially in those areas where her coal deposits were close to the sea. This facilitated the export of coal and manufactured products to overseas markets and simultaneously permitted the import of raw materials and foodstuffs. Such concentrations included the Lancashire industrial area, centered on Manchester and the port of Liverpool (Merseyside) which specialized in cotton textiles; Southern Wales which emphasized steel and metal fabrication; the Northeast coast, centered on Newcastle and Middlesbrough (Tyneside and Teeside), a region concentrating on steel, heavy metal fabrication, machinery and shipbuilding; and finally, the Scottish Lowlands with its focus on steel and shipbuilding centered at Glasgow (Clydeside). All of these industries were export oriented, and their long-term success resulted from scale, agglomeration and urbanization economies. Early entrepreneurial initiative, labor skills and technological innovation also played significant roles. However, over time, these industries found themselves extremely vulnerable to cyclical and secular changes. By the late 1920s, Britain had lost a large share of her overseas markets for coal and heavy industrial products. When industrial technology had changed, her industries had not responded effectively to those shifts; her coal reserves were depleted; the remaining deposits were high cost, and the demand for coal suffered from competition with petroleum and natural gas. Thus from the 1920s onward, spurred on by the prolonged coal strike of 1926, the export sectors (coal, steel, ships and textiles) and the regions in which these activities were concentrated stagnated. The Depression of the 1930s brought severe unemployment to the traditional export-oriented mining and industrial areas. In contrast, those areas that specialized in the so-called growth industries, for which demand had not drastically declined, or in the production of consumer goods, all with different locational requirements than the traditional industries, fared far better. For example, unemployment levels in London in the depths of the Depression ranged from 10 to 12%. These values were less than one-third of the average for Clydeside and South Wales.

The two regions that were only marginally affected by the economic crisis and recovered most rapidly were Greater London in the Southeast and the Midlands. Unlike Paris, which is the dominant political and economic center in a highly centralized state, London, in a less centralized system, plays a more modest role. Nevertheless, it is the seat of governmental power and an international metropolis. It is also the nation's chief commercial and financial center and the focus of the British transport network. Its sheer population size makes it the country's leading market, so that the industries that developed there involved the production of consumer-oriented goods and high value-added products like precision engineering equipment.

The Midlands, situated north and northwest of London, was the only other region that profited from the decline of the older industrial areas and from the changing demands of national and international markets. Its location, in fact, was more central to domestic markets than London. The Midlands has had a long industrial tradition. Its industrial development was based upon local low-cost coal supplies, a skilled labor force and presumably entrepreneurial initiative and innovation. On the negative side, the region's inland position is less advantageous for export than that of the coastal industrial areas, but that disadvantage is more than offset by the nature of its products. The region has a highly diversified industrial base which makes it less sensitive to cyclical and secular fluctuations. Its products are in the high value-added sectors for which transport expenditures are a relatively small fraction of total assembly, production and distribution costs. These include precision engineering products such as motor vehicles and machine tools. In addition, the region manufactures a vast assortment of consumer durables for which there has been an increasing demand on the domestic and international scene. This short review has laid the foundation for my discussion of regional planning in the United Kingdom.

Britain was the first western nation to recognize the significance of regional inequalities in economic development as a major national problem. Even this brief summation of her experience reveals many of the difficulties that arise in advanced western societies as they attempt to redress spatial imbalances. However, it must be stressed at the outset that, aside from the highlands of Scotland and perhaps the Southwest, the problem regions of the United Kingdom fall basically into that category which I have termed "depressed." These are areas that formerly had a high socio-economic level of development but must now be classified as "unhealthy" as a result of changed economic circumstances. In my review, to this point, of development experiences only the northern and northeastern regions of France fitted this description. Such areas presumably already possess a social and economic infrastructure. Unlike the underdeveloped regions where infrastructure expenditures are necessarily high, these depressed areas require infrastructure funding mainly for modernization and expansion. They are also endowed with a literate labor force holding residual skills that could be recaptured by vocational retraining programs. Finally, and of considerable importance, these problem regions often possess entrepreneurial talent that could conceivably be utilized for the revitalization of their economies.

The late 1920s and 1930s comprised the first period of development experience in Britain. Most of the traditional mining and industrial areas noted above failed to recover from the acute postwar depression. With no previous experience to guide it, the Conservative government encouraged unemployed miners and industrial employees to move to somewhat more prosperous areas elsewhere after retraining. Limited relocation allowances were made available for such moves. Brown calculates that by 1938, 150,000 relocated workers remained in their new homes; this occurred despite the enormous impact of the Depression of the 1930s.[11] Nevertheless, most students of "the regional problem" were convinced that this initial program was misdirected. It drained the depressed areas of pools of young skilled workers and failed to contribute to a high and stable national employment level. With the shift to a coalition government under MacDonald in 1934, the development policy was shifted to one of "bringing the work to the workers." The government designated four "Special Areas": South Wales, Northeast England, West Cumberland and the Clydeside area of Scotland. All of these regions had unemployment levels averaging 40% or more of their labor force in the height of the Depression. However, perhaps because their unemployment levels were not as high, the major urban centers of these regions which could have acted as growth points were excluded from the incentive plan. That program called for the building of advance factories with state funds, loans at low interest rates and tax relief. In retrospect, the investments and their results were extremely modest considering the severity of the problem. Nevertheless, as noted by Sundquist,[12] "The techniques and methods of developing the areas had been tested; they had been proven effective, at least to some degree, and a body of experience had been built."

Probably the most significant event in regional development planning in

Britain during the prewar era was the publication in 1939 of the recommendations of the Royal Commission on the Distribution of the Industrial Population. The "Barlow" report was clearly the basis for postwar development programs, and its goals still continue to be the major objectives of regional policy in the United Kingdom. Using social, economic and strategic criteria the Commission advocated two courses of action. First, it recommended the decentralization of economic activity, especially industry, from congested urban agglomerations like London. Second, it argued that the chief regional goal should be a "balanced" distribution of industry which would require, over time, its dispersal into the problem regions. With regard to the latter recommendation, the report also argued for a policy of industrial diversification[13] in the depressed regions. It was argued by the Commission that the traditional industries in those areas were far too susceptible to cyclical and secular fluctuations of employment. The prescribed cure was concerted action to achieve a greater variety of industries which should result in a more consistent demand for their products. I should add that business, labor and government all apparently agreed with the goals recommended in the Barlow report.

As might be expected, however, there was no immediate consensus on the methods needed to achieve those objectives. The Distribution of Industry Act of 1945 and the Town and Country Planning Act of 1947 were the enabling legislative measures for postwar regional planning in Britain. True there have been shifts in policy during the era since 1947 depending upon the party then in power, but the basic design has been reasonably consistent.

There are two major complementary facets of British regional policies which are aptly termed "the carrot and the stick" or incentives and controls. The incentives are available to various categories of depressed or stagnant regions. Four such types of areas were delimited using a single criterion–the level of unemployment. The minimum base was an unemployment rate of at least 4.5%.

(1) Development areas–large depressed regions including their key urban centers or growth points. The intent, however, was not a growth pole strategy. Various types of incentives were available, with no locational selectivity, in these regions.

(2) Special development areas–truly depressed regions with the heaviest levels of unemployment. Typically they were former coal mining districts. Such areas are eligible for higher subsidy levels.

(3) Intemediate areas–sometimes termed the "grey" areas. These are restricted zones that are eligible for much lower subsidy levels.

(4) "New towns"–eligible for infrastructure subsidies.

There are currently five "Development Areas" in the United Kingdom (parts of which are designated as Special Development Areas); these include: Scotland, Wales, the North, and portions of the North West and the South West. The "Special Development Areas" comprise all of Northern Ireland (dominated by Belfast); much of the western Scottish lowlands (especially

Glasgow in Clydeside); West Cumberland (whose key urban center is Carlisle), and Tyneside and Teeside (particularly Middlesbrough and Newcastle) in the North; Merseyside (whose focus is Liverpool); and interior South Wales. As for the "Intermediate Areas," these include the North West (particularly Manchester); Yorkshire and Humberside; and a small portion of the South West. It should be noted that in most instances the "assisted" regions contain one or two major urban centers. However, these potential growth poles are not granted special subsidies over and above those availabe in their hinterlands. I should add that the various types of "assisted" areas were chosen mainly because of their high unemployment levels, but other factors also played a role, for this dynamic locational policy is sensitive to economic developments. Areas can be added or removed and their classification can change. Other factors taken into consideration in classifying areas include their general employment structures, male and female activity rates and potential population change (here anticipated net losses through out-migration are evaluated). The "assisted" areas now cover above two-thirds of the land area of the United Kingdom and contain roughly 40% of the nation's population.

There are now 28 new towns in Britain with a total estimated population in 1977 of 800,000 (21 in England, 2 in Wales and 5 in Scotland). These centers are located either in the "assisted" areas or in the overspill regions adjoining major conurbations. Other than the normal subsidies that may be available in their regions, the new towns receive priority in state subsidized infrastructure expenditures. Some consider the new towns to be de facto growth points.

As of 1977, the regional incentives available in the United Kingdom were among the most attractive in Western Europe. They now account for nearly 2% of the annual expenditures of the central government or roughly (at current exchange rates) $800,000,000. Given the state of economic health in Britain in recent years, this sum constitutes a major resource allocation for regional development. The major financial incentives taken the form of grants, loans and regional employment premiums.

There are two key type of grants: one supports building construction and the other subsidizes expenditures for plants and equipment. Currently, the building grants in the Development Areas and the Intermediate Areas are 20% of investments, while the comparable level is 22% in the Special Development Areas. Equipment grants (plant and machinery) are not available in the Intermediate Areas. These are fixed at a level of 20% in the Development Areas and double that value in the Special Development Areas. Smaller grants are also available to aid in the relocation of industrial plants. Similar sums are granted to service and research activities to encourage their movement out of such conurbations as London and Birmingham to the "assisted areas." The size of the grant depends on the number of workers moved and the number of additional employees created by the move. Here success has been minimal as far as the depressed regions are concerned, but the movement of offices to the overspill regions has relieved some of the congestion problems of the metropolises.

Loans at subsidized reduced interest rates are available for new industrial plants and for the expansion and modernization of older plants in the "assisted" areas. Here an element of selectivity exists, for the size of the loans is apparently negotiable.

The last of the major incentives is the Regional Employment Premium. This subsidy began in 1967 and was to be phased out by 1974. The Labor Government not only extended this program but doubled the amount paid per worker. These incentives which, of course, favor labor-intensive rather than capital-intensive industries are a continuing annual subsidy for plants in the "assisted" areas and their value for an employer is estimated to be roughly 8% of the wage bill.

Other less significant incentives include government expenditures for the preparation of industrial sites and for infrastructure improvements, advance factory construction for rental or purchase, occupational training grants and assorted tax relief measures.

British regional policy includes the strongest and most varied operational disincentives of any major development effort in the West. The key element of the control program is the Industrial Development Certificate which is now required for all industrial buildings of more than 5,000 square feet. These were originally designed to control industrial growth in London and in Birmingham (in the West Midlands). In fact, all new projects in the United Kingdom require such a certificate, but only in the megalopolis stretching between those urban centers is approval controlled. Thus it is estimated that 95% of *all* requests are approved. To secure a certificate for an industrial plant to be located within London or Birmingham, an applicant would have to demonstrate that the project was "in the public interest." "Mere convenience or extra cost cannot be accepted as grounds for granting permits."[14] However, a firm denied a certificate for construction of a plant in London, may be able to obtain one for a site in its broader region (the South East), if it can demonstrate that it could not operate at a profit in the "assisted" areas. I should add that the policy is reasonably flexible; thus individual cases are decided on their merits. The degree of rigor in enforcement appears to vary with national economic circumstances at different time periods. While persuasion is legitimate, political and economic pressures are apparently uncommon and rarely successful. Concessions have been made in the national interest, especially when an applicant agrees to put one facility in the depressed regions and another in the South East or the West Midlands.

In 1964 office building controls were adopted for the South East (including London), East Anglia and the West Midlands (Birmingham). However, this program must be classified as an intra-regional effort rather than an inter-regional shift. Finally, a decentralization of Government offices has been underway since the 1960s. From 1963 to 1974 some 26,000 positions were dispersed from London, with 65% going to be "assisted" areas. Plans now call for the dispersal of an additional 31,000 jobs between 1974 and 1984, again mainly to the Development Areas.[15]

Given the notably high levels of investment in the depressed regions by various government agencies, (Britain has no central regional planning authority comparable to the Cassa in Italy or DATAR in France), what have been the achievements of regional planning in Britain? The OECD has estimated that about 560,000 manufacturing posts were created in the "assisted" areas from 1945 to 1971. Of these, roughly half originated in the South East and East Anglia, 80,000 in the Midlands and 90,000 from abroad. In contrast, OECD also calculates that reductions in pre-existing jobs in this sector, within the 5 major depressed regions of Britain between 1951 and 1971, totaled 1,300,000 places. Thus the inward movements, cited above, replaced about half of the losses.[16] Moore and Rhodes, using a variant of the shift-share technique, have estimated that employment in the depressed regions was 220,000 more than those areas might have expected, if it were not for the use of regional policy measures between 1963 and 1970.[17]

It also appears that the drift of population to the South East, one of the major concerns of regional planning in Britain, has been sharply reduced. However, as Sundquist[18] notes, this diminution is, in a sense, a "statistical artifact," for there had simultaneously been considerable growth in the adjacent counties of East Anglia and the South West (all within commuting range of London). Brown, too, has argued that without the use of regional policy instruments the migratory flow to the South East might have been far greater.[19] Nevertheless, it must be emphasized that the growth industries and the major service sectors are still disproportionately located in the South East and the Midlands. Unemployment data and per capita income values for 1975-76 years are presented in Table 60. These materials clearly indicate that spatial imbalances persist after 40 years of experimentation with regional policy instruments. The performance of the British economy in 1976 was quite poor as is illustrated by the data in Table 60. Thus, the average unemployment rate for the nation was 5.8% or double the value for 1972. Note the striking regional disparities in unemployment levels. Clearly, the 4.5% criterion used earlier to delimit the depressed regions is no longer really appropriate, for only the value for the South East was below that level. If we take the national average unemployment percent as our base, only Yorkshire-Humberside, the East Midlands, East Anglia and the South East were below the 5.8% value. The per capita income data for 1975 are also revealing, for only the South East exceeded the national average. Scotland and the North had improved notably since 1973, presumably reflecting the impact of the North Sea oil boom. Despite this relative improvement, these regions still lagged behind the more prosperous areas of the South. In contrast, Northern Ireland has had the most dismal record related, of course, to the prolonged internal strife in that troubled region. Since 1976, there has been a modest economic upturn in the United Kingdom based mainly on greater wage and price stability and the easing of labor unrest.

In retrospect, despite the lengthy experience of Britain in regional planning and the major infusion of funds into the problem areas, regional socio-economic disparities still persist. Of course, those spatial imbalances are far less than was true for Italy. Students of British regional policy agree that

Table 60. Unemployment Ratios and Per Capita Income by Planning Region in the United Kingdom for 1975-76

Region	1976[a] Unemployment (% of labor force)	1975[b] Per capita income (UK = 100)
Northern Ireland	10.3	77.4
Scotland	7.0	96.0
Wales	7.4	88.1
England	5.4	101.9
North	7.0	94.2
Yorkshire and Humberside	5.6	96.3
North West	7.0	96.3
West Midlands	5.9	96.8
East Midlands	4.8	96.6
East Anglia	4.9	92.6
South West	6.4	94.8
South East	4.2	112.6
Greater London	–	123.8
Rest of South East	–	104.5
United Kingdom	5.8	100.0

[a] *Abstract of Regional Statistics*, 1977, (London: Central Statistical Office, 1977), p. 96.
[b] Ibid, p. 175.

her development program has not been sufficiently selective in terms of sectoral priorities, nor have there truly been selective locational measures. In the first instance, there has been a failure to foster the growth of the propulsive industries. By this I mean, insufficient priority has been given to those industries that typically have strong backward and forward linkages with greater potential multiplier effects. Similarly, there has never been a coordinated attempt to foster a selective growth center policy in the United Kingdom. Lastly, in my view, no regional policy can truly be successful if it is not an integral part of national spatial planning for economic development. Despite the panoply of regional policy measures in force in Britain, no real spatial strategy exists.

As for the depressed regions, they still operate under severe handicaps. In particular their location is still peripheral to the national market. With the entry of Britain into the EEC, it is obvious that the South East is far closer to the center of gravity of that huge potential market than the "assisted areas." Thus it appears inevitable that Greater London and its immediate hinterland will continue to prosper. For the development regions, the future is unfortunately far less certain.

The United States

The experience of this country[20] in regional development planning is drastically different from those of Spain, France, the United Kingdom and Italy. Ours is a highly urbanized and industrialized society whose real per capita gross national product is among the world's highest. Yet our "average" conceals disgraceful rural and urban poverty as well as striking regional income inequalities whose resolution would require a massive allocation of national resources and certainly would stretch our technical skills to the limit. We could, if we so desired, build on the regional planning experiences of Western Europe.[21] However, to do so would require that our society make a major national commitment to the solution of the inequities in our system or, to use Lyndon Johnson's terminology, declare a war on poverty. To date, I would contend that there is minimal evidence of such a commitment.

Regional planning in the United States began on a very modest basis with the creation of the TVA (Tennessee Valley Authority) in 1933. This project was the first significant direct intervention of the Federal government in regional economic development. I should note that a number of regional development plans were proposed by the National Resources Planning Board in the 1930s but never implemented. In the case of the TVA, this planning region, at the outset, was truly backward with per capita incomes that were less than half of the national average. While most of its population was engaged in subsistence agriculture, the region did possess a small aluminum plant, iron foundries as well as several low-wage lumber and textile works. The development of this river basin authority was sparked by the creation of several multi-purpose high-water dams. Their purpose was the improvement of navigation along the river, flood control and the provision of cheap power to the Valley for industrial purposes and for rural electrification. With time, power requirements escalated mainly as a result of the development of manufacturing. As a result, a number of thermo-electric plants were built fueled by low quality-low cost strip-mined coal. In the postwar era, energy has also been derived from nuclear facilities. The Authority has always taken an active role in conservation. While reforestation and the prevention of soil erosion have traditionally been major concerns of the Agency, now thermal pollution and general environmental protection come under its umbrella. Insofar as I am aware, no meaningful cost-benefit analysis has been attempted to measure the economic impact of this river-basin authority. However, employment opportunities have clearly expanded and unemployment levels are now well below those of surrounding regions.

Before I discuss postwar regional development programs, it is necessary to treat the nature of the various types of "distressed areas" in the United States. The Public Works and Economic Development Act of 1965, unquestionably the most important planning legislation enacted in the postwar era, recognized five major criteria that would determine the eligibility of regions for assistance by the Federal Government. The grounds for eligibility are as follows: (1) substantial unemployment (6% or more was the chosen critical level), (2) persistent unemployment, (3) the prospect of a sudden rise in unemployment, (4) low median family income (40% or less of the national average) and (5) high net out-migration. High unemployment and low income were clearly the common denominators for the delimitation of these distressed areas. Economists would argue that low income is the key problem of underdeveloped areas, while high unemployment is typically associated with technological change and structural and locational shifts in demand. I should also note that two supplementary criteria were ultimately added by EDA to this list. The first is logical in that all Indian Reservations were included, while the requirement that each state have at least one designated distressed area was obviously politically motivated.

Chinitz, in a now famous paper written in 1967,[22] developed a series of model regions that are an appropriate scheme for discussing the problem areas of the United States. My analysis here builds on his treatment supplemented by Cameron's[23] modification of the Chinitz terminology. Both authors argued that the following 6 major types of distressed areas could be identified in the United States (as of the late 1960s):

(1) High Income–Fast Growth Regions,
(2) Mature Industrial Regions,
(3) Moderately Depressed–Rural Regions,
(4) Poor Rural Regions,
(5) Poor Mountain Industrial–Mining Regions,
(6) The Inner City Ghetto.

While some of the socio-economic criteria used to delimit distressed areas might suggest that these Type 1 urbanized-industrial areas are thriving and healthy, in fact, they must still be classified as problem regions because of their pockets of persistent unemployment. In-migration, Chinitz argues, had been far more rapid in these areas than the growth of employment opportunities. In other cases, over-reliance on one key industry has led to economic difficulties, when demands for its products have sharply declined. Parts of southern California are commonly cited as examples of such areas, and the aerospace industry is used as an example of a sector with declining demand and reduced job opportunities. I suspect, given the hindsight of the 10 years since Chinitz wrote this definitive article, that the inclusion of this category within the group of distressed areas is clearly questionable in 1978.

The mature-industrial group is a replication of the history of the depressed regions of the United Kingdom and northeastern France. These are areas where the industrial structure is basically the same as that of the turn of the century. Here secular decline has hit traditional industries such as steel in Pittsburgh and textiles in New England. The social and economic infrastructure of these areas is technologically outdated and labor skills and organization are narrow and commonly inflexible.

As for Type 3, the moderately depressed rural areas, Chinitz would include in it the Upper Great Lakes and Northern New England. In these regions, unemployment is relatively high, and income is moderately below national averages. The difficulties of these areas have been caused by the decline of employment opportunities in extractive industries (agriculture, mining, lumbering and fishing). Their population is declining because of heavy rural to urban migration. Yet the urban centers in these regions cannot provide sufficient job opportunities for those who are no longer needed in the extractive sector.

The poor-rural regions category defined by Chinitz is another anomaly. He describes this type of distressed region as having high rates of unemployment and low incomes. As of 1967, these areas were confined essentially to the southern segments of the nation, where in his terms "the classic characteristics of underdevelopment prevail." The areas included were characterized by declining agricultural employment and minimal incomes for those still engaged in this sector. Certainly, these conditions still apply to many segments of the South, but industrialization in the past decade has made major inroads in this region. The new industries of the South have been located both in the non-metropolitan and the metropolitan regions. Thus, new employment opportunities now exist in areas where agriculture was formerly dominant. While, in general, low wages prevail, incomes are far higher than in the pre-industrial era. Unfortunately, by design or by chance, the regions of the South which are predominantly black have attracted far less industrial activity than those populated by whites. Over the decades net out-migration has been relatively heavy. The directional bias was clearly to the urban centers of the Northeast. That migration has now slowed considerably, checked by the intervening opportunities of growing southern urban centers like Atlanta. In the sizeable poor-rural areas that remain, the few urban centers that exist tend to be small and scattered; urban services, particularly for the non-white population, are still totally inadequate.

The fifth category comprises the core region of Appalachia stretching from eastern Kentucky through portions of West Virginia, Virginia, the Carolinas and Tennessee. It will be discussed in connection with my treatment of the Appalachian Regional Commission. The reasons for the inclusion of the Indian Reservations as problem areas is self-evident, for their poverty is a national disgrace.

Finally, the last of the distressed areas and unquestionably our most urgent and pressing societal problem is, of course, the inner city ghetto. It too will be treated separately in my discussion of ongoing regional problems.

Aside from the TVA and several other river basin projects, there was no additional organized program for assisting problem regions until 1961 when the ARA (the Area Redevelopment Administration) was established to provide loans and grants chiefly for industrial development in distressed areas. Commercial establishments could also obtain aid but normally at a reduced scale. However, that program was insufficiently funded and staffed. Thus the passage of the Appalachian Regional Development Act in 1965, coupled with the Public Works and Development Act, marked the take-off point for regional development planning in the United States.

The mission of the EDA (Economic Development Administration) in cooperation with the states was to help them to revitalize their distressed areas. It did so by planning and subsidizing regional economic development. The Public Works and Economic Development Act which created EDA recognized under Title IV the following major types of geographical regions that could be eligible for financial assistance: (1) Redevelopment Areas which comprise single counties, labor market areas, Indian Reservations or municipalities below a designated size, and (2) Economic Development Districts which are composed of multi-county areas; such units are needed because individual redevelopment areas may not be viable units for regional planning. These districts comprise groups of adjacent counties with three basic characteristics: adequate size for effective planning, at least one redevelopment area, and normally one development center whose size should be sufficient to stimulate the economic growth necessary for the alleviation of "distress" in their economic development districts. The criteria for eligibility for aid were noted above. I should stress that no redevelopment area or economic development district need meet *all* of those criteria.

Over time, the EDA issued operational guidelines for the selection of development centers (both in redevelopment areas and economic development districts). These guidelines sound as if they were directly derived from the criteria recommended in the growth pole literature. Milkman and his coauthors covered the EDA standards and its early experience in great detail.[24] In summary form these are as follows: (1) the population of a development center may range from 15,000 to 250,000; (2) strong econmic linkages should exist between the development centers and their hinterlands; (3) their labor force should be well trained; and (4) unemployment should be well below the level in their district or development area. Perhaps the most important criterion is that the economy of these centers should be diversified and healthier than that of their surrounding areas with a sound potential for rapid growth. The role of these growth centers is to provide public and commercial services to their hinterlands, stem the migration to larger metropolitan centers and provide jobs for the unemployed and low income residents of their tributary areas. Once a development center is identified, it must develop a comprehensive plan (with Federal funding and technical assistance). This plan must include a development strategy not only for the centers, themselves, but also for their tributary districts or areas. Agency-funded studies have generally been inconclusive as to the success of this growth center strategy. However, academic critics of their findings have argued that the apparent lack of success of this strategy may, in part, be attributable to the small size of the majority of the designated growth centers. Despite this counter-argument, there appears to be minimal agreement in the literature as to the optimal size for such growth poles. Nevertheless, this focus still appears to be central to EDA strategy. By 1977, 333 growth centers had been identified (or selected); presumably they are all eligible for financial aid in 1978.

Lest the reader be led astray by my extended discussion of growth center policy under the EDA, I should note that by far the great bulk of that agency's funds has been allocated for "local public works" or, in other terms, economic and social infrastructure. Clearly, growth centers were aided by these expenditures, but far too small a fraction of Federal funds has been spent for direct and indirect subsidies for industrial and commercial development.

Title V of the Public Works and Economic Development Act of 1965 provides for Multi-State Economic Development Regions and Commissions. These are the problem areas which extend across state boundaries and require macro-scale solutions. Currently, aside from Appalachia, which is covered by separate legislation, there are 8 such economic development regions: (1) New England, (2) the Coastal Plains of the Southeast, (3) the Ozarks, (4) the Upper Great Lakes, (5) the so-called Four Corners which contain portions of Colorado, Utah, New Mexico and Arizona, (6) the Old West, (7) the Southwest Border and (8) the Pacific Northwest.

The redevelopment areas and the economic development districts are eligible for grants and loans for infrastructure development, industrial and commercial loans, loan guarantees for working capital and technical assistance. Loans for industrial development can cover up to 65% of the total cost of land, buildings, machinery and equipment. As for the economic development regions, they have the responsibility to develop long-range economic plans, and they are eligible for supplementary grants and technical assistance.

The EDA was apparently destined for elimination during the Nixon years, but it was given renewed life during successive administrations. Most recently, the Public Works and Development Act of 1977 has provided long-term funding. EDA would now appear to be the major regional planning agency in the nation for the immediate future. However, less than 10% of its funds have been allocated to urban areas, so its role in resolving the nation's urban crisis has been minimal.

The Appalachian Regional Development Act of 1965 was a response to the widely publicized poverty of that region. Chinitz considered Appalachia to be a special variant of the poor-rural region. The inhabitants of this area are white, poor, and unemployed. The long history of coal mining in Appalachia has left a heritage of ravaged land and of streams and underground water supplies polluted by acid mine drainage. Until the energy crisis of the mid-1970s, many of the miners were unemployed or underemployed and generally unsuited for alternative employment opportunities. The Appalachian Regional Commission was created to assist the region in meeting its special problems and in promoting its economic development. The Act explicitly calls for the pursuit of a growth point strategy. Its initial focus was on improving the region's infrastructure, particularly its highways. More recently, however, there has been a major reorientation toward the projects which emphasize human resources. Hansen states that "this approach would seem to be a milestone on the road from place-orientation towards approaches recognizing that the welfare of people is, or should be, the principal objective of regional policy."[25]

No treatment of regional policy for the distressed areas of the United

States can be complete without a discussion of our most urgent regional problem: the inner city ghetto. The incidence of poverty in central cities in 1970 was well over twice that in what the U.S. Census terms the "Metropolitan Ring." According to Hansen, about one-quarter of the metropolitan black population in the U.S. was below the poverty-line in 1969, the comparable value for the white inhabitants of the same areas was only 7%. If one were to contrast the current (1978) unemployment rates for black youth and for whites in the same age bracket, the results would be truly appalling. Metropolitan governments have clearly failed to respond to the problems of the ghetto, nor have they been able to treat the other problems of central cities such as economic obsolescence and decay resulting, in large measure, from the rapid suburbanization of our metropolitan population and economic activity. More recently, we have witnessed a significant movement of people and jobs beyond the limits of our SMSA's (Standard Metropolitan Statistical Areas). Our urban problems are clearly complicated by the fact that cities do not have the financial resources to respond effectively to the demands of their poverty-striken residents. It is here that the Federal government must play the key role. It is the only administrative entity that has the monetary base to cope with the problems of our cities. However, the experience to date of the willingness of Washington to assume that responsibility is discouraging.

The United States, like many other developed nations in the West, has attempted to develop a system of measures to subsidize the growth of its depressed regions. Yet our commitment to the elimination of regional socio-economic disparities has been far weaker than that of many western European countries that are far less wealthy than we are. The regional development legislation of the mid-1960s and even the new Public Works and Economic Development Act of 1977 were strongly influenced by our free enterprise system. Like Italy, a decade earlier, we believed that public works projects would in turn stimulate private investment in our distressed areas. True, a share of the development funds has been channeled into loans and loan guarantees for new industrial and commercial initiatives, but the types of incentives and disincentives that are now commonplace in Britain, France, Sweden and the Netherlands are yet to be adopted in this country. We certainly have much to learn from their experience, but even more important there is the urgent need to make a major national commitment to regional development and particularly to the solution of the pressing problems of our cities. Private initiative and local financial efforts can never resolve these national problems. The necessary resources are clearly available, but the willingness to encumber those funds is far from realization. It was argued earlier that in the cases of Italy and the United Kingdom there was a need for a national spatial strategy, yet it must be conceded that elements of such policy already exist in those countries. In contrast, in the United States we are far from a consensus, even in principle, on the nature of a meaningful development program that might solve the problems of our distressed areas.

THE SIGNIFICANCE OF THE ITALIAN EXPERIENCE

Having reviewed the development of regional planning in Spain, France, Britain and the United States, I can return to the question first posed in the introduction to this chapter. What can the United States and other western market-oriented societies learn from the Italian experience? I should preface my answer by stressing that in the Italian case that assessment must be both positive and negative.

Ideally too, I should like to present the reader with a neat package conveniently subdivided into two parts: the first labeled, "successes," and the second termed, "failures." Of course, such an arbitrary classification is not possible, for there are no such sharp distinctions. Instead, I plan to couch my analysis in terms of the viable options for regional development planning and assess the Italian experience.

Incentives

I have already discussed in previous chapters the evolution of the industrial incentive program in Italy. It is fair to argue that the incentives available for investors in the Mezzogiorno are as good or in most instances more attractive than those offered by regional development authorities elsewhere. As in Britain, France, the Netherlands and West Germany, there has been a general shift from loans to direct grants in Italy. Tax breaks too vary enormously from country to country, but here again the Italian inducements are the most attractive.

The only instance where Italy's subsidies have failed to match those of her competitors has been in the case of employment premiums. For example, in Britain and France there are now special subsidies linked to every new job created in their development zones. The regional employment premiums in the United Kingdom, at least, do not appear to have a time limit (they are paid annually). Their size is now roughly equivalent to the sums available for direct grants. In the debate on the 1976 development legislation in Italy, it was suggested that financial incentives be tied directly to the number of jobs created, but that proposal was ultimately defeated. However, the provision providing for the absorption of social security, medical assistance and accident insurance by the state for all firms in the South is a form of employment subsidy. Those social insurance rebates, it should be recalled, were designed to redress the reductions in regional wage differentials that had resulted from nation-wide labor contracts in 1968.

The failure in most regional development programs to provide subsidies for working capital is a clear deficiency. It is most discouraging for small and medium-sized firms who need funds for day to day operations which are often unobtainable on the open credit market. Here Italian regional development policy could be innovative and take this initiative which would encourage local entrepreneurship.

Controls and Disincentives

One of the major tools of regional policy that has proven to be effective in Britain, France and the Netherlands is the use of controls and disincentives. These have been particularly effective where there are major concentrations of population, industry, commerce and government offices in one metropolitan area, normally the capital.

Such measures can take several forms: (1) licensing policies like the Industrial Development Certificate used in Britain, particularly in Greater London; (2) punitive or penalty taxes such as the French levy on new industrial plants, on the expansion of existing manufacturing facilities, and on new tertiary activities in Paris. There can also be combinations of both measures as has been the case in the Randstad area (Rotterdam, the Hague, Utrecht and Amsterdam) of the Netherlands. The advantage of these controls and disincentives is that the rigor of their enforcement can be responsive to changing economic circumstances (both by sector and by location). The West Midlands which now has a higher unemployment rate than the average for the United Kingdom is a case in point. There is, of course, the ever-present danger that entrepreneurs faced with such controls or disincentives may decide to shelve their expansion plans. The impact of such negative decisions could, in the case of a large facility such as an integrated automobile plant, result in slower national economic growth. Thus, there is evidently the need for flexibility in the application of any regional policy measure. Persuasion can and should be part of the planning process.

The controls and disincentives used in Britain and France have not been completely successful, for there has been a tendency for new establishments to be sited in the peripheral regions circling their capital. Thus, although such policies may have helped to reduce congestion, in numerous cases, they have not brought new employment opportunities to the problem regions. The move to the periphery has been less true for branch plants of major corporations typically headquartered in the Paris, London or Randstad areas than for new initiatives. Here market proximity may dictate the locational decision, and a move to "the provinces" may well be economically rational.

As noted in Chapter 3 and 5, disincentives or controls have not been features of Italian regional policy. The only measure which has really had regional clout is the requirement that state-controlled firms locate the major share of their investments in the South. This is, of course, of major significance, for, as indicated earlier, these state concerns have been the chief force in the industrial development of the South. Until recently there was no such requirement for private firms. Persuasion was used, but the response to such efforts was minimal. With the passage of the development legislation of 1971 and 1976, disincentives and controls were finally authorized. Originally the government had asked for a penalty tax on employers who expanded their employment in the congested areas. In fact, one plan called for a levy of 1,000,000 lire (c. $1,600) per employee. That proposal was ultimately shelved, and the current program requires that all state-controlled firms must submit their five-year investment plans to CIPE. In the case of private firms, whose assets exceed $8,000,000, they too must submit their plans to CIPE where the investment in a project exceeds $11,000,000. Negative decisions are to be rendered if the project does not fit the overall national plan, or if the chosen areas are congested. State firms must adhere to these decisions, while private firms which ignore the recommendations or, for that matter, fail to report their new investment plans could theoretically be subject to fines of 25% of their investment costs. Thus the program, on paper at least, includes both controls and disincentives, but insofar as I am aware the control policy with respect to private investments has *not* been enforced. The only negative decision that has come to light was the refusal of CIPE to authorize Alfa Romeo (a state establishment) to build a second major automobile plant near its headquarters in Milan.

In sum, the Italian regional development authorities now have the power to use controls and disincentives to discourage new or expanded industrial facilities. Those measures are evidently applied in the state sector, but for private enterprise the controls have not been enforced. I should also note that western European experience indicates that where these regional policy instruments have been rigorously enforced; the authorities, in the areas where growth has slowed or even stabilized, have argued that controls had throttled their economies. Whether these complaints are justified or not, some fine tuning of the control and disincentives measures is certainly needed to make these regional policy tools more effective. In Italy, the business community has been one of the bulwarks of the Christian Democrat Party which has held power since the last war. Industrial interests appear to have blocked any rigorous enforcement of the disincentive and control legislation. I should also add that perhaps because of the pressure of the business community of northern Italy, the development laws have always included provisions for helping "the depressed areas of the Centro-Nord." It is perfectly valid to argue that there are several truly impoverished areas in northern Italy. However, the territorial delimitation of such regions is so broad that literally only the largest cities like Turin, Milan, Genoa, Bologna and Florence are specifically excluded from assistance. Thus, theoretically a plant located on the periphery of those metropolises might be eligible for support. Of course, the sums available for aid to the depressed areas of the Center and the North are quite limited and direct grants are specifically excluded, so that expenditures there have been minimal. However, southern advocates would argue that the mere existence of such provisions is potentially harmful to the economic development of the Mezzogiorno.

Returning to the question of disincentives and controls, I see no evidence in Italy of any direct attempts to control the location of tertiary activity. However, with the shift of authority to the "regioni," I am sure that there will be a growth of the bureaucracy in the provinces. Whether there will be a parallel decline in government personnel in Rome is doubtful.

Location Policy

My review of the evolution of the industrial location program for the Mezzogiorno has indicated that a growth pole policy was adopted at a very early stage; yet, over time, that policy of concentrated development in a limited number of centers has been diluted as a result of political pressure. The proliferation of areas and nuclei of industrial development has clearly impaired the effectiveness of what was an avowed location program. I noted earlier that Italian regional planners have argued that the approval of such a vast number of areas and nuclei was simply a result of political bargaining for votes in the South. They also contended that there had never been any real intention of providing meaningful support for all of these zones. I can accept their arguments, but there is no question that expenditures for infrastructure improvement in these areas and nuclei have squandered scarce resources and vitiated meaningful support for a limited number of growth centers. Allen's argument that the current location policy is one of "diffuse concentration" is not supported by the ILSES findings that "the gradation of support on the basis of location is not a meaningful incentive for spatially differentiated investment within the Mezzogiorno."

Thus I would argue that the Italian model with respect to location policy should *not* be emulated by regional planners elsewhere. However, I should reiterate that despite the absence of meaningful locational incentives, market forces have produced a concentration of investment in a remarkably limited number of industrial zones. Hence, as noted earlier, by a self-selection process, both state and private firms have chosen to concentrate their resources in a few favorable locations rather than disperse their investments in response to the spatial distribution of areas and nuclei. There is one option that I believe should be adopted from the British experience. Given the new North-South dichotomy discussed in Chapter 11, it might be fruitful for CIPE to adopt a policy of graded locational incentives for different regions *within* the Mezzogiorno; one that would be comparable to the graduated scale of support given the Special and the Intermediate Areas of the United Kingdom. Above all, there is the need for the adoption of a national spatial strategy. For the moment, no such plan is in prospect.

Sectoral Policy

My review of the literature appears to indicate that other than the Italian case no regional planning program in market-oriented economies has a specific policy providing for graduated incentives based on industrial sectors. A number of programs do pay obeisance to "growth" sectors, and, in fact, the negotiation process may provide more favorable incentives for such industries, but their sectoral policy is not specific. Here again, the Italians have taken the initiative, for they have had a system of graduated incentives since the late 1950s.

However, there has been an ongoing debate within the Italian planning community with regard to the types of industries that should be favored. It was agreed, from the start, that the further growth of the "traditional" southern industries should be discouraged. It was from that point that a cleavage occurred. Some argued that, given the high level of unemployment in the South, labor-intensive industries should be favored, while others contended that the "growth sectors" should receive priority. These, almost invariably, were capital-intensive. The experience of the first two decades of the development effort, that has been reviewed at length in this book, indicated that, in fact, the capital-intensive industries received by far the largest share of the loans and grants. The result, of course, was the limited growth of manufacturing employment between 1951 and 1971.

I should note at this point that regional policy measures in western Europe have typically attracted capital-intensive industries, so that the Italian case is no exception to the rule. These new and modern facilities were typically branches of northern or foreign firms. Local capital, local entrepreneurial skills and local markets played no role in their operations. Their raw materials were imported, while their products were destined primarily for northern or foreign markets. As noted earlier, these plants had minimal multiplier effects because there were no significant backward or forward linkages with other plants in the Mezzogiorno. They clearly did not contribute to any appreciable degree to the solution of unemployment and underemployment problems in the region.

In contrast, there is no question that supporting labor-intensive industries tends to relieve unemployment. Andersson reports that a Norwegian econometric study has demonstrated that labor premiums are 3 to 15 times more effective than capital subsidies as regional policy tools.[26] Increasingly development programs in western Europe have adopted this instrument. Witness the British and the French examples. I would voice one reservation. Is such an investment program an appropriate regional policy measure when labor costs are rising so rapidly in Italy, given the fact that she must compete on the open market with textile producers in the Far East (where production costs are so much lower) or with petrochemical producers in the OPEC nations where costs are subsidized? Essentially we are talking about a region unprotected by tariff barriers which must ultimately become not only an integral part of the Italian economy but must eventually be integrated with the economy of the European Community. This being the case, the adoption of a labor intensity criterion makes little sense. It is far more important, in the long term perspective, for the Mezzogiorno to attract modern-innovative-high technology industries that might prosper under free market conditions, particularly those industries that have strong backward and forward linkages. Probably the key missing element in the industrial evolution of southern Italy, during the development era, has been its failure to attract small and medium sized satellite plants as contrasted with the case of the Milan and Turin areas. Industrial hierarchies like those are still missing in most of the South and their absence reflects the immaturity of southern industrial development.

Conclusions

As I reflect on the Italian development experience and its significance for regional planning in western societies, my judgments are necessarily mixed. However, I suspect that the same type of assessment would result from an evaluation of the regional policy programs of each of the countries that I have examined as well as others that were excluded from my analysis.

It is abundantly clear that in many ways Italy has been far more adventurous in her regional policy measures than most of her collaborators in the OECD. Admittedly, the frequent shifts in the form and value of her incentive program must be a constant frustration for potential investors who are attempting to assess the costs and benefits of her incentives. However, the willingness to experiment with regional policy tools, abandoning some, while retaining and reinforcing others, has given the Italian program remarkable flexibility. One example was the approval of the "contrazzione programmata" (program contracts) in 1971. These were designed as instruments whereby entrepreneurs could be consulted so as to tie investments to national planning objectives. These program contracts are also used as a method for bargaining as to the level of incentives that are made available to large investors. The *flexibility* of Italian regional policy measures *should probably be emulated in other planning programs.*

It should be remembered too that Italian planners have been treating a truly underdeveloped region which comprises 43% of the nation's territory and 37% of her population. No other advanced western society faces a comparable problem. This contrasts markedly with the French and British examples where both depressed and underdeveloped regions must be dealt with under the same umbrella. The case could be made that it may be relatively easier to design regional policy measures for depressed areas than it is for populous underdeveloped territories.

The instruments used by the Italian government to improve socio-economic conditions in the Mezzogiorno have, in general, been commendable, but as we have seen, there are obvious shortcomings. All too often, the rules are bent or even broken, so that in fact that the great bulk of all requests are approved regardless of location, sector or scale. Then too, investments, particularly since the early 1960s, have been largely devoted to manufacturing and infrastructure, while the service sector has been essentially ignored. Development experience elsewhere indicates that services and the "quaternary" sectors offer significant potential for regional development.

Most regional policy measures, and the Italian case is no departure from the norm, rely on financial incentives, yet as Allen and Stevenson have noted "incentives per se are passive."[27] Greater effort, they argue, needs to be devoted to vocational training programs, for the experience of northern and foreign firms has indicated that there is a clear deficiency of worker skills in the South.[28] I noted earlier that one of the problems of industrial development in southern Italy was the shortage of capable entrepreneurs. Most of the new modern firms in the Mezzogiorno are subsidiaries of state enterprises or branches of private firms headquartered in the North or abroad. Managerial personnel are typically drawn from northern Italy. Thus there are minimal (in house) opportunities for the training of local entrepreneurial talent. Management programs do exist, but they need to be strengthened and diffused throughout the region.

I have stressed that it is the state sector that must ultimately bear the brunt of the development effort in the South, but typically these state-controlled plants are large and capital-intensive. To date such enterprises have had minimal repercussive or multiplier effects. Much more attention must be devoted to the small and medium sized firms, and this was the obvious intent of the 1971 and 1976 legislation.

As noted previously, southern Italy has witnessed a remarkable rural to urban movement during the past two decades. That process has carried with it the typical economic and social problems that we associate with urbanization. Thus like the North, southern cities are plagued with poor housing, slums, pollution, congestion and a general deficiency of urban services. Urban planning is still in the formative stage in Italy and even more so in the Mezzogiorno. So far minimal efforts have been made to cope with these problems.

Given rural depopulation in so many of the villages of the interior highlands, those who remain (perhaps 10% of the southern population) cannot be ignored. They too require financial assistance to sustain themselves in their villages, or if necessary, relocation allowances to permit them to move to nearby urban centers. The intense out-migration from the rural areas of the Mezzogiorno has clearly been destructive to the social and family fabric of many villages. Nevertheless, I strongly disapprove of that aspect of the new legislation which provides subsidies for the industrial development of the areas that have suffered severe depopulation. However, I suspect that market forces will ultimately deter attempts to industrialize the "osso" (the mountain spine).

Finally, I must reemphasize the extraordinary role of political forces in regional policy decisions in the South. Equity versus efficiency are recurring issues in all regional development programs, but in no developed nation that I have studied has the political element played such a decisive role as it has in Italy.

NOTES

[1] *The World Bank Report on the Spanish Economy: The Economic Development of Spain* (Madrid: 1962), J. R. Lasuen, "Regional Income Inequalities and the Problems of Growth in Spain," *Papers and Proceedings of the Regional Science Association*, Vol. 8, 1962, pp. 169–188; H. W. Richardson, "Regional Problems in Spain," *Urban Studies*, Vol. 8, No. 1, February 1971, pp. 39–54; J. R. Lasuen, "Spain's Regional Growth," *Public Policy and Regional Economic Development: The Experience of Nine Western Countries*, N. Hansen, ed. (Cambridge: Ballinger, 1974). pp. 235–269; H. W. Richardson, *Regional Development Policy and Planning in Spain*, (Lexington: Westmead, 1975);

OECD *Regional Problems and Policies in OECD Countries* (Paris: OECD, 1976), Vols. 1 and 2; ———. *Regional Policies: The Current Outlook*, (Paris: OECD, 1977).

[2]Lasuen, 1974, op. cit. and Richardson, 1975, op. cit.

[3]Lasuen, 1974, ibid, p. 257.

[4]N. Hansen, *French Regional Planning* (Bloomington: Indiana University Press, 1968); K. Allen and M. C. MacLennon, *Regional Problems and Policies in Italy and France*, University of Glasgow Social and Economic Studies (London: George Allen & Unwin, 1969); DATAR (La Délegation a' l Aménagement du Territoire et et 'l Action Régionale), *Regional Investment Incentives* (Paris: 1972); R. Prud 'homme, "Regional Economic Policy in France, 1962-1972" in Hansen, 1974, op. cit., pp. 33-63; J. Sundquist, *Dispersing Population: What America Can Learn from Europe*, (Washington: Brookings, 1975), pp. 91-145; OECD, 1976, op. cit.; ———, 1977, op. cit.

[5]G. Tagliacarne, "Le Regioni Forti e le Regioni Deboli della Communità Allargata," *Informazioni SVIMEZ*, Vol. 27, No. 2, January 31, 1974, pp. 108-120.

[6]J. F. Gravier, *Paris et le Désert Francais* (Paris: Portulan, 1947).

[7]MacLennon, op. cit., p. 198.

[8]Sundquist, op. cit., pp. 124-126.

[9]G. McCrone, *Regional Policy in Britain*, University of Glasgow Social and Economic Studies (London: George Allen & Unwin, 1969); M. Chisholm and G. Manners, eds., *Spatial Problems in the British Economy* (Cambridge: Cambridge University Press, 1971); A. G. Brown, *The Framework of Regional Economies in the United Kingdom*, National Institute of Economic and Social Research, Economic and Social Studies, No. 27 (Cambridge: Cambridge University Press, 1972); G. McCrone, "The Location of Economic Activity in the United Kingdom," *Cities, Regions and Public Policy*, G. Cameron and L. Wingo, ed. (Edinburgh: Oliver & Boyd, 1973); G. Cameron, *Regional Economic Policy in the United Kingdom*, in N. Hansen, 1974, op. cit., pp. 65-102; M. Sant, *Industrial Movement and Regional Development: The British Case* (Oxford: Pergamon, 1974); Sundquist, 1975, op. cit., pp. 37-90; OECD, 1976, op. cit.; OECD, 1977, op. cit.

[10]Most of the statistics cited in this review of British experience were drawn from the *Abstract of Regional Statistics for the United Kingdom* for 1975 to 1977 and from OECD, 1976 and 1977, op. cit.

[11]Brown, 1972, op. cit., p. 281.

[12]Sundquist, 1975, op. cit., p. 40.

[13]A. Rodgers, "Some Aspects of Industrial Diversification in the United States," *Economic Geography*, Vol. 33, No. 1, January 1958, pp. 16-30.

[14]Sundquist, 1975, op. cit., p. 55.

[15]Ibid., p. 79.

[16]OECD, 1976, op. cit., p. 36.

[17]B. Moore and J. Rhodes, "Evaluating the Effects of British Regional Policy," *Economic Journal*, March 1973, p. 99.

[18]Sundquist, op. cit., p. 67.

[19]Brown, op. cit., pp. 336-37.

[20]B. Chinitz, "The Regional Problem in the U.S." in E. A. G. Robinson, ed., *Backward Areas in Advanced Countries* (New York: St. Martins, 1969); G. Cameron, *Regional Economic Development: The Federal Role* (Baltimore: Johns Hopkins, 1970); J. Cumberland, *Regional Development Experience and Prospects in the United States of America* (Paris, The Hague: Mouton, 1971); M. Newman, *The Political Economy of Appalachia: A Case Study in Regional Integration* (Lexington: D. C. Heath, 1972); J. Pickard, "U.S. Metropolitan Growth and Expansion: 1970-2000 with Population Projections," in S. M. Mazie, *Population Distribution and Policy*, Vol. 5, U.S. Commission

on Population Growth and the American Future (Washington: 1972), pp. 135–145; L. Wingo, "Issues in a National Urban Strategy for the U.S.," *Urban Studies*, Vol. 9, No. 1, 1972, pp. 3–27; R. Milkman et al., *Alleviating Economic Distress*, (Lexington: D. C. Heath, 1972); N. Hansen, "Regional Policy in the U.S.," in Hansen, ed., 1974, op. cit., pp. 271–304; Sundquist, 1975, op. cit., pp. 1–36, 239–281.

[21]This is the major thrust of the Brooking's study on *Dispersing Population* by Sundquist.

[22]B. Chinitz, "The Regional Policy in the USA," in E. A. G. Robinson, ed., *Backward Areas in Advanced Countries* (London: Macmillan, 1969), pp. 52–61.

[23]G. C. Cameron, "The Regional Problem in the United States: Some Reflections on a Viable Federal Strategy," *Regional Studies*, Vol. 2, pp. 207–220.

[24]Milkman, 1972, op. cit., pp. 129–131.

[25]Hansen, 1974, op. cit., p. 293.

[26]A. Andersson, "Regional Economic Policy in Sweden," Hansen, 1974, op. cit., p. 211.

[27]Allen and Stevenson, 1974, op. cit., p. 211.

[28]"Foreign Investments in the Mezzogiorno," *IASM NOTIZIE*, Vol. VI, 1973 and *Per un Rilancio della Politica di Industrializzazione del Mezzogiorno* (Rome: Confindustria, 1971).

AFTERWORD

The economic and social gap between northern and southern Italy has still not narrowed despite more than two decades of major development subsidies. Progress has certainly been registered, but considering the hopes of many southerners the results have fallen far short of their expectations. Some have argued that the development program's goals were set too high, while others have reasoned that the passage of one generation was far too short in time to have expected anticipated outcomes. Still others, not only skeptics from the North but also such well informed foreign observers as Vera Lutz,[1] have reasoned that "forcing" industry into unsuitable locations in the Mezzogiorno may have retarded the overall growth of the Italian economy as did past tariff policies. I cannot accept the last contention. Clearly, the success of the new Fiat plants in the South, the remarkable performance of the Taranto steel works, and the general accomplishments of a series of new petro-chemical plants, among other undertakings, in southern Italy belie that analysis. Perhaps the most telling counterarguments are those of Professor Pescatore (the former head of the Cassa) and Saraceno[2] who stressed the undeniably rapid economic growth of the South viewed as an integral unit. In fact, the annual growth rates of southern Italy, in terms of gross domestic product, have been far higher than those of the other "underdeveloped regions" in the Mediterranean Basin and have also been significantly greater than those of several of the "developed nations" of northwestern Europe.

There is clearly an urgent need to change the southerner's image of the development effort, so as to convince him that the program is not merely a political charade, whose visible result was "cathedrals in the desert." Nevertheless, it must be admitted that the inauguration of many new industrial facilities in the Mezzogiorno has not necessarily resulted in significant socio-economic transformations of the regions where the plants were sited. On the contrary, in a number of cases unforseen hardships have been engendered. In such instances there was a sharp increase in the number of relatively low-skilled workers hired during the construction phase, followed by layoffs and ultimate replacement, at a reduced level, by managerial personnel, engineers and foremen from the North and by young graduates of local vocational programs in the South. This pattern is decidedly different from the experience in France and Britain. The construction workers, many of whom had shifted from agricultural pursuits, could no longer find employment. Instead, they either joined the unemployment rolls which pay bare subsistence benefits, or shifted to low income, marginal tertiary activities. This has been proven by the well documented cases of Taranto, Brindisi, Gela and Pomigliano D'Arco (Alfa-Sud). To take another example, the failure of highly skilled workers returning from abroad to find job opportunities in southern manufacturing plants has clearly been disillusioning. Inflation, often greater than the southern Italian norm, particularly the cost of housing and food has been frustrating for those employed in the new factories. The provision of new housing and services has also fallen behind demand, and environmental pollution is now a serious problem, particularly in the regions where the new petro-chemical plants have been located.

The resulting loss of faith in the southern development effort is understandable. If these perceptions can be changed, perhaps the "meridionalisti" (southerners), worker-manager-entrepreneur alike, can be instilled with a much needed sense of confidence. There is a particular necessity to encourage local entrepreneurial initiative, for so many of the middle and upper management level personnel were originally from the North, and so large a fraction of the state and private establishments are branch plants of firms whose headquarters are in Rome, the North or abroad.

The stimulation of trust in the economic effectiveness of the development program is also vital for northern Italians. This responsibility is officially a function of the Institute for the Promotion of Industry in the Mezzogiorno, but the support of economic journals like *Il Mondo* is also crucial. It is in the North that private investment funds and technical know-how exist. Barring a deeper recession, those moneys and skills could be channelled to the Mezzogiorno. Here the adoption and *enforcement* of controls or disincentives, whether they be industrial certificates as in Britain or punitive taxation, could conceivably play major roles in the dispersal of new industry to the South. However, as I have indicated elsewhere,[3] it is the state-controlled firms which must continue to bear the brunt of the burden.

Finally, having lived and studied in Italy for long periods of time during the past two decades, I was impressed by the changes seen in the South. It is true that southern cities are far from matching the socio-economic milieu of Milan, Turin, Bologna, Florence and Rome. Yet, despite the continuing pressure of population and the persistent widespread poverty, life for many has unquestionably improved. In contrast, it is valid to note that progress has been locationally selective, and it is also accurate to say that some areas like the "osso" (the mountain spine) have clearly retrogressed. Then too, some segments of the population have prospered far more than others; thus rural-urban differentials are still great, and gross intra-urban inequities persist. In a sense, Naples and Palermo almost replicate the worst problems in the United States. Nevertheless, the general tenor of life in the South is still far better than it was in the 1950s. I suspect that the central problem for many southerners is not the issue of relative improvement in their standards of living, but that which they justly view as failure of the South to match the socio-economic levels of northern Italy.

The catching-up thesis is not just an illusory dream, for the virtual isolation of much of the Mezzogiorno has been eliminated. No different than in the United States and Western Europe, the southerners, watching television and movies or reading magazines, are now aware of the "good life" in the North. They want an equitable share of the nation's economic product. Thus the Italian southerner and his counterpart in the underdeveloped regions or urban ghettos of the market oriented economies of America and Europe is now demanding decent housing, schools, hospitals and medical services. He also wants an improvement in his diet, so that meat and dairy products are no longer unattainable luxuries. Like the relatively more fortunate Italian northerner and his counterpart in America and Europe, he too wants a new car. Regretfully, most of these goals are still far from realization.

NOTES

[1]V. Lutz, *Italy: A Study in Economic Development* (London: Oxford University Press, 1962), pp. 141–152.

[2]Notes from interview with Professor G. Pescatore in 1976; P. Saraceno, *La Mancata Unificazione Economica Italiana a Cento Anni dell' Unificazione Politica*, SVIMEZ (Rome: Giuffrè, 1961).

[3]Rodgers, 1976, op. cit., pp. 47–81.

INDEX

Entries refer to Italy unless otherwise indicated. Italian place names omitted to avoid repetition.